# International Conference on Materials, Processing and Product Engineering 2015 (MPPE2015)

AA006416

IOP Conference Series: Materials Science and Engineering
Volume 119

Leoben, Austria
3 – 5 November 2015

**Editors:**

**Sven Eck**
**Reinhold Ebner**
**Andreas Ludwig**

ISBN: 978-1-5108-2276-4
ISSN: 1757-8981

**Printed from e-media with permission by:**

Curran Associates, Inc.
57 Morehouse Lane
Red Hook, NY 12571

**Some format issues inherent in the e-media version may also appear in this print version.**



Printed by Curran Associates, Inc. (2016)

For permission requests, please contact the Institute of Physics
at the address below.

Institute of Physics
Dirac House, Temple Back
Bristol BS1 6BE UK

Phone: 44 1 17 929 7481
Fax: 44 1 17 920 0979

techtracking@iop.org

**Additional copies of this publication are available from:**

Curran Associates, Inc.
57 Morehouse Lane
Red Hook, NY 12571 USA
Phone: 845-758-0400
Fax: 845-758-2633
Email: curran@proceedings.com
Web: www.proceedings.com

# Table of contents

## Volume 119

**International Conference on Materials, Processing and Product Engineering 2015 (MPPE2015)**
**3–5 November 2015, Leoben, Austria**

**Accepted papers received: 7 March 2016**
**Published online: 14 April 2016**

**Preface**



**Papers**

IOP Conf. Series: Materials Science and Engineering **119** (2016) 011001 doi:10.1088/1757-899X/119/1/011001

# Preface: International Conference on Materials Processing and Product Engineering (MPPE 2015), *3-5 November 2015, Leoben, Austria*

**S Eck[1], A Ludwig[2] and R Ebner[1]**
Editors, *International Conference on Materials, Processing and Product Engineering*, in *IOP Conference Series: Materials Science and Engineering*
[1] Materials Center Leoben Forschung GmbH, Leoben, Austria
[2] Chair of Modelling and Simulation of Metallurgical Processes, Montanuniversitaet Leoben, Austria

E-mail: sven.eck@mcl.at

The Materials Center Leoben Forschung GmbH (MCL) and the Austrian Society for Metallurgy and Materials (ASMET) have jointly organized the "First International Conference on Materials, Processing and Product Engineering - MPPE" held in the Congress Center of Leoben from Nov. 3rd to 5th, 2015.

The main topic of the conference was to present the state of the art in fundamental knowledge and technological expertise enabling and stimulating the development of knowledge based innovations across the entire product value chain starting from the material synthesis, proceeding with the processing of products and concluding with their in-service behaviour until the end of the product life time.

Hence, the International Conference on Materials, Processing and Product Engineering MPPE 2015 focused on the core regions of industrial production chains, covering topics such as

- Development and characterization of materials;
- Materials synthesis and processing;
- Dimensioning of components including integrated materials and process modelling;
- Behaviour of materials and components during service.

The scientific and technological goal was to present the state of the art of theoretical, experimental and numerical techniques and their combinations that are capable of improving the competitiveness of modern production facilities.

The following renowned scientists and representatives from industry (list in alphabetic order) gave corresponding key note lectures at the conference:

- Prof. Peter Groche, Technical University Darmstadt, Germany
- Prof. Nenad Gubeljak, University of Maribor, Slovenia
- Dr. Matthias Hutter, Fraunhofer IZM Berlin, Germany
- Prof. Ernst Kozeschnik, Technical University Vienna, Austria
- Prof. Andrej Ruban, KTH Royal Institute of Technology, Sweden
- Dr. Martin Schloffer, MTU Aero Engines AG, Munich, Germany
- Dr. Martin Stockinger, Böhler Schmiedetechnik, Kapfenberg, Austria
- Prof. Ephraim Suhir, Portland State University, United States

During the two and a half day conference in Leoben the more than 120 participants coming from 14 different countries used the opportunity for extended discussions with their colleagues.

In addition to the key note presentations more than 60 contributions from scientists from all over the world including a number of results from projects performed under the Austrian COMET funding programme in frame of the COMET K2 Center on "Materials, Processing and Product Engineering (MPPE)" were presented. 31 written contributions were accepted for publication via a peer reviewing process during and after the conference.

The conference organizers gratefully acknowledge financial support by the Austrian Federal Government (in particular from Bundesministerium für Verkehr, Innovation und Technologie and Bundesministerium für Wissenschaft, Forschung und Wirtschaft) represented by Österreichische Forschungsförderungsgesellschaft mbH and the Styrian and the Tyrolean Provincial Government, represented by Steirische Wirtschaftsförderungsgesellschaft mbH and Standortagentur Tirol, within the framework of the COMET Funding Programme.

The editors:
Dr. Sven Eck
Professor Reinhold Ebner
Professor Andreas Ludwig

# Peer review statement

All papers published in this volume of *IOP Conference Series: Materials Science and Engineering* have been peer reviewed through processes administered by the proceedings Editors. Reviews were conducted by expert referees to the professional and scientific standards expected of a proceedings journal published by IOP Publishing.

# Investigation of the influence of the chemical composition of HSLA steel grades on the microstructure homogeneity during hot rolling in continuous rolling mills using a fast layer model

**M Schmidtchen[1], A Rimnac[2], P Warczok[3], E Kozeschnik[3, 4], C Bernhard[5], S Bragin[2], R Kawalla[1], B Linzer[2]**

[1]TU Bergakademie Freiberg, Institute of Metal Forming, Bernhard-v-Cotta-Str. 4, D-09599 Freiberg
[2]Primetals Technologies Austria GmbH, MT C&R TI ESP, Linz
[3]Institute of Materials Science and Technology, Vienna University of Technology, Getreidemarkt 9, 1060 Vienna, Austria
[4]Materials Center Leoben Forschung GmbH, Roseggerstraße 12, 8700 Leoben, Austria
[5]MU Leoben, Lehrstuhl für Eisen- und Stahlmetallurgie, 8700 Leoben, Austria

E-mail: matthias.schmidtchen@imf.tu-freiberg.de

**Abstract.** The newly developed LaySiMS simulation tool provides new insight for inhomogeneous material flow and microstructure evolution in an endless strip production (ESP) plant. A deepened understanding of the influence of inhomogeneities in initial material state, temperature profile and material flow and their impact on the finished product can be reached e.g. by allowing for variable layer thickness distributions in the roll gap. Coupling temperature, deformation work and work hardening/recrystallization phenomena accounts for covering important effects in the roll gap. The underlying concept of the LaySiMS approach will be outlined and new insight gained regarding microstructural evolution, shear and inhomogeneous stress and strain states in the roll gap as well as local residual stresses will be presented. For the case of thin slab casting and direct rolling (TSDR) the interrelation of inhomogeneous initial state, micro structure evolution and dissolution state of micro alloying elements within the roughing section of an ESP line will be discussed. Special emphasis is put on the influence of the local chemical composition arising from direct charging on through-thickness homogeneity of the final product. It is concluded that, due to the specific combination of large reductions in the high reduction mills (HRM) and the highly inhomogeneous inverse temperature profile, the ESP-concept provides great opportunities for homogenizing the microstructure across the strip thickness.

## 1. Introduction

During the past three decades simulation of metal forming processes has become a necessary tool for quality insurance, automation and the development of new materials technologies. Especially the

demands of thermo-mechanical treatments considering the interaction of process entities such as temperature, stress and strain state on the one hand and the dependencies of microstructure developments during hot rolling on the other require numerical tools which are able to describe these complexly interrelated properties accordingly. In hot rolling applications the material flow is simulated with different approaches and models. Overviews of the concepts, differing in the computing effort necessary and used modelling depth, are given in [1-3].
One of the key issues is a proper choice of the modelling approach for stress and strain state, which can range from involvement of all components of the strain and strain rate tensor to the assumption of parallelepipedic deformation. Nowadays Finite Element methods, boundary approaches or slab theory are used. As a rule of thumb the accuracy and complexity of the obtained strain rate and stress tensor scales insures the computational effort to be done. Depending on the field of application the used modeling approach usually lies somewhere in between these methods [2].
To combine the advantages of a higher modelling depth with a fast end flexible solution procedure an improved layer model, based on slab theory has been developed [4-7]. The model was first developed and tested as a symmetric model and is now extended to general asymmetric rolling conditions. Thereby asymmetries in relation to roll gap geometry, tribology, initial material state and temperature distribution are taken into consideration. Residual stresses are also available.
In the following a short overview of the main features and models included is given. For selected examples of hot rolling tasks a comparison to Finite Element computations is shown. In relation to rolling schedules for TSDR or ESP-concepts the effect of an inhomogeneous temperature distribution on local chemical compositions and strain states is discussed.

## 2. Model description

The basis of the new general N-layer asymmetric rolling model "LaySiMS" was described earlier in [4-6]. Regarding the inhomogeneous deformation state in the roll gap as well as the temperature gradient, the strain, strain rate and the hardening and softening, which are inhomogeneous across the thickness, can be calculated. The simulation model consists of four sub models: material flow model, mechanical properties, thermal model and microstructure model, see Fig. 1. The assumptions and simplifications of the layer model are summarized in the Figure as well. In Fig. 1 the calculated quantities are also indicated. The model of material flow is based on an extended strip theory and simplified thermal computations [5, 6]. The roll gap is divided into N – layers along the y-direction (direction perpendicular to the rolling direction). Within the computation, $h_i$ are the thicknesses, $\vartheta_i$ the temperature, $\sigma_{Fi}$ the flow stress and $v_{wxi}$ the velocity of each individual layer. Interfaces between the layers as well as the friction coefficients $m_i$ are numbered beginning at the top of the strip. The roll gap length is given by $L_d$. Back and forward tension is indicated with $\sigma_R$ and $\sigma_V$. At the neutral line $x_{Ni}$ the flow speed of the material is equivalent to the horizontal speed of the rolls $v_{Wxi}$ or a value in between, fulfilling the requirements of mass balance. Within the model dissimilar radii of the work rolls $R_1$ and $R_2$ were used. Flattening of the work rolls is considered using the approach of Hitchcock [1, 2].
The changes in thickness of each layer will be followed individually along the rolling direction. Due to the possibility of a high number of individual layers, inhomogeneous deformation states can be approximated using visioplasticity methods. Incompressibility is guaranteed due to the application of volume constancy in all layers, starting with corresponding velocities for each individual layer in the neutral line. The basic equations for plastic and elastic deformation of the layers were derived in a similar way as described in [1, 2]. The yield criterion of v. Mises and an analytic description of the roll gap geometry were used. In addition, there is a need for an additional semi empirical equation, which describes the thickness variation of each individual layer during the deformation process. For this purpose, the model of Zhang for two layers [6] was modified for an application in N-layer configurations with arbitrary hardness and thickness distributions. The layer thickness model depends on ratios of local yield stresses of each layer, friction, layer thickness ratio as well as the strain history of the individual layer.

The material flow model is based on the data of the stress-strain-behaviour of the used material as function of the deformation, deformation rate, and temperature. Within the model a mean yield stress of the layers is used, which is determined by a recursive approach using a linear rule of mixing to compute the resulting rolling force and torque. For each individual layer the yield stress can be determined from different yield stress approaches, which are semi-empirical models after Hensel and Spittel [5], Li et. al. [9], mean flow stress approaches of Misaka-Minami [13] or phenomenological models e.g. after Engberg et. al. [16].

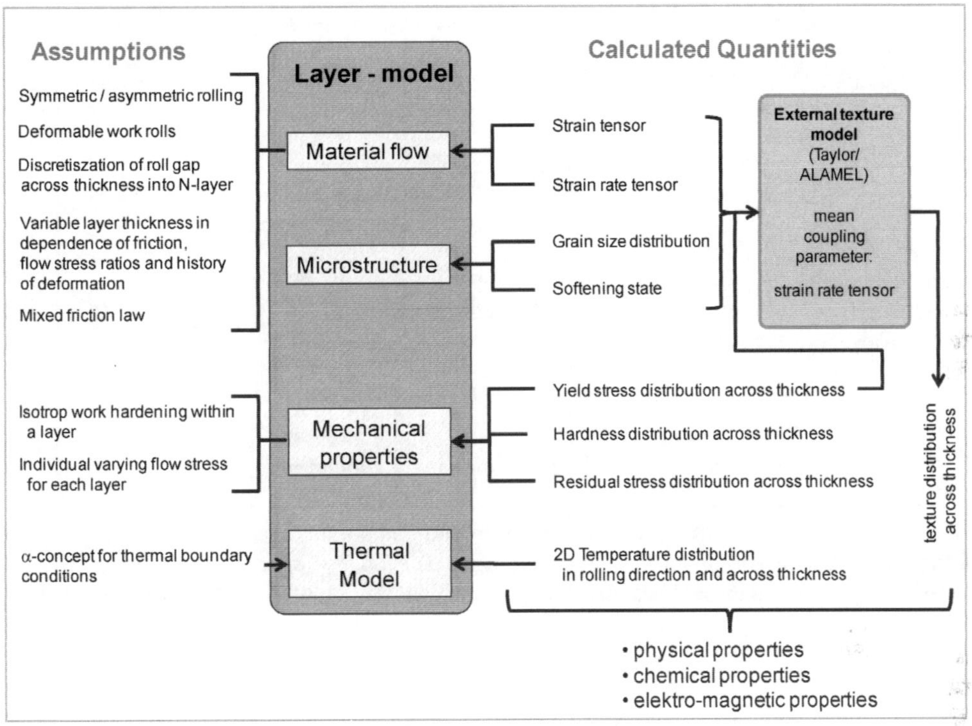


**Figure 1.** Description of the model [17].

The dynamic and static softening is included in the simplest version using a modified JMAK-Model which was described in [8-13]. The model takes into consideration the retarding effect of microalloying elements. In the initial model approach described by [14] the dissolute contend was predicted using solubility products. For a more detailed investigation of the substructure development in steel grades an approach based on Engberg and Lissel [16] was also implemented. The computational results given here are based on a version of LaySiMS using thermodynamical computations based on MATCALC [15]. Therefore an interface between LaySiMS and MATCALC was implemented which allows for automated computations of the dissolution state or precipitation kinetic for real chemical compositions.

During computation, an equivalent plastic strain is required. It is given by the actual deformation state itself and two additional terms arising from shear at the entrance region $\Delta\varphi_{vorE\,i}$ [1, 2, 6, 7] and from the prestrain $\Delta\varphi_{eff\,i}$ of the former deformation step. The temperature evolution was computed using a quasi 1D-heat conduction equation for each layer taking convective terms into consideration [5, 6]. In general, all thermal equations are coupled to deformation by a dissipative term. The mechanical properties across the thickness are calculated based on the description of the material flow and the flow curves of the material.

Due to the assumptions and simplifications made a stiff set of first order differential equations has to be solved, Fig. 2, which strongly depends on several internal parameters especially the individual position of the neutral line $x_{Ni}$ for each layer. In general, a numerical solution has to be found by imposing an iteration for finding the neutral zone position $x_{Ni}$. The program package implemented is based on the computer algebra system MATHEMATICA.

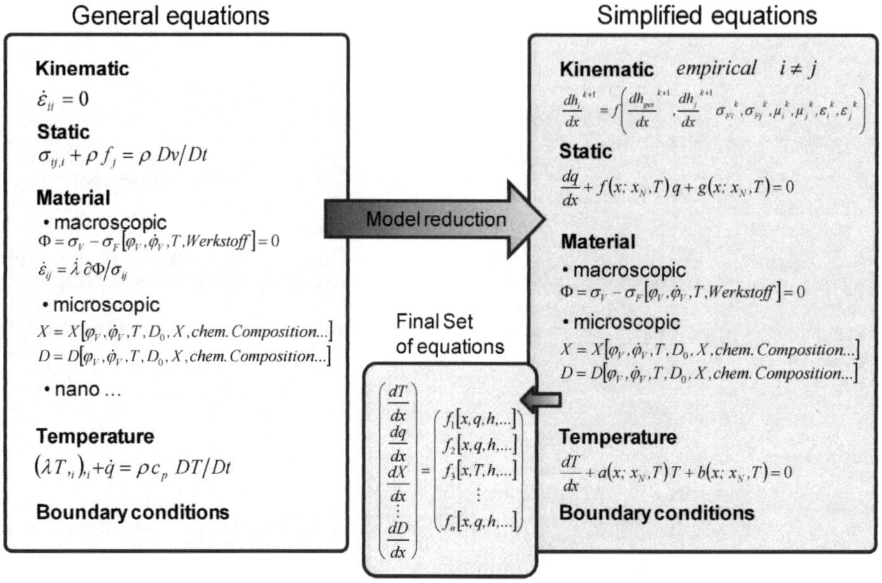


**Figure 2.** Scheme for complexity reduction of the full model towards the simplified layer-based model with minor sacrifice of accuracy and large reduction of computational effort [17].

The final result will be obtained within an iterated approach: First, a numerical solution of the thermal mechanical problem including the layer thickness distribution is obtained using the simplified homogeneous strain state and position of neutral line of the classical slab theory. The plastic deformation starts and ends at the same position for each layer [17]. After determining the new strain state and the improved position of the neutral line, the real material flow, the temperature distributions as well as the microstructure evolution due to dynamic softening are calculated in a second step. The third step includes the computation of the final stress and strain state from the condition of local constancy of volume flux for each layer, using the predicted strain, strain rate and temperature distributions which results in a shift of the initial positions were the plastic deformation starts or ends, mostly toward the centre of the roll gap. To obtain the shifted positions, residual stresses for each layer have to be introduced at the entrance and exit of the roll gap. Starting from the neutral line, the volume constancy conditions for each layer together with a force balance give a set of equations from which the residual stresses can be deduced.
Due to the simplified structure of the equations the whole solution for one deformation step can be found within a fraction of seconds, depending on the number of layers used for discretization [6, 7]. A detailed description of the model as well as several comperative studies will be published elsewhere [7].

## 3. Application to hot rolling

Within an extensive comparative study the new layer model was tested in relation to analytical solutions, Finite Element computations and experimental investigations [7]. Temperature distribution, microstructure evolution and strain state agree well with results from generally accepted numerical and analytical methods as well as experiments. The accuracy of the microstructure evolution is well

IOP Conf. Series: Materials Science and Engineering **119** (2016) 012001 doi:10.1088/1757-899X/119/1/012001

satisfying but, of course, depends strongly on the applied constitutive model. In the mentioned investigations the accuracy checks were performed with microalloyed HSLA steel grades. Further experimental trials were performed and compared to the model for a more general assessment of the performance of LaySiMS. As an example for the accuracy of the predicted strain state in Fig. 4 the tracking lines of the knots of an FE computation and the layer distribution computed by LaySiMS are given. The tracking lines agree well but the computational time of LaySiMS was more than thousand times faster than MSC.MARC which was used for the FEM trials.

The application of LaySiMS to different rolling technologies gave a clear understanding of the local material behavior during hot rolling. Due to the model approach the material can be followed along fixed tracking lines through the whole rolling mill without any intermediate homogenization or other additional simplifications. The differences are clearly visible in Fig. 6 for similar rolling schedules, which differ in the initial temperature distribution only. Due to a high equivalent strain within the first rolling steps and an inhomogeneous temperature distribution with largest temperature values in the strip core a strain distribution with increased homogeneity is reached. Fig. 7 shows the homogenization of strain distribution across the strip thickness for rolling conditions equivalent to the roughing section of an ESP mill. The temperature distribution corresponds to Fig. 6a.

In both cases of Tab. 1 and Fig. 3 and 5 three forming steps were necessary to reduce the microstructural inhomogeneity. In these cases a change in dissolution state is more significant for a softening behaviour than for the mean yield stress. Due to the retarding effect of the dissolved micro alloying elements, a static or metadynamic softening during the transport between the first stands can be used to avoid unwanted grain growth caused by the high temperatures at this process stage.

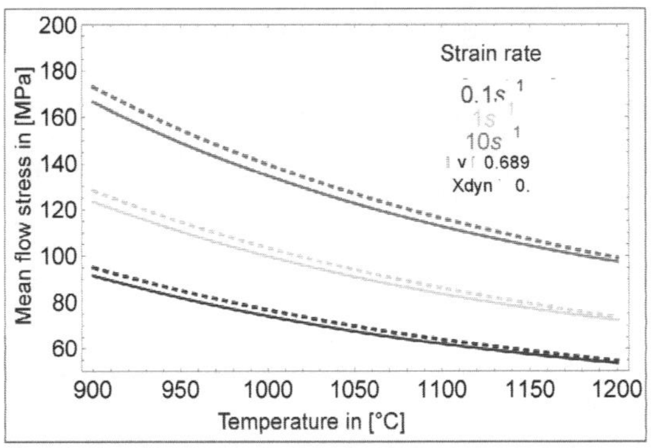


**Figure 3.** Influence of dissolution state on mean flow stress after Miska-Minami [13] (dashed lines: nominal composition given in Tab.1).

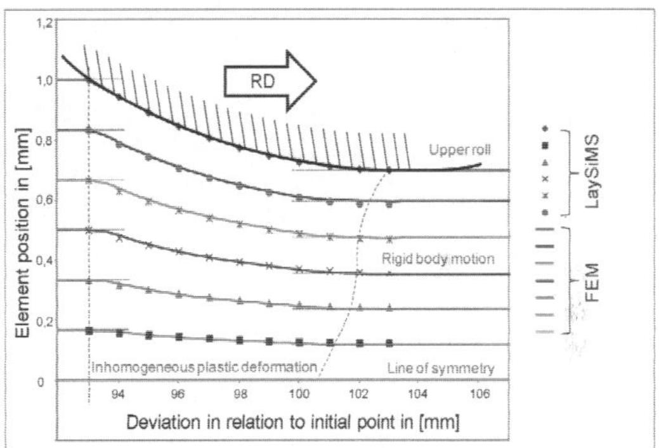


**Figure 4:** Comparison between the track lines of FE knots and LaySiMS layer computation (RD- rolling direction) [17].

**Table 1**. Influence of chemical composition on recrystallization kinetics (model approach after [14]), equilibrium dissolution state from MATCALC computations

| **Nominal chemical composition** | **Equilibrium dissolution state** |
|---|---|
| Dissolved content of microallying elements in S355 | |
| **{wtNb, wtTi, wtV, wtMo}** | **{wtNb, wtTi, wtV, wtMo }** |
| {0.0423, 0.0028, 0.0336, 0.097} | {0.00213004, 0.000046024, 0.0335575, 0.096999} |

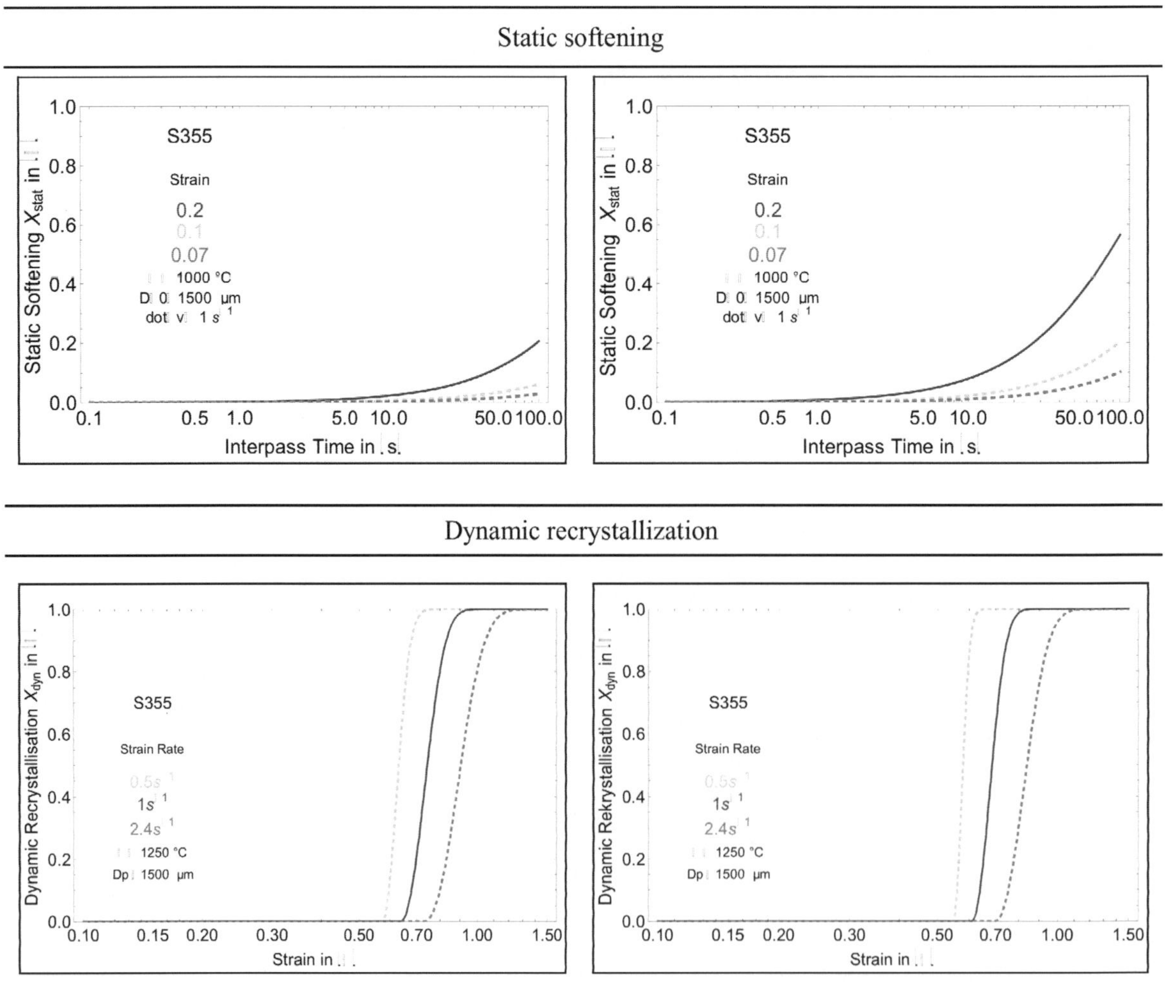


**Figure 5.** Softening kinetics of HSLA S355 and their dependence on chemical composition (model approach of [14])

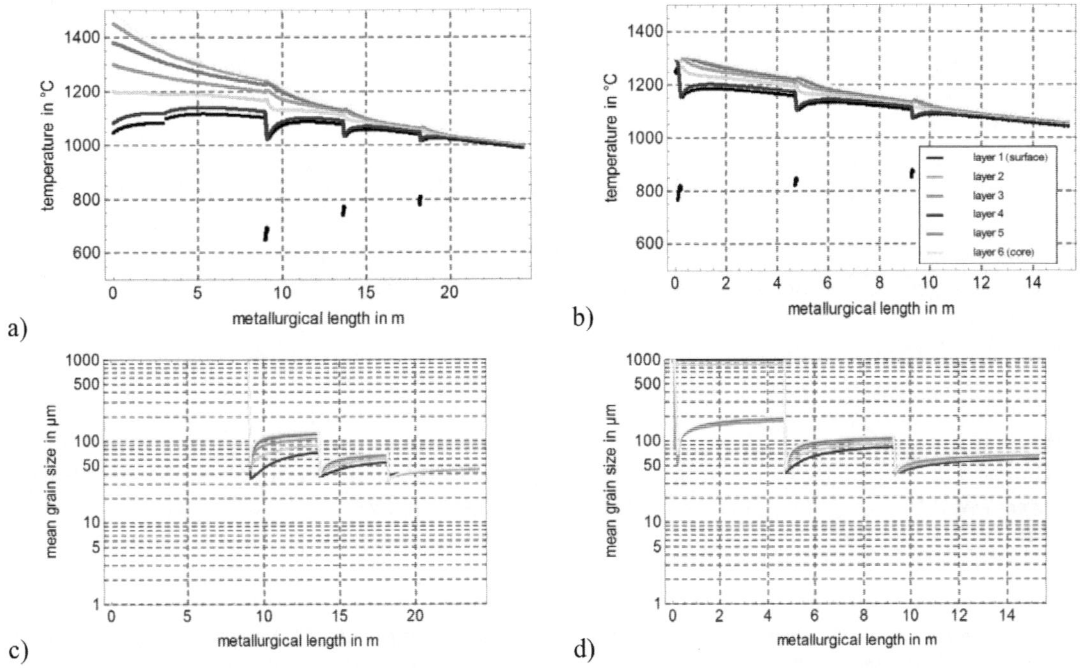


**Figure 6.** Temperature and mean grain size distribution of S355 along the metallurgical length in the layers after the roughing section with different initial temperature distributions: left (a,c) - inhomogeneous, right (b,d) - homogeneous; for computation an symmetric 11 layer configuration were used [17].

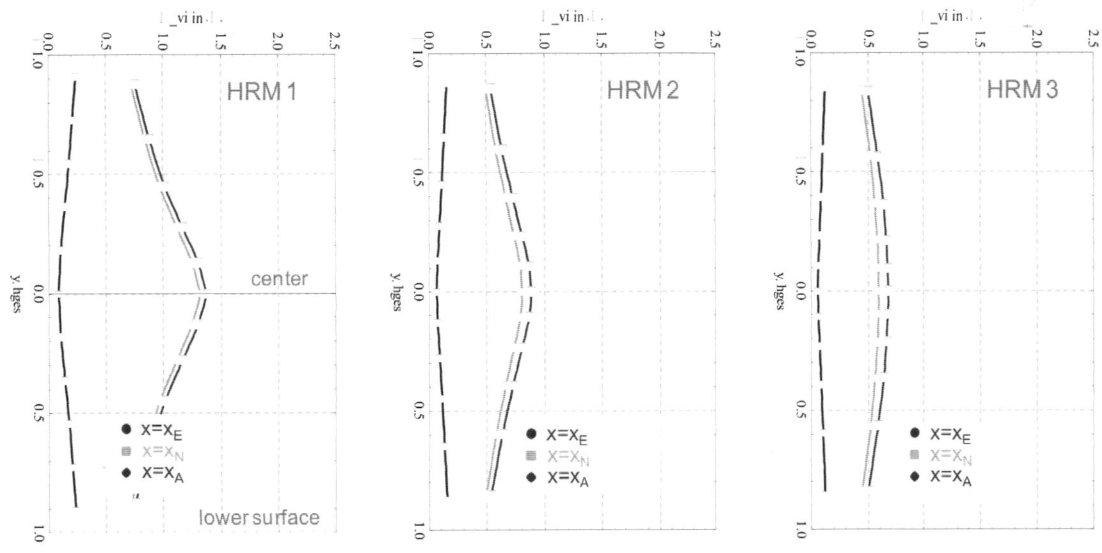


**Figure 7.** Homogenization of equivalent strain distributions during rolling in a roughing section with inhomogeneous initial temperature distributions (computation with a symmetric 11 layer configuration; $x_E$...exit region, $x_A$...entrance region, $x_N$...neutral line).

## 4. Conclusions

For accomplishing accurate and thorough analysis of hot rolling processes within reasonable times a large modelling depth at a minimal computational time is demanded when the focus is on gaining reliable information regarding inhomogeneous material evolution. To meet these concurring requirements, a fast simulation method for hot rolling was developed. The computation of the local strain state under consideration of the material evolution processes was performed using the Mathematica-based program package *LaySiMS*. Depending on the number of layers, an inhomogeneous strain state in the roll gap can be approximated with sufficient accuracy at reasonable computation times. Further refinement of results can be achieved by either increasing the number of equidistant layers or introducing layers of different initial heights for local refinement in regions of highest interest accounting for a higher resolution in areas of significant local gradients. A comparison of computational results calculated with the layer model with those resulting e.g. from FE calculations correspond well. Hence, the benefit achieved by using the LaySiMS–system is a dramatically reduced computational time by keeping the necessary modelling depth and accuracy.

## 5. Acknowledgments.

The Authors have to thank DFG as well as FFG Austria for financial support within *SPP1204* resp. *FFG-SimMoStahl*.

## 6. References

[1] Dahl W, Kopp R, Pawelski O, Pankert R 1993 *Umformtechnik, Plastomechanik und Werkstoffkunde* in German (Verlag Stahleisen Düsseldorf & Springer Verlag,)
[2] Lenard J G, Pietrzyk M, Cser L 1999 *Mathematical and Physical Simulation of the Properties of Hot Rolled Products* (Elsevier Ltd.)
[3] Kawalla R, Schmidtchen M 2013 *Mater. Sci. Forum* **762** 22
[4] Schmidtchen M, Kawalla R 2008 *Proc. Int. Conf. MEFORM* in German (TU Bergakademie Freiberg) 245
[5] Schmidtchen M, Kawalla R 2010 *Steel Research Int.* No. 9 special edition Proc. Int. Conf. Metal Forming (Toyohashi, Japan) 230
[6] Schmidtchen M, Senger B, Kawalla R, Seidel J, Ernst O, Seefeldt M, Decroos K 2013 *Proc. Int. Conf. MEFORM* in german (TU Bergakademie Freiberg) 278
[7] Schmidtchen M Multiscale simulations of non-isothermal roll bonding and rolling of metal matrix composites, to be published
[8] Uranga P, López B, Rodriguez-Ibabe J M 2009 *Mater. Sci. Technol.* **25** No. 9 1147
[9] Liu Q Li Z, Tang G,. Tian Z, Siciliano F 2010 *J. Iron Steel Research Int.* **17**, No. 1 70
[10] Siciliano F, Leduc L L 2005 *Mater Sci.Forum* **500-501** 221
[11] Uranga P, Fernández A I, López B, Rodriguez-Ibabe J M 2003 *Material Science and Engineering* **A345** 319
[12] Siciliano F, Jona J J 2000 *Metallurg. Mater. Trans.* **31A** 511
[13] Minami K, Siciliano F, Maccagno T M, Jonas JJ 1996 *ISIJ International* **36** No.12 1507
[14] Rodriguez-Ibabe J M 2007 *Thin Slab Direct Rolling of Microalloyed Steels* (TransTech Publications LTD, Switzerland)
[15] Kozeschnik E, Buchmayr B 2001 *Diffusion and Phase Transformation* **5** 349
[16] Engberg G, Lissel L 2008 *Steel Research Int.* **79** No. 1, 47
[17] Schmidtchen M, Kawalla R, Rimnac A 2014 *Mater Sci. Forum* **783-786** 2194

# Modelling and simulation of randomly oriented carbon fibre-reinforced composites under thermal load

**R Treffler[1], J Fröschl[1], K Drechsler[2] and E Ladstätter[2]**

[1] BMW Group, Knorrstrasse 147, 80788 Munich, Germany

[2] Institute for Carbon Composites, Department of Mechanical Engineering, Technische Universität München, Munich, Germany

E-mail: rene.treffler@bmw.de

**Abstract.** Carbon fibre-reinforced sheet moulding compounds (CF-SMC) already exhibit a complex material behaviour under uniaxial loads due to the random orientation of the fibres in the matrix resin. Mature material models for metallic materials are generally not transferable. This paper proposes an approach for modelling the fatigue behaviour of CF-SMC based on extensive static and cyclic tests using low cost secondary carbon fibres (SCF). The main focus is on describing the stiffness degradation considering the dynamic modulus of the material. Influence factors such as temperature, orientation, rate dependence and specimen thickness were additionally considered.

## 1. Introduction

In the last decade, metallic materials have been successively replaced by fibre-reinforced polymer composites in the automotive industry due to further potential in weight reduction. However, the mechanical and thermal behaviour of fibre-reinforced plastics is so diverse and complex that todays knowledge is far from complete. Most of current simulations of composite structures are based on isotropic material models [1] and anisotropy is not usually considered. Moreover, Young's modulus reductions due to cyclic and thermal loads are mostly not sufficiently taken into consideration.

A current research project at BMW AG deals with the process chain including planning, design, engineering and manufacturing automotive components made of carbon fibre reinforced sheet moulding compounds (SMC) based on a vinylester resin using low cost carbon fibre (CF) recyclate. Sheet moulding compounds (SMC) are ready to mould fibre reinforced polymers on the basis of thermoset resins which are primarily used in hollow section components. The reinforcement can be achieved through randomly distributed glass or carbon fibres 5 to 50 mm in length. Hence, from the macroscopic point of view, the material has isotropic mechanical properties. But locally the structure of the material is anisotropic which is additionally influenced by the flowing motion during the compression moulding process. Despite the increasing demand, the limiting factor for a breakthrough of CF in automotive mass production from todays point of view is still the high cost of the production of virgin CF. An option for reducing the cost of CF-SMC is to utilize the increasing market of secondary carbon fibres (SCF). The easiest way represents the use of fibres from waste cuttings [2].

The aim of the present study is to provide an approach for a quantitative prediction of the mechanical behaviour of sheet moulding compounds using randomly oriented secondary carbon fibres (SCF-SMC) under thermal loads of up to 150 °C. Further influence factors such as specimen thickness, rate dependence and fibre orientation are taken into account. The developed material models are based on test results of mechanical experiments including tensile, compression and fatigue tests at various temperatures. In particular, stiffness degradations at several stress ratios and tension levels are considered in the model approach which is used as a failure criterion for the fatigue behaviour of the composite.

## 2. Experimental

### *2.1. Test-specimen production and experimental procedure*

To determine the mechanical behaviour of SCF-SMC a number of tests were carried out on plain specimen at 23 °C and at 150 °C. The samples for all tests were machined from moulded plates of 2, 4, 6 and 8 mm thickness using a CNC milling machine. For each test specimen geometry and experimental procedure were defined by following its corresponding ASTM standard. Samples were cut out in two perpendicular directions (0°, 90°) to investigate the anisotropy of the material. All tests were carried out in a laboratory environment of 50 % relative humidity.

### *2.2. Monotonic tensile and compression tests*

Monotonic tensile and compression tests were performed using a Zwick/Roell materials testing machine with an optical extensometer for strain measurement. Young's modulus and ultimate tensile stresses (UTS) were determined at different temperatures (23 °C, 50 °C, 80 °C, 110 °C and 150 °C) performing at least two isothermal tensile tests for each temperature level. For data presentation average values of the generated results were taken. All monotonic tensile and compression tests were carried out at a strain rate of 0.1 %/s. Preliminary tests with different strain rates have shown that the material properties are not rate dependent.

Compression tests were carried out according ASTM D695 to determine the ultimate compressive strength (UCS) of the material at room temperature and at 150 °C. To prevent local buckling appropriate test fixtures were utilised.

### *2.3. Fatigue testing*

All fatigue tests were conducted at different stress levels using a computer-controlled MTS servo-hydraulic testing machine under load control. For fatigue tests under pulsating tensile stresses ($R$ = 0.1) and alternating load ($R$ = -1) the different stress levels were obtained by various percentages of the ultimate tensile strength (UTS). For fatigue tests under pure compressive loading ($R = -\infty$) stress levels were defined by taking the ultimate compressive strength (UCS) as a reference. At each stress level at least two test specimens were tested and average values for the generated results were taken for data presentation. Test frequency is adapted to each stress level but remains constant during the test and reaches values between 5 and 10 Hz. The duration of the fatigue tests were limited by restricting the number of cycles to a maximum number of 2 x $10^6$.

In order to analyse the material behaviour of different loading conditions fatigue tests at several $R$-values ($R$ = 0.1, -1, $-\infty$) were determined. The $R$-ratio is defined as the ratio between the minimum and maximum cyclic stress [3]. Figure 1 shows the annotation in form of a Haigh diagram which is defined by the main parameters mean cyclic stress, $\sigma_m$, cyclic amplitude, $\sigma_a$, and $R$-ratio. Haigh diagrams can be obtained from stress-life-diagrams (S-N-curves) of the predefined $R$-ratios and approximate the fatigue life of the material under loading patterns for which no experimental data exists [3]. Haigh diagrams of materials are usually implemented in commercial post-processing software to estimate the fatigue life of a structure.

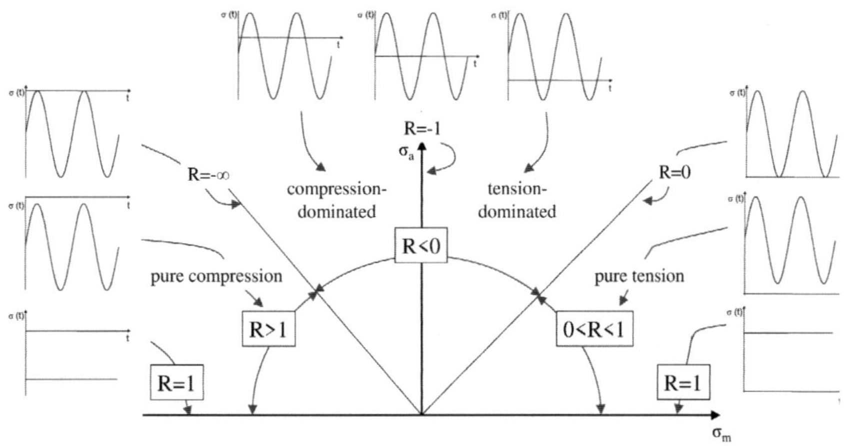


**Figure 1.** Haigh diagram ($\sigma_m$-$\sigma_a$-plane) [3]

## 3. Data reduction

*3.1. Hysteresis measurements*

During cyclic tests, hysteresis curves can be measured continuously which form the basis of evaluating the stiffness degradation. Studies such as [4] revealed the change in hysteresis curves within a fatigue test using polymer composites. This effect is the result of material damage and visco-elastic effects and is shown schematically in Figure 2 (a). Three moduli, the tangent modulus $E_T$ at $\epsilon_0$, the secant modulus $E_S$ and the dynamic modulus $E_{dyn}$ can be determined from the hysteresis-curve (see Figure 2 (b)) and can be calculated according Figure 2 (c) [4, 5].

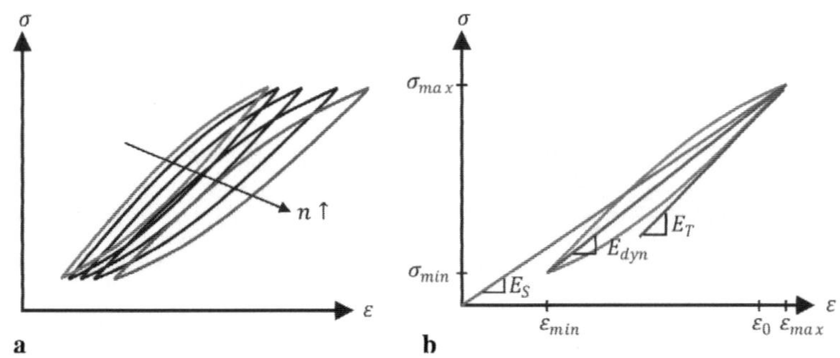


$$E_T = \left.\frac{d\sigma}{d\epsilon}\right|_{\epsilon_0}$$
$$E_S = \frac{\sigma_{max}}{\epsilon_{max}}$$
$$E_{dyn} = \left|\frac{\sigma_{max} - \sigma_{min}}{\epsilon_{max} - \epsilon_{min}}\right|$$

**Figure 2.** Hysteresis measurements, (a) changes in hysteresis curves as a function of load cycles $n$, (b) definitions of different stiffness moduli, (c) equations of defined stiffness moduli [4]

The most suitable property for the evaluation of the stiffness degradation is the dynamic modulus $E_{dyn}$ as it describes the instantaneous material reaction at a specified cycle number. Compared to the secant modulus $E_S$, $E_{dyn}$ only characterises stiffness degradation due to material damage and not degradation because of visco-elastic effects. The advantage over the tangent modulus $E_T$ is that it is averaging the whole hysteresis loop [4].

*3.2. Stiffness degradation*

In the following the terms stiffness, elastic modulus and stiffness degradation are related to the dynamic modulus $E_{dyn}$. According to Paepegem [6] the stiffness degradation of the majority of fibre-reinforced composite materials can be devided into three distinctive stages (see Figure 3):

- initial decrease (stage 1): significant decrease of 2 - 5 % which is dominated by transverse matrix cracks
- intermediate region (stage 2): approximately linear reduction (1 - 5 %) in a wide range of load cycles due to an increasing number of fibre/matrix separations
- final region (stage 3): progressive stiffness decay in abrupt steps until final failure because of fibre fractures

Figure 3 illustrates a typical stiffness degradation curve of a carbon fibre reinforced sheet moulding compound under pulsating tensile stresses ($R = 0.1$).

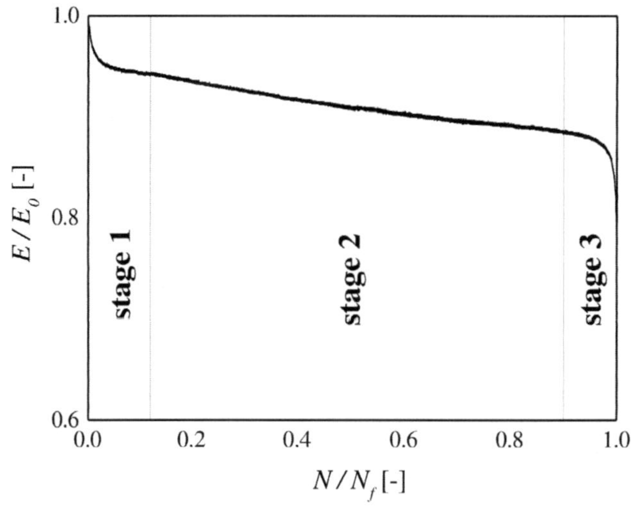


**Figure 3.** Three stages of stiffness degradation of fibre-reinforced composite materials [6]

In order to assure a better comparability of the results the stiffness at a certain point of fatigue life $E$ is referred to the undamaged longitudinal stiffness $E_0$ and the number of cycles $N$ are standardised to the number of cycles at failure $N_f$.

## 4. Results and discussion

*4.1. Monotonic tensile testing*

The stress-strain behaviour of the examined carbon fibre-reinforced sheet moulding compound (SCF-SMC) under thermal load is summarised in Figure 5. As tensile properties are mainly fibre-dominated [4], SCF-SMC exhibits a pronounced linear-elastic response with a slight non-linear behaviour and brittle fracture properties. In [7] similar observations were observed on quasi-isotropic woven plies. The reason for the observed slightly non-linear behaviour is that fibre bundles which are not oriented in tensile direction (0°) during loading show a localized plastic behaviour. Nevertheless this behaviour is limited by the 0° bundles and plasticity cannot occur [7]. Figure 4 shows two scanning eletron microscope (SEM) images of a fractured specimen under monotonic tensile loading.

**Figure 4.** Scanning electron microscope (SEM) images of fractured specimen under monotonic tensile loading

At higher temperatures, the ductile behaviour of the vinylester resin is enhanced but the material response is still primarily fibre-dominated [8]. However, a significant reduction in ultimate tensile strength (UTS) and elastic modulus can be observed (see Figure 5(b)). This effect is even more noticeable when service temperature is higher than the material glass transition temperature region (110 - 120 °C). According to [9] differences in the thermal expansion coefficients of the contacting materials (carbon fibre, vinylester resin) are responsible for the resulting micromechanical stresses in the fibre/matrix interface. It is expected that this effect is significantly enhanced due to large angle differences between fibre bundles in randomly oriented fibre-reinforced polymers such as SCF-SMC as carbon fibres have different thermal expansion coefficients in directions parallel and perpendicular to the fibres. Additionally, pyrolysis of the matrix material occurs when the glass transition temperature is achieved. As a result high temperatures force the primary valency bonds in large molecules to break down [9].

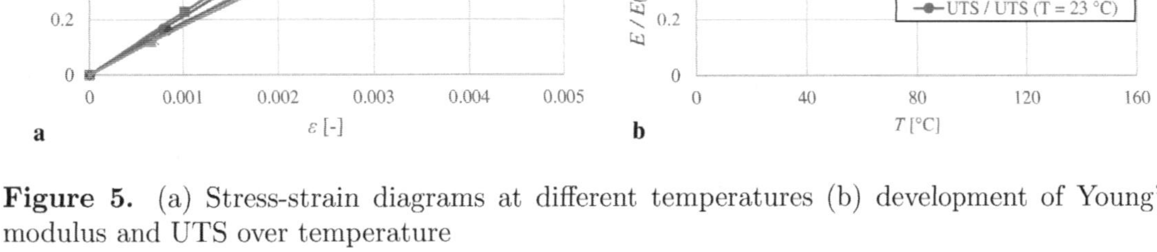


**Figure 5.** (a) Stress-strain diagrams at different temperatures (b) development of Young's modulus and UTS over temperature

Figure 6 illustrates mean value and standard deviation of Young's modulus over specimen thickness. Similar curves can be determined for UTS and UCS. As a result it can be noted that the mechanical properties are not significantly influenced by the specimen thickness. Furthermore, the scatter of the experimentally determined data is very present in specimen with low thickness. This observation is indicative for a heterogeneous material behaviour. By increasing the test volume, for example through enlarging specimen width or thickness, the variance of the results decreases. The material now behaves more homogeneous.

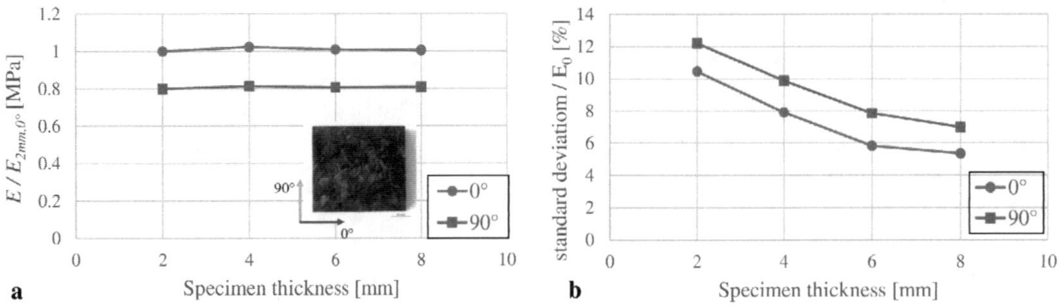


**Figure 6.** (a) Impact of sample thickness on mean value and (b) standard deviation of Young's modulus

Figure 6(a) also indicates the dependence on the orientation of the machined samples. Comparing the tensile modulus of specimen which were cut out in two perpendicular directions (0°, 90°) the examined material shows a degree of anisotropy of 15 - 20 %.

### *4.2. Fatigue testing*

Figure 7 shows the determined degradation curves of the examined SCF-SMC at temperatures 23 °C and 150 °C and several $R$-values ($R = 0.1$, -1, $-\infty$). The results of the fatigue tests can be summarised as follows:

- Different stress levels $\sigma_{max}$ do not influence the degradation process.
- At room temperature ($T = 23$ °C) SCF-SMC under pure pulsating tensile or compressive loading do not have a significant stiffness decrease. The test specimens display a sudden fracture behaviour. Under alternating load the material exhibits a remarkable stiffness degradation, particularly in the second stage.
- Under thermal load of 150 °C the $R$-ratio has no more influence on the stiffness degradation. In comparison to room temperature the degradation behaviour is more pronounced.

In following studies the observed phenomena will be investigated further through detailed material analyses. The current hypothesis is that a lower stiffness degradation ($R = 0.1$ and $R = -\infty$) indicates a fibre-dominant damage behaviour. Furthermore, it is assumed that load spectra with combined tensile and compressive loading cause more damage in the matrix system and result in a more pronounced stiffness behaviour of the material.

As already described, the material behaviour of a fibre-reinforced composite under thermal load is dominated by the polymer. Therefore, compared to room temperature the stiffness decay is more pronounced and not influenced by the stress ratio.

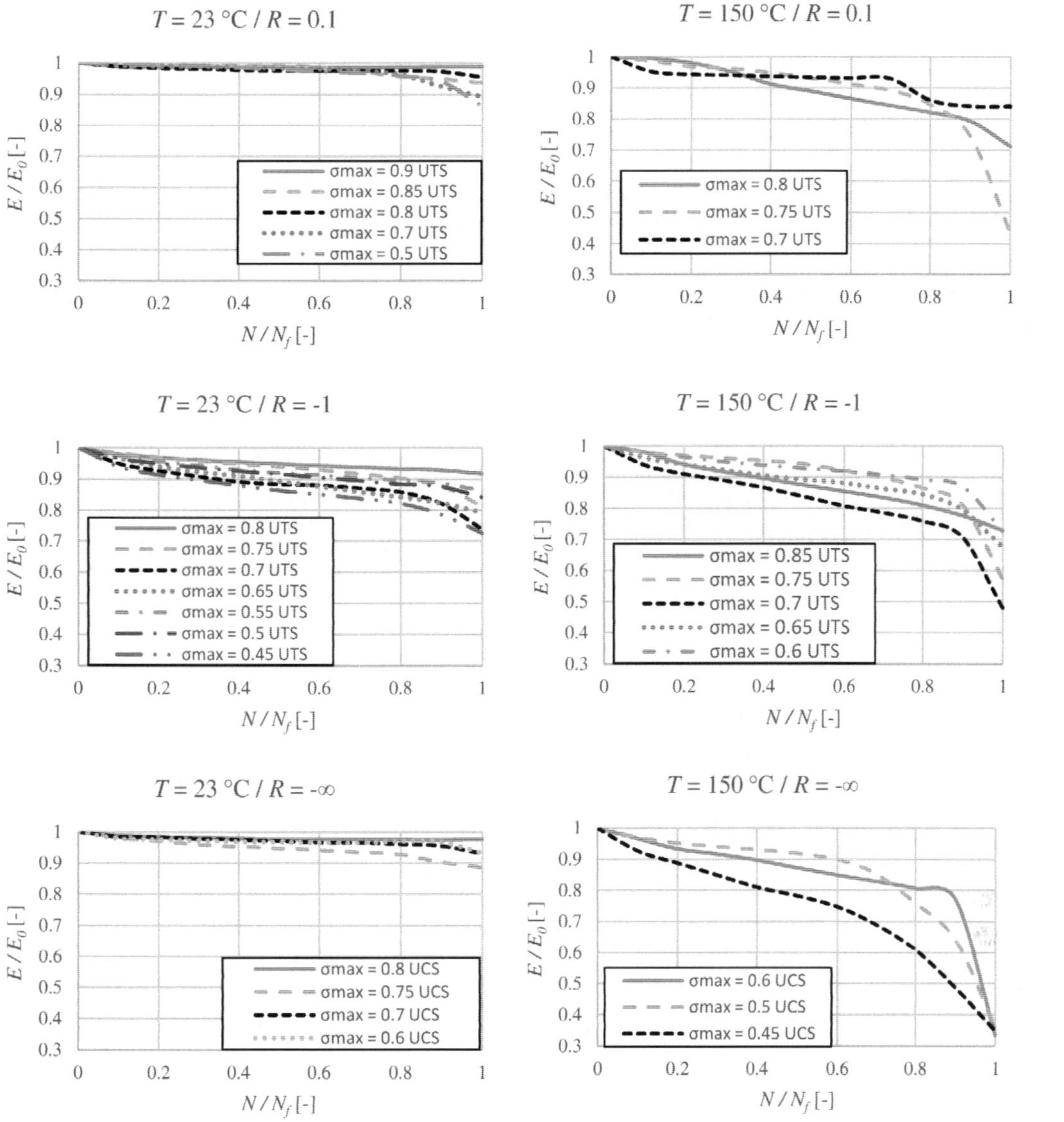


**Figure 7.** Stiffness degradation curves of SCF-SMC at various stress ratios ($R = 0.1$, -1, $-\infty$) and temperatures (23 °C, 150 °C)

## 5. Modelling approach

Figure 8 illustrates the stiffness degradation approximation in respect to $R$-ratio and temperature ((a) room temperature, (b) 150 °C). Each symbol corresponds to an examined stress ratio and is representing mean values of the determined stiffness at certain points of fatigue life (10 %, 20 %, ..., 100 %). As seen above, at room temperature stiffness degradation depends on stress ratio. Additonal fatigue tests at $R = -2$ and $R = -0.5$ confirm this behaviour. Under cyclic loading with combined tensile and compressive rates a significant stiffness reduction is noticed. The two observed damage mechanisms are each approximated by a cubic trial function according to Equation (1) with the parameters a, b and c describing the three stages of stiffness degradation, respectively. In Figure 8 these models are illustrated by master curve lines.

$$f\left(\frac{N}{N_f}\right) = a\left(\frac{N}{N_f}\right)^3 + b\left(\frac{N}{N_f}\right)^2 + c\left(\frac{N}{N_f}\right) + 1 \quad (1)$$

$N/N_f$ describes the number of cycles related to the total number of cycles at fracture for the current load sequence of a spectrum loading. Thus, the stiffness degradation rate for each value of a rainflow-counting analysis can be obtained and accumulated for the current load sequence.

As a result, this process provides the possibility to consider stress redistribution due to local stiffness degradation in the fatigue assessment of the composite. Stress conditions which are situated in between the two calibrated master curves illustrated in Figure 8(a) are taken into account by linear interpolation of the model parameters $a$, $b$ and $c$.

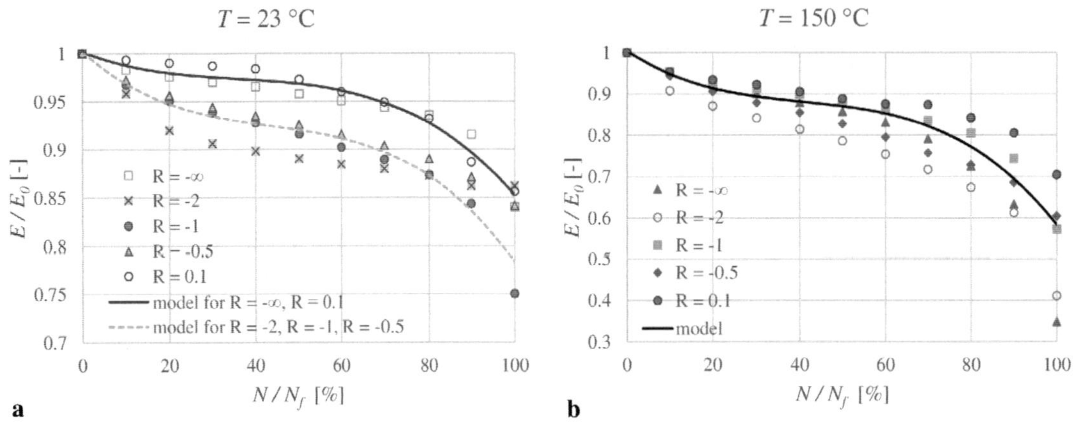


**Figure 8.** Models of stiffness degradation behaviour of SCF-SMC in respect to different $R$-ratios at (a) room temperature and (b) 150 °C illustrated by master curve lines

At 150 °C stiffness degradation is not influenced by stress ratio. Therefore, only one master curve is obtained and approximated by Equation (1) (see Figure 8(b)).

## 6. Method of iterative stiffness adaption

In this project the method of iterative stiffness adaption [10] is transferred to the class of randomly oriented polymer composite materials. Based on a linear viscoelastic Maxwell model the first analysis is performed. As a result the stress tensor $\sigma(\mathbf{x}, t)$ is gained which is the input for the damage evaluation. To consider the multi-axiality in the next step the stress tensor can be modified. In the damage evaluation a nonlinear damage accumulation on basis of S-N-curves, Haigh-diagrams and stiffness degradation curves is performed. Thus, temperature, R-ratio and mean stresses are considered in the assessment.

If the local stiffness changes by a predefined value, a new stress analysis is performed. Therefore, the stiffness matrix is adapted by a damage tensor according to Paepegem [11] using a scalar damage variable $d$ which is a measure for the degradation of the longitudinal stiffness. Finally, the iterations are stopped when the calculated stiffness reaches a value below the required stiffness of the structure. Thus, a quantative assessment of the fatigue behaviour in form of cycle number is achieved. In Figure 9 the method of iterative stiffness adaption is shown schematically.

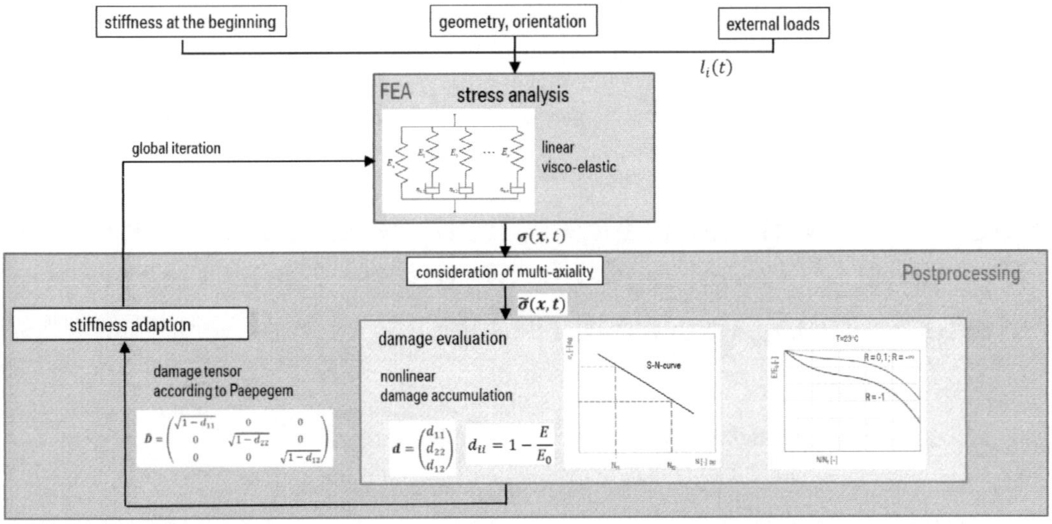


**Figure 9.** Schematic illustration of the method of iterative stiffness adaption

## 7. Conclusions

An approach for describing the stiffness degradation of randomly oriented carbon fibre-reinforced sheet moulding compounds using low cost secondary carbon fibres (SCF-SMC) was carried out in this paper. For this purpose monotonic tensile and cyclic fatigue tests were performed at different temperatures and resulting data, especially the dynamic modulus has been evaluated. Influence factors such as orientation, rate dependence and frequency were considered. At room temperature the examined material does not have a significant stiffness decrease under pure pulsating tensile or compressive loading whereas under alternating load a remarkable stiffness degradation is noticed. At 150 °C the stiffness behaviour does not depend on the R-ratio but exhibits substantially more degradation compared to room temperature. On basis of the evaluated data stiffness degradation models have been developed. These models were approximated by cubic trial functions containing temperature and $R$-ratio dependent material parameters describing the three stages of stiffness degradation. The approach is the basis for an iterative stiffness adaption method which provides the possibility to consider stress redistribution in the fatigue assessment of composites.

## References

[1] Oldenbo M 2004 *Anisotropy and non-linear effects in SMC Composites* Ph.D. thesis Luluea University of Technology
[2] Palmer J, Savage L, Ghita O R and Evans K E 2010 *Composites Part A: Applied Science and Manufacturing* **41**(9) 1232–37
[3] Vassilopoulos A P, Manshadi B D and Keller T 2010 *International Journal of Fatigue* **32**(4) 659–69
[4] Pinter G, Ladstätter E, Billinger W and Lang R W 2006 *International Journal of Fatigue* **28**(10) 1277–83
[5] Talreja R 2000 Fatigue in polymer matrix composites *Comprehensive Composite Materials* Volume 2: Polymer Matrix Composites ed Kelly A and Zweben C (Amsterdam: Elsevier) pp 529–52
[6] Paepegem W V and Degrieck J 2002 *International Journal of Fatigue* **24**(7) 747–62
[7] Vieille B and Taleb L 2011 *Composites Science and Technology* **71**(7) 998–1007
[8] Albouy W, Vieille B and Taleb L 2014 *Composites: Part A* **67** 22–36
[9] Schürmann H 2007 *Konstruieren mit Faser-Kunststoff-Verbunden* (Berlin, Heidelberg: Springer-Verlag)
[10] Talreja R and Singh C 2012 Damage and failure of composite materials (Cambridge University Press) pp 1–8
[11] Paepegem W V 2002 *Development and finite element implementation of a damage model for fatigue of fibre-reinforced polymers* Ph.D. thesis Ghent University

# Investigation of metallurgical phenomena related to process and product development by means of High Temperature Confocal Scanning Laser Microscopy

**U Diéguez-Salgado[1], S Michelic[1] and C Bernhard[1]**

[1]Chair of Ferrous Metallurgy, Montanuniversitaet Leoben, Franz Josef Straße 18, 8700 Leoben, Austria

E-Mail: Uxia.Dieguez-Salgado@unileoben.ac.at

**Abstract.** An increased interest for high temperature metallurgical processes appeared during the last decades, in order to achieve the high quality requirements in steel products. A defined steel cleanness and microstructure essentially influence the final product quality. The high temperatures involved in metallurgical processes and the lack of in situ observations do not only complicate the verification of simulation model predictions but also make significant conclusions regarding the industrial processes difficult. For that reason, new tools and techniques are necessary to develop. By combining the advances of a laser, confocal optics and an infrared image furnace, the High Temperature Confocal Scanning Laser Microscopy (HTCSLM) is a strong tool which enables high temperature in situ observations of different metallurgical phenomena. Next to solidification processes and phase transformations also the behavior of inclusions at different interfaces in the system steel-slag-refractory can be observed. The present study focuses on the aspects of inclusion agglomeration in the liquid steel and the inclusion behavior at the steel/refractory interface in two different steel grades. Out of the obtained experimental data, attraction forces are calculated and compared. This information provides an important basis for a better understanding of inclusion behavior in industrial processes and the therewith related process optimization, like for example the clogging phenomenon during continuous casting.

## 1. Introduction

The final quality of a steel product is always closely connected to its metallurgical production process. Different possibilities and procedures during secondary metallurgy as well as during continuous casting exist in order to ensure the demands of customers regarding mechanical properties, corrosion resistance and surface quality. A parameter which significantly influences steel quality is the so called steel cleanness. The latter is defined through the number, size, chemical composition, distribution and morphology of non-metallic inclusions in the steel matrix. Due to the necessity of deoxidation, the formation of inclusions cannot be avoided completely. In order to ensure a possibly low content of inclusions in the steel and to enable a specific inclusion modification, their behavior in the system steel-slag-refractory needs to be well understood.

Figure 1 gives an overview on possible reactions and interactions of inclusions at different interfaces. Non-metallic inclusions collide and agglomerate in the liquid steel in order to form larger particles; if the particles are large enough, they are transported by flotation to the liquid steel/slag interface due to the density difference between the liquid steel and inclusion, by bubble attachment or by fluid transport in the metallurgical vessel and removed from the steel to the slag by dissolution. Some small inclusions could be transported to the liquid steel/refractory interface and adhering to the wall refractory and others could stay suspended in the liquid steel and pass onto the next stage of the steelmaking process. [1,2]

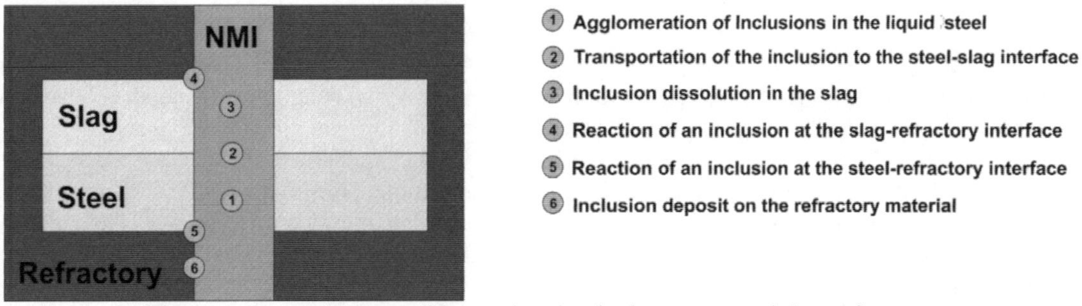


**Figure 1.** Overview on possible reaction sites in the system steel-slag–refractory.

The present study gives an insight in the possibilities of in-situ observation concerning the behavior of non-metallic inclusions at different interfaces between steel, slag and refractory material and indicates the practical relevance for process and therefore product improvements in steelmaking.

## 2. Theoretical background

In the continuous casting process, the submerged entry nozzle (SEN), plays a vital role in order to ensure stable casting conditions. The clogging phenomenon – characterized by a substantial reduction of the inner section of the nozzle caused by material build up – can lead to a disruption of normal casting operations and also produces an uneven flow of molten steel in the mold, finally leading to the formation of inclusion clusters which deteriorate the product quality. The agglomerates found in the nozzle often primarily consist of $Al_2O_3$ particles. [3–8] Next to the principal source of particles [9] causing clogging, the adhesion between particles and the refractory material as well as the attraction between two particles are points of interest.

### *2.1 Adhesion between non-metallic inclusions and liquid steel*

Several research groups dealt with this phenomenon in the past. Singh [7,10–12] divided the deposition process into three stages: (i) The particles are transported from the bulk melt to the refractory surface and come in contact with the nozzle refractory (ii) The inclusions adhere to the refractory surface and (iii) inclusions adhere to each other and sinter together to form a network. The characteristics of the system liquid steel/refractory, such as wettability, composition and surface tension will significantly influence the attachment step.

Sasai et al. [13,14] conclude in their investigation that the agglomeration force between alumina particles in Al deoxidized molten steel is not caused by the van der Waals force but the cavity bridge force created due to the fact that alumina particles are unlikely to be wetted by molten steel. In non-wetting conditions, when two spherical non-metallic inclusions/particles of radii $r_1$ and $r_2$ approach, the steel in contact area may be expelled, leaving an empty neck (filled with gases and vapours) of concave lens shape. The adhesion force is bigger than the buoyancy force and also the flow resistance that acts on the particles, so that the particles will remain at the interface. [10,11,13–15]

This attractive force is the sum of the surface tension force and the force that results from the pressure drop across the liquid-gas interface. [13,15] The balance of forces can be written as follows (The force balance is calculated in the thinnest point of the neck):

$$F_{adh} = 2\pi R_2 \sigma + \pi R_2^2 \Delta P \quad (1)$$

where $\sigma$= surface tension of molten steel, $\Delta P$= difference pressure and $R_2$ = Curvature radius (geometrical parameter) [10,11,13–16]

Applying Equation 1, Figure 2 is calculated. The attraction force is estimated assuming the case that non-metallic inclusions (radii 2,5µm) attach to different refractory walls at the melt depth of 1 m and with 1,8N/m surface tension of molten steel. Three different situations are plotted in Figure 2:

1) The adhesion force between alumina particles and an alumina refractory wall.
2) The adhesion force between alumina particles and a zirconia refractory wall.
3) The adhesion force between liquid calcium aluminate inclusions, from Ca-treated steel, and an alumina refractory wall.

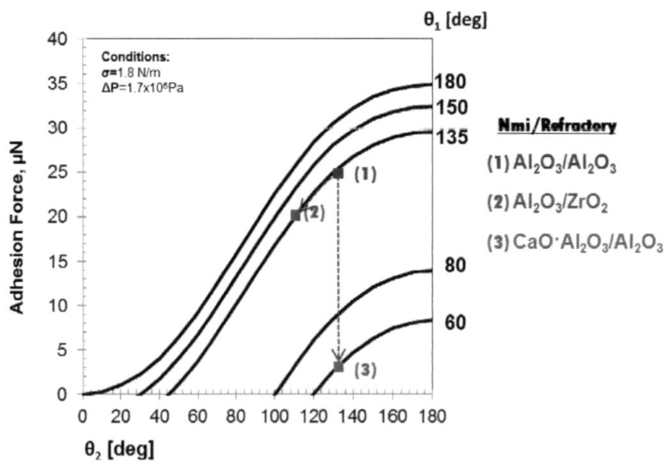


**Figure 2.** Relation between the contact angles and adhesion force. ($\theta_1$ is the contact angle of the non-metallic inclusions with the molten steel and $\theta_2$ is the contact angle of the refractory material with molten steel, the value of the contact angles used are found in [17]).

The adhesion force between alumina non-metallic inclusions and an alumina refractory wall amounts to 25µN (1).

a) The decrease of the contact angle between the molten steel and the refractory leads to a lower adhesion force (2). However, according to a work by Eustatopoulos [5,18], it appears to be impossible to find a refractory-nozzle material with a low contact angle (below 60°-40°) and coincident low reactivity.
b) Assuming a particle from liquid calcium-aluminate instead of solid alumina results in a decrease of the contact angle between the molten steel and the non-metallic inclusions and consequently a significant decrease of the adhesion force (3).

As described in Equation 1, the attraction force is influenced by surface tension of the molten steel, the difference pressure between the molten steel and the gas cavity and the 'shape' of the cavity which depends on the non-metallic inclusions size, morphology and the wetting behaviour.

### *2.2. Attraction between non-metallic inclusions*

The non-wettability behavior of the majority of non-metallic inclusions for liquid iron generates a deformation of the liquid surface, which is called meniscus. If two particles approach one-another, the meniscus in between them is further depressed or drawn downward. [19] This change, which is called capillary attraction, can lead to a difference of capillary pressure between the inside and the outside area of the pair, which will push the two bodies towards each other. [20] The capillary attraction is believed to be the source of the attraction between alumina inclusions at free surfaces and at the surface of gas bubbles inside the liquid steel melt. [19]

The strength of the capillary attraction force for inclusion particle pairs in molten steel, which was found to be much different for particles with different morphology, can be listed as follows in ascending order [21]: liquid/liquid pair < liquid/semi-liquid pair < semi-liquid/semi-liquid pair < liquid/solid pair < semi-liquid/solid pair and finally solid/solid pair as the strongest. The latter sequence is only valid if only the effect of the particle morphology is considered. But the attraction between particles is also strongly affected by other parameters like for example by the particle size, the contact angle of a particle with steel melt, the particle density, the shape and the interfacial tension [22].

## 3. High Temperature Confocal Scanning Laser Microscopy (HT-CSLM)

By combining the advances of a laser, confocal optics and an infrared image furnace, the high temperature Confocal Scanning Laser Microscopy (HT-CSLM) is a strong tool which enables high temperature in situ observation of metallurgical phenomena as phase transformations in the steel [23,24] (e.g. the peritectic reaction or the nucleation of acicular ferrite), austenite grain growth [25], inclusion agglomeration in the liquid steel [26] or inclusion dissolution in a slag phase [27,28], as well as reactions at the steel/refractory interface. The obtained experimental data can be used for a quantitative evaluation of dissolution rates, attraction forces or diffusion coefficients.

The present work focusses on the in situ observation of non-metallic inclusions behavior at high temperature at liquid steel and liquid steel/refractory interfaces by means of High Temperature Confocal Scanning Laser Microscopy.

### *3.1. Set up.*

A Confocal Scanning Laser Microscope (VL2000DX from Lasertec) and a high temperature furnace (SVF17-SP from Yonekura) are used for all investigations. The decisive advantage of HT-CSLM in contrast to conventional microscopes or experimental facilities is the possibility of in-situ observations at temperatures up to 1700 °C in combination with a very good image quality. This is primarily possible due to the fact, that a laser is used as a light source. A schematic view of the experimental set-up in the high temperature furnace which shows an elliptic, gold coated inner contour as well as detailed configuration data of the set-up can be found elsewhere [29]. The halogen lamp is situated in the bottom focal point of the furnace. The sample holder with the crucible and sample inside is located in the top focal point of the ellipse. The temperature is measured with a thermocouple fixed at the bottom side of the sample holder. High-purity argon with a flow rate of 150 $cm^3$/min ensures a neutral atmosphere in the furnace. Additionally, the oxygen content in the furnace is measured for every experiment.

### *3.2. Experimental procedure.*

Two different experiments are performed:

- Inclusion agglomeration in the liquid steel: a steel disc with approximately 600μm thickness is placed in an $Al_2O_3$ crucible and subsequently set up on the sample holder in the ellipsoidal furnace (see Figure 3(a)). The specimen is heated up following the heating cycle given in Figure 4. An alumina disc is placed under the steel disc in order to avoid the attachment of the steel disc onto the crucible. When the liquidus temperature of the steel is reached, inclusions are starting to emerge from the bulk to the steel surface and their agglomeration behavior can be studied. In the present study a Ca-treated steel and a ULC steel are used (composition given in Table 1) in order to investigate the effect of particle morphology into the agglomeration behavior.
- Inclusion behavior at the steel/refractory interface: An exogenous alumina particle (400μm ECD) is placed on the top of a steel disc with approximately 600μm thickness, which is placed in an $Al_2O_3$ crucible (see Figure 3(b)). An alumina disc is placed below the steel disc to avoid the steel sticking onto the crucible. The specimen is heated up following also the heating cycle given in Figure 4. When the liquidus temperature is reached, inclusions are starting to emerge from the bulk to the steel surface and their behavior at the steel/refractory interface is investigated. In the present study the behavior of nonmetallic inclusions at the interface between alumina and a Ca-treated steel (see composition in Table 1) is observed.

**Tabla 1.** Investigated steels [wt.-%].

| Steel | C | Si | Mn | Al | S | Ca |
|---|---|---|---|---|---|---|
| **Ca-treated** | 0.15 | 0.02 | 1.07 | 0.04 | 0.004 | 0.0012 |
| **ULC** | 0,002 | 0,001 | 0.074 | 0,020 | 0,008 | 0,000 |

(a)

(b)

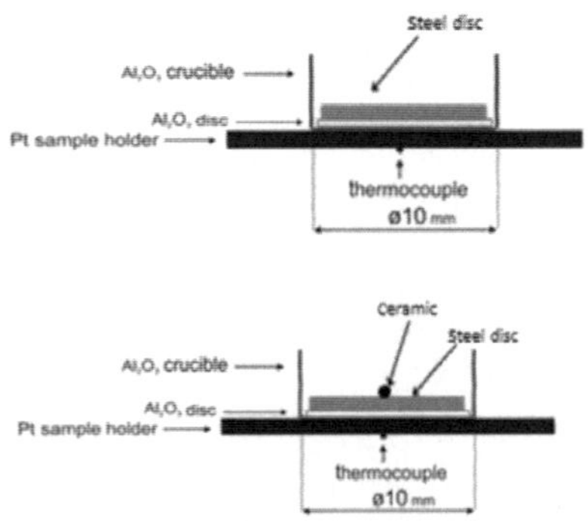


**Figure 3.** Experimental set up. (a) Inclusions agglomeration experiments. (b) Inclusion behavior at the steel/refractory interface.

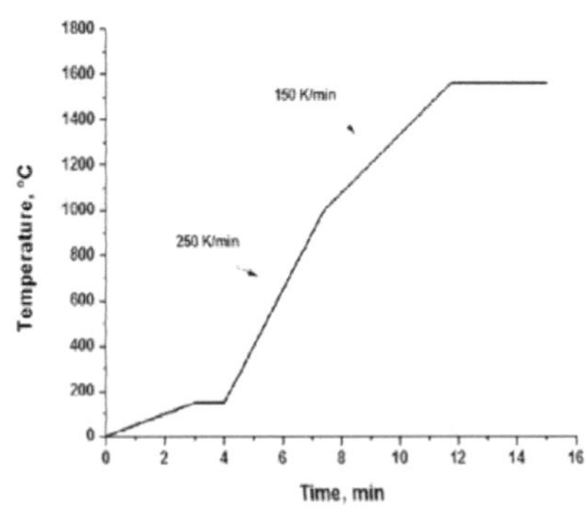


**Figure 4.** Time-temperature profile of the experiments.

A posteriori examination of the sample at room temperature is carried out using a Scanning Electron Microscope with Energy Dispersive XRays (SEM/EDS). In order to characterize (composition, shape, size) the inclusions observed during the experiment, a fast cooling of the sample is important [26].

For the quantitative evaluation of the experimental results, an approach of Yin et al. [20,21] is used. They described and compared the magnitude of the attraction force from measured acceleration of particles with a certain mass based on in-situ observations in the confocal microscope. The latter approach was also used of many other research groups. [12,19,30–32] Details about the evaluation procedure can be found in [20,26].

## 4. Results

### *4.1. Inclusion agglomeration on the molten steel.*

Inclusion behavior in the liquid steel was studied in different steel grades (composition given in Table 1) also involving different inclusion types. Regarding the Ca-treated steel, results of inclusion agglomeration experiments and in-situ images displaying the inclusion movement and the subsequently calculated attraction forces have already been published previously [26]. It was found, that as soon as a large part of the sample surface melted, many inclusion particles emerged from the interior of molten steel bath and floated on the melt surface. The shape of the observed inclusions was mostly observed to be semi-liquid with irregular shape or liquid with globular shape.

Representative examples of detected inclusions analyzed in the SEM after the experiment and the corresponding chemical composition can also be found in a previous publication [26]. Generally, semi-liquid and liquid inclusions do not attract each other.

The contact angle of solid non-metallic inclusions and steel melt usually amounts to more than 90°. The capillary force decreases with decreasing contact angle and smaller density of the particle. The reduction of the contact angle and the density for liquid as well as semi-liquid particles results in smaller capillary forces compared with solid particles in agreement with [21]. Nevertheless, some interaction between the inclusions is observed close to the solidification front. The investigation of the trajectory path and the subsequent calculation of the attraction force for two semi-liquid inclusions showed that the tendency of agglomeration for liquid as well as for semi-liquid inclusions is smaller than for solid inclusions. According to literature [21] the attraction force between two of semi-liquid CA60 particles is approximately $10^{-16}$ -$10^{-15}$ N, which agrees well with results published previously [26].

In contrast, regarding the ULC steel, intense agglomeration, cluster formation and a large number of interactions between particles were observed during the whole experiment. A SEM-analysis of the inclusions after the in-situ experiments proved that solid, irregular shaped $Al_2O_3$ is the predominant inclusion type in this steel.

The experiments showed that alumina inclusions strongly attract to each other; the inclusions grow quickly by agglomeration which is again in good agreement Yin et al. [20,21].

From Figure 5, the calculated attraction force for two solid, irregular shaped alumina inclusion is in the range $10^{-14}$ -$10^{-15}$ N; for a pair of semi-irregular and semi-liquid inclusion the attraction force amounts to between $10^{-15}$ -$10^{-16}$ N. Therefore, the tendency of agglomeration for solid inclusions is higher than for the semi-liquid inclusions. This is again in good agreement with published by Yin et al. [21].

Possible explanations for these observed differences regarding the agglomeration tendency could be:

a) The wettability behavior: Solid inclusions are less wetted than liquid inclusions, this leads to a meniscus formation around the inclusions and therefore the capillary force appears.
b) The velocity of the inclusions: Liquid particles seem to move slower than solid ones and hence the calculated attraction force is lower.
c) The density of the particles.

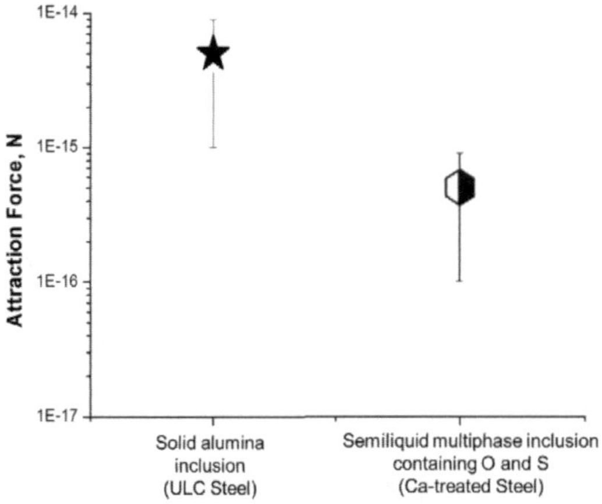


**Figure 5.** Comparison of attraction force between two same inclusion type for different composition/morphology and shapes.

*4.2. Inclusion behavior at the steel/refractory interface.*

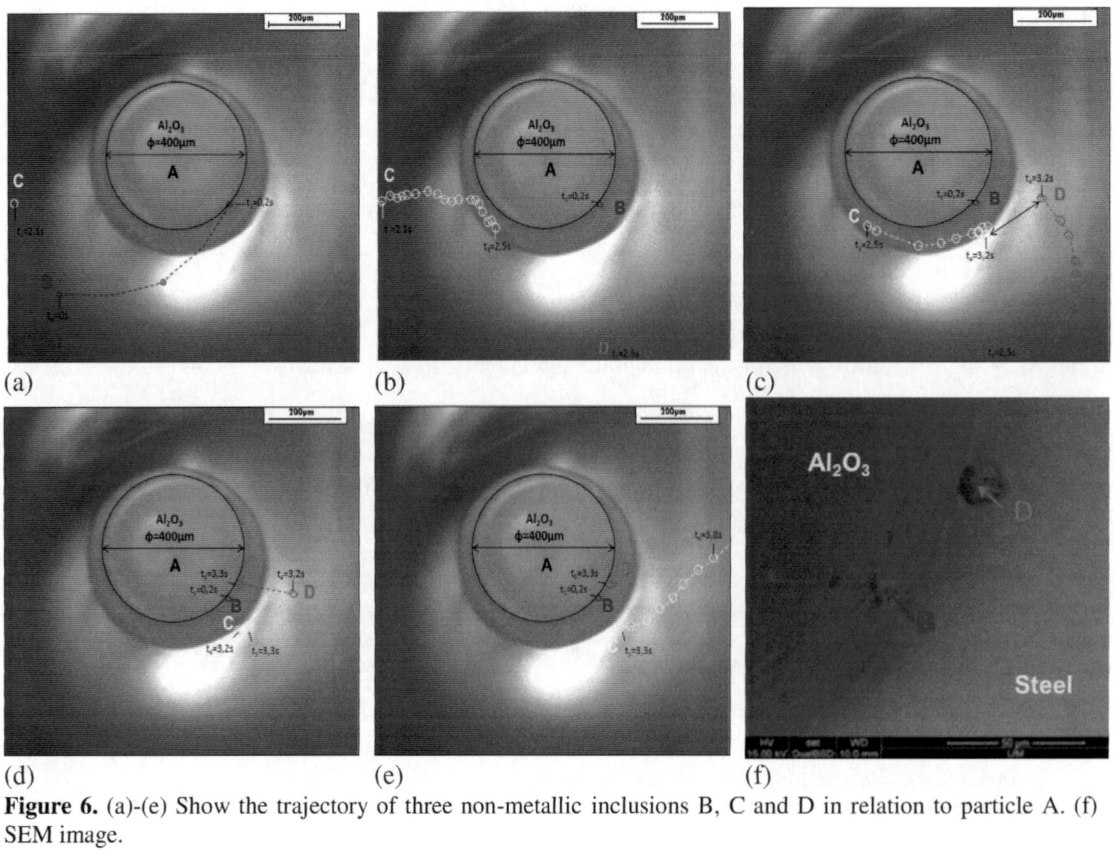


**Figure 6.** (a)-(e) Show the trajectory of three non-metallic inclusions B, C and D in relation to particle A. (f) SEM image.

In a first step, applying Equation 1, the attraction force is calculated modifying the size of one of the particles. The results show that for a 5μm ECD particle the attraction force is constant as soon as the second particle is up to 100μm ECD. For this reason, a 400μm ECD alumina particle is placed on the top of a steel disc in order to observe the inclusions behavior at a flat refractory/steel surface.

From the agglomeration experiments with Ca-treated steel (see section 4.1) is observed that the tendency of agglomeration between semi-irregular semi-liquid inclusions is low. In this section, the behavior of the inclusions from Ca-treated steel at a liquid steel/alumina interface is investigated.

When the Ca-treated steel was melted around the alumina particle, some non-metallic inclusions float up. Different behavior of the particles in relation with the big alumina particle was observed. In Figure 6 (a) a particle B rapidly attaches to the large alumina particle (Particle A), after 2 seconds a globular particle C appears. The trajectory of particle C is observed (see Figure 6 (b)-(e)), it moves towards particle A, but instead of attaching, particle C keeps moving close to A and after some time, it moves away. In Figure 6 (b) a particle D is observed; after 0.8 seconds particle D attaches to A (see Figure 6 (c) and (d)). SEM-analysis of the inclusions after the in-situ experiments proved that inclusions B and D are attached to the alumina surface (See Figure 6 (f)).

The non-metallic inclusions B, C and D have different shapes. Globular particles like C are expected to be liquid and to show a low interaction at the steel/alumina interface as it is observed during the experiment, whereas B and D are semi-irregular semiliquid inclusions which show higher interaction at the steel/alumina interface.

The alumina sphere added to the experiment is non-wetted by the molten steel. The contact angle of B, C and D with the molten steel depends on the inclusions morphology and leads to different deformations of the liquid steel surface. During the experiments, it is observed that non wetting inclusions, like B and D, tend to attach to the alumina surface. However, the better wetted inclusions, like C, approach to the non-wetted surface but they don't attach.

The different morphology and wetting behavior of these non-metallic inclusions may be the cause of for the different behavior observed at the steel/alumina interface.

## 5. Conclusions

Based on the in-situ studies using Laser Scanning Confocal Microscopy the following conclusions regarding the behavior of non-metallic inclusions in the liquid steel and at the steel/refractory interface can be drawn:

- In general, the observation results confirm the common understanding that the wetting angle of liquid iron on non-metallic inclusions or on refractory materials is the dominating influencing factor for adhesion and agglomeration phenomena.
- For this reason, the tendency of agglomeration between solid inclusions (usually with a poor wettability for liquid iron) is higher than for semi-liquid inclusions (usually with a good wettability for liquid iron). Besides this, shape and density of particles are minor influencing parameters.
- The calculated attraction forces between the particles are in good agreement with literature.
- The same conditions apply to the behaviour of particles at refractory/steel interface.
- Future work will comprise the investigation of a wide number of different steel grades with their associated inclusion population in contact with commercial refractory materials for submerged entry nozzles and slide plates. The results will further on be validated in up-scaled, close-to-process experiments and finally help to specify criteria for the development of clogging inert refractory materials.

**Acknowledgements**

Financial support by the Austrian Federal Government and the Styrian Provincial Government, represented by Österreichische Forschungsförderungsgesellschaft mbH and by Steirische Wirtschaftsförderungsgesellschaft mbH within the research activities of the K2 Competence Centre on "Integrated Research in Materials, Processing and Product Engineering", operated by the Materials Center Leoben Forschung GmbH in the framework of the Austrian COMET Competence Centre Programme, is gratefully acknowledged.

**References**

[1] Braun T B, Elliott J F and Flemings M C 1979 *Metall. Trans. B* **10B** 171.
[2] Zhang L, Pluschkell W and Thomas B G 2002 *85th Steelmaking Conference* **85** 463.
[3] Rackers K G and Thomas B G 1995 *78th Steelmaking Conference* 723.
[4] V Vermeulen. B Coletti, B Blanpain, P Wollants and J Vleugels 2002 *ISIJ Int.* **42** 1234.
[5] Nadif M, Burty M, Soulard H, Boher M, Pusse C, Lehmann J and Meyer F 2004 *IISI Study on Clean Steel.*
[6] Wilson F G, Heeson M J and Rawson J D W 1988 *Technical Steel Research.*
[7] Singh S N 1974 *Metall.Trans.* **5** 2165.
[8] Ogibayashi S 1995 *Taikabutsu Overseas* **15**.
[9] Bernhard C, Xia G, Egger M, Pissenberger A and Michelic S 2012 *AISTech.*
[10] Uemura K, Takahashi M, Koyama S and Nitta M 1992 *ISIJ Int.* **32** 150.
[11] Kawecka-Cebula E, Kalicka Z and Wypartowicz J 2006 *Arch. of Metall. and Mater.* **51**.
[12] Aneziris C G, Schroeder C, Emmel M, Schmidt G, Heller H P and Berek H 2013 *Metall. Trans. B* **44B** 954.
[13] Sasai K 2014 *ISIJ Int.* **54** 2780.
[14] Sasai K and Mizukami Y 2001 *ISI Int.* **41** 1331.
[15] Zheng L, Malfliet A, Wollants P, Blanpain B and M Guo 2015 *ICS.*
[16] Mizoguchi T, Ueshima Y, Sugiyama M and Mizukami K 2013 *ISIJ Int.* **53** 639.
[17] Raiber K, Hammerschmid P and Janke D 1995 *ISIJ Int.* **35** 380.
[18] Eustathopoulos N and Drevet B 1994 *J. Phys. III* **4** 1985.
[19] Vantilt S, Coletti B, Blanpain B, Fransaer J, Wollants P and Sridhar S 2004 *ISIJ Int.* **44** 1.
[20] Yin H, Shibate H, Emi T and Suzuki M 1997 *ISIJ Int.* **37** 936.
[21] Yin H, Shibate H, Emi T and Suzuki M 1997 *ISIJ Int.* **37** 946.
[22] Nakajima K and Mizoguchi S 2001 *Metall. and Mater. Trans. B* **32B** 629.
[23] Loder D, Michelic S K and Bernhard C 2014 *MS&T.*
[24] Loder D, Michelic S K, Mayerhofer A, Bernhard C and Dippenaar R J 2014 *Junior Euromat.*
[25] Fuchs N, Krajewski P and Bernhard C 2015 *BHM* **160** 214.
[26] Michelic S K , Diéguez Salgado U and Bernhard C 2015 *LMPC* .
[27] Feichtinger S, Michelic S K, Kang Y, Bernhard C and Jantzen C 2014 *J. Am. Ceram. Soc.* **97** 316.
[28] Michelic S, Goriupp J, Feichtinger S, Kang Y, Bernhard C and Schenk J 2015 *Steel Research Int.* **86**.
[29] Bernhard C, Schider S, Sormann A, Xia G and Ilie S 2011 *BHM* **156** 161.
[30] Coletti B, Vantilt S, Blanpain B and Sridhar S 2003 *Metall. and Mater. Trans. B* **34 B** 533.
[31] Kimura S, Nakajima K and Mizoguchi S 2001 *Metall. and Mater. Trans. B* **32**.
[32] Wikström J, Nakajima K, Shibata H, Tilliander A and Jönsson P 2008 *I&S* **35** 589.

# Simulation of thermos-solutal convection induced macrosegregation in a Sn-10%Pb alloy benchmark during columnar solidification

**Y Zheng[1], M Wu[1,2], A Kharicha[1,2], A Ludwig[1]**

[1]Chair of Modelling and Simulation of Metallurgical Processes, Montanuniversitaet Leoben, Austria
[2]Christian Doppler Lab for Advanced Simulation of Solidification and Melting, Montanuniversitaet Leoben, Austria

E-mail: menghuai.wu@unileoben.ac.at

**Abstract:** In order to investigate the effect of thermo-solutal convection on the formation of macrosegregation during columnar solidification, simulations with a liquid-columnar two phase model were carried out on a 2D rectangular benchmark of Sn-10%Pb alloy. The solidification direction in the benchmark is unidirectional: (1) downwards from top to bottom or (2) upwards from bottom to top. Thermal expansion coefficient, solutal expansion coefficient and liquid diffusion coefficient of the melt are found to be key factors influencing the final macrosegregation. The segregation range and distribution are also strongly influenced by the benchmark configurations, e.g. the solidifying direction (upwards or downwards) and boundary conditions, et al. The global macrosegregation range increases with the velocity magnitude of the melt during the process of solidification.

## 1 Introduction

Macrosegregation is the inhomogeneous solute distribution at the scale of a casting and it is very sensitive to thermal-solutal convection [1]. It is important to understand the macrosegregation, and be able to model and control it because macrosegregation is usually detrimental for castings. Macrosegregation is mainly due to a combination of microsegregation and relative motion between liquid phase and solid phase [2-5]. The relative motion during the solidification is induced by different flow phenomena, for example, equiaxed sedimentation, thermo-solutal convection or feeding flow. In this study it is dedicated to illustrate the delicate mechanism of macrosegregation during solidification with the presence of thermal-solutal convection. This is also an important topic in the annual solidification course in Swiss Federal Institute of Technology of Lausanne (EPFL) [4, 6]. There are lots of works carried out on lateral solidification benchmark [7-11] while the investigation on the vertical-directional solidification is not sufficient. In order to investigate the effect of thermo-solutal convection on the formation of macrosegregation during vertical-unidirectional solidification, a series of simulations with a liquid-columnar two phase model were carried out on 2D rectangular benchmark of Sn-10%Pb alloy. The aim of this paper is to explore the effect of thermo-solutal convection on macrosegregation in detail.

## 2. Numerical model and simulation settings

Details of the two-phase columnar solidification model are described by Wu and Ludwig [12, 13]. The main assumptions are:

a) The two phases in the model are the liquid melt and the solidifying columnar dendrite trunks. The morphology of the columnar dendrite trunks is approximated by step-wise growing cylinders positioned in a staggered arrangement with constant primary dendritic arm spacing, $\lambda_1$.
b) The columnar trunks grow from the casting surface when constitutional undercooling exists and the columnar tip front is tracked explicitly.
c) The liquid-to-solid mass transfer (solidification/melting) rate, $M_{\ell s}$, is calculated as a function of the growth velocity of the columnar trunks, $v_{R_c}$, which is governed by diffusion of the solute in the interdendritic melt surrounding each cylindrical trunk.
d) Thermodynamic equilibrium is assumed at the liquid-solid interface. The concentration difference ($c_\ell^* - c_\ell$) is the driving force for the growth of columnar trunks.
e) Volume-averaged concentrations for each phases ($c_\ell, c_c$) are solved by global species conservation equations. To evaluate macrosegregation, a mixture concentration is defined: $c_{mix}$ = $(c_\ell \rho_\ell f_\ell + c_c \rho_c f_c)/(\rho_\ell f_\ell + \rho_c f_c)$.
f) A linearized binary Sn-Pb phase diagram with a constant solute redistribution coefficient *k* and a constant liquidus slope *m* is used.
g) Interdendritic flow resistance in the mushy zone is calculated via a permeability law according to the Blake-Kozeny [14]. Solidification shrinkage is not accounted for and thermos-solutal convection is modelled with the Boussinesq approach.

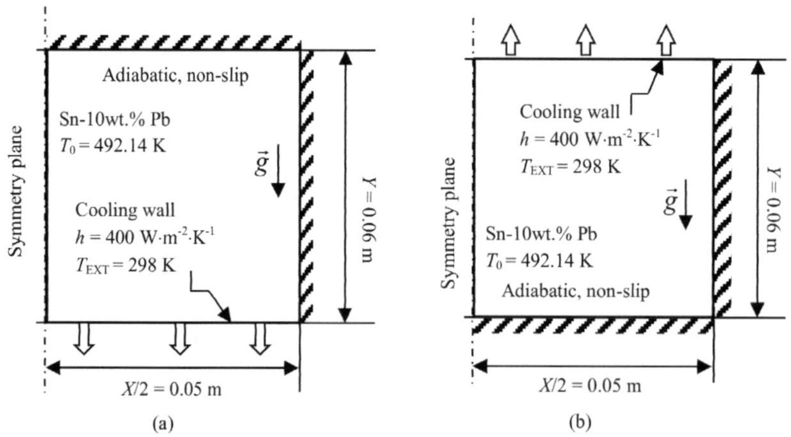


**Figure 1**. Benchmark geometry with boundary and initial conditions.

In this paper, a 'macrosegregation index",

$$c^{index} = \frac{c_{mix} - c_0}{c_0} \times 100 \tag{1}$$

and a "global macrosegregation intensity (GMI)" following reference [15],

$$\text{GMI} = \frac{1}{V_{domain}} \cdot \iiint_{V_{domain}} |c^{index}| dV \tag{2}$$

are used to qualify the macrosegregation and global severity of macrosegregation.

The benchmark simulations are carried out in 2D as shown in Figure 1, which is derived from the one proposed by Bellet et al [16], labelled with boundary and initial conditions. Due to the symmetry, half of the benchmark is modelled and a panel symmetry boundary condition is defined at the left boundary. In order to investigate the influence of the direction of gravity, the heat is extracted on the wall bottom (Figure 1(a)) or top (Figure 1(b)). In order to study the effects of different thermo-solutal parameter on the melt convection and final macrosegregation, Sn-10wt. %Pb alloys is chose. Sn-Pb alloy system is a typical eutectic system and it has been investigated for a long time [17-19]. Furthermore, the thermal dynamic and physical properties are ready to obtain, and they can be found in references [17, 21-23].

There are two cases. Case 1, let the benchmark grow upwards from the bottom. Because the density of lead is much higher than tin, the solute enriched melt is heavier and have the tendency to flow downward. Case 2, let the benchmark grow downwards from top. The important features in the solidification process, such as growing direction, expand coefficient, flow convection, et al, can be investigated from the 2 benchmark cases.

Columnar solidification model is realized numerically with a control-volume based finite volume method through ANSYS FLUENT software version 14.5. Both the liquid and solid share a single pressure field P. The pressure correction equation is obtained from the sum of the normalized mass continuity equations using an extended SIMPLE algorithm [20]. For each time step, 60 iterations are adopted to decrease the normalized residual of concentration, flow quantities and pressure below $10^{-4}$ and enthalpy quantities below $10^{-7}$. The time steps used impact the accuracy and reliability of the numerical results. Due to the complexity of the coupling, there is no formulation to determine the optimal $\Delta t$. It must be determined empirically by testing simulations. Initial time step $10^{-3}$ s is set. The iterations are run to the end of solidification at about 500 s and each simulation would take 2 days or so to complete the simulation on the parallel EVA cluster (2.6 GHz, 8 cores).

## 3. Simulation results

### *3.1 Case 1*

This case considers the solidification of Sn-10wt.%Pb alloy from the bottom. The start of cooling immediately establishes a relatively stable temperature gradient. As soon as the temperature drops below the liquidus, the melt is undercooled, and solidification ($M_{fc} > 0$) begins Figure 2 (a.1). Since gravity force due to both thermal and solutal expansion points downwards, theoretically no flow should occur. Probably due to the numerical reason (inaccuracy or instability), a weak flow with the velocity magnitude of $10^{-6}$ m/s (Figure 2(c.1)) is induced. This weak flow leads to a weak segregation, $c^{\text{index}} \sim 0.28$. A relatively strong negative segregation at the bottom is obtained. This is due to diffusion of the solute in the interdendritic melt. It is known that segregation is originated from solute partitioning at the solidification interface. Given a control volume is isolated from its neighboring volume, there is no convectional and diffusional mass exchange, and the solute partitioning can only cause microsegregation. Thus no macrosegregation would occur under this hypocritical condition. Here diffusional mass exchange is present as well. Therefore, there is mass exchange among neighboring control volumes through diffusion. Pb element is rejected into the interdendritic melt at the solid-liquid interface, causing a concentration gradient in the interdendritic melt. Diffusion in the interdendritic melt leads to formation of the negative segregation in the casting surface at the bottom at the initial solidification stage (Figure 2(d.1)). This negative segregation is slightly strengthened during the rest solidification (Figure 2(d.2), (d.3)). The top positive segregation layer forms in the same mechanism, i.e. diffusion of solute element in the interdendritic melt. In the end of the solidification, the columnar phase fraction has a very narrow range from 0.75 to 0.78, indicating that the solidification is a very steady one. The velocity during the solidification process is relatively small (at the magnitude of $10^{-3}$ m/s), and the final macrosegregation change for Case 1 is about -2.3 ~0.87.

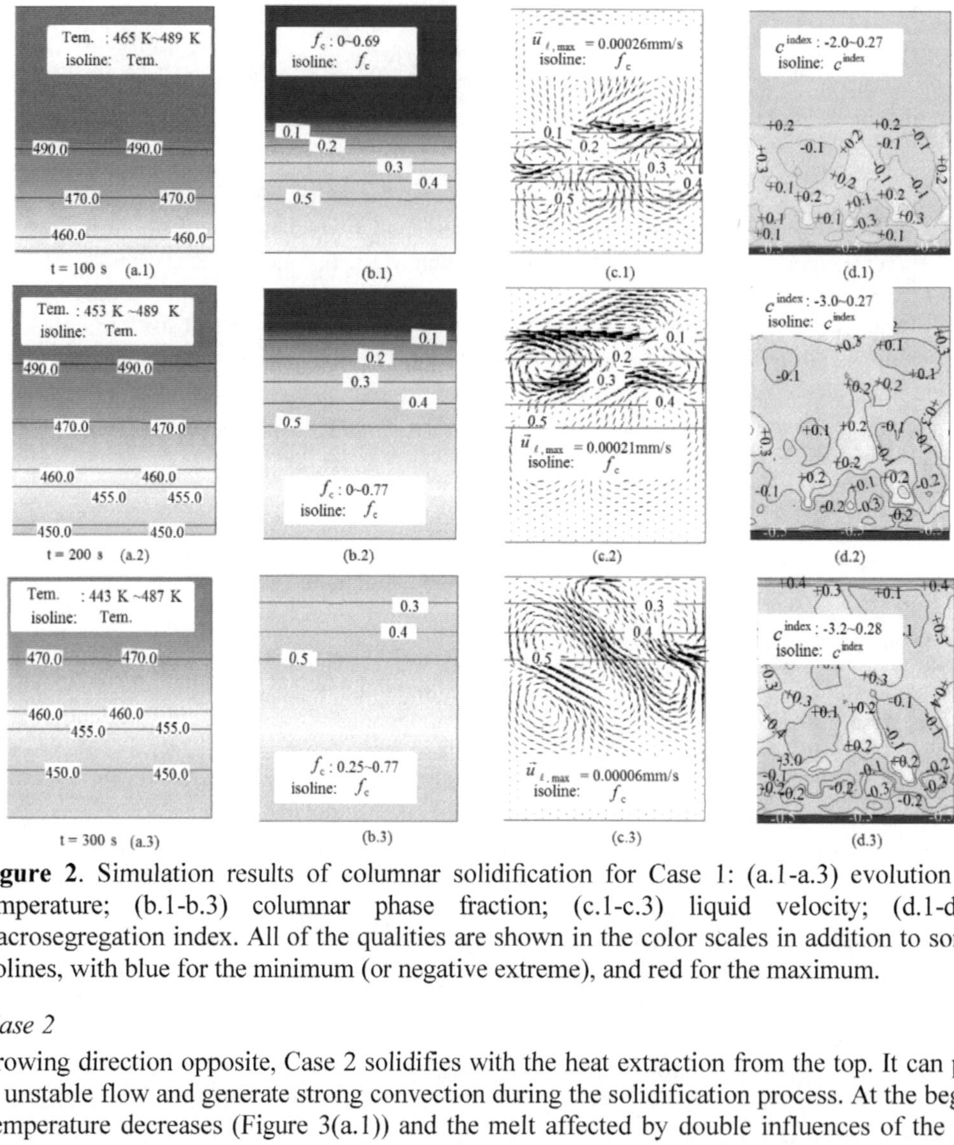

**Figure 2**. Simulation results of columnar solidification for Case 1: (a.1-a.3) evolution of temperature; (b.1-b.3) columnar phase fraction; (c.1-c.3) liquid velocity; (d.1-d.3) macrosegregation index. All of the qualities are shown in the color scales in addition to some isolines, with blue for the minimum (or negative extreme), and red for the maximum.

*3.2 Case 2*

Growing direction opposite, Case 2 solidifies with the heat extraction from the top. It can produce some unstable flow and generate strong convection during the solidification process. At the beginning, the temperature decreases (Figure 3(a.1)) and the melt affected by double influences of the thermal and solutal expansions starts to move downwards. A strong convection forms instantly after the mass transfer begins and the convection becomes more intensively as the phase transfer continues (Figure 3(c.1) and (c.2)). The convections in the melt varied acutely and decompose to more convection cells (vortex), taking on the feature of typical unstable flow and occupying the whole bulk liquid region. Since the permeability effect exists between liquid and columnar phase in the mushy zone, those connections are confined and driven to those area with the liquid fraction more than 0.7, which is shown in Figure 3(c.3). As a result, the velocity in the remained areas decreases to $10^{-4}$ m/s or less, which is ignorable and have no significant influence on the other scalars, such as temperature, species, pressure, et al. the phase fraction evolve gradually from 0 to 0.90 up to 300 s (Figure 3(b.1) - (d.2)).

Accordingly, the variations of macrosegregation are shown in Figure 3(d.1), (d.2) and (d.3). When the solidification starts, as is shown in Figure 3(d.1), channels form in top region adjacent to the chill wall due to the unstable flow, but they are not well pronounced. This macrosegregation is in accordance with the weak flow pattern at the beginning. When the solid-liquid interface moves downwards, some channel segregation become significant. The liquid region has relatively homogenous species distribution without any stratification due to mixing effect of strong convections (Figure 3(d.2) and (d.3)), which stem from the instability of melt flow. The content of Pb becomes larger in the liquid phase as the solidification proceeds. As a consequence a severe positive macrosegregation region ($c^{\text{index}} \sim 124$) remains in the bottom when the solidification is completed. When the solidification time exceed about 400 s, the velocity in the whole domain decrease to $10^{-4}$ m/s or less, which is ignorable and have no significant influence on the other scalars, such as temperature, species, pressure, et al. The final GMI is about 39 and the global macrosegregation fluctuation in Case 2 would range from -58.2 to 156.0. The columnar phase distribution is also not homogeneous, ranging from 0.13 to 0.91. The rest melt solidifies as eutectic.

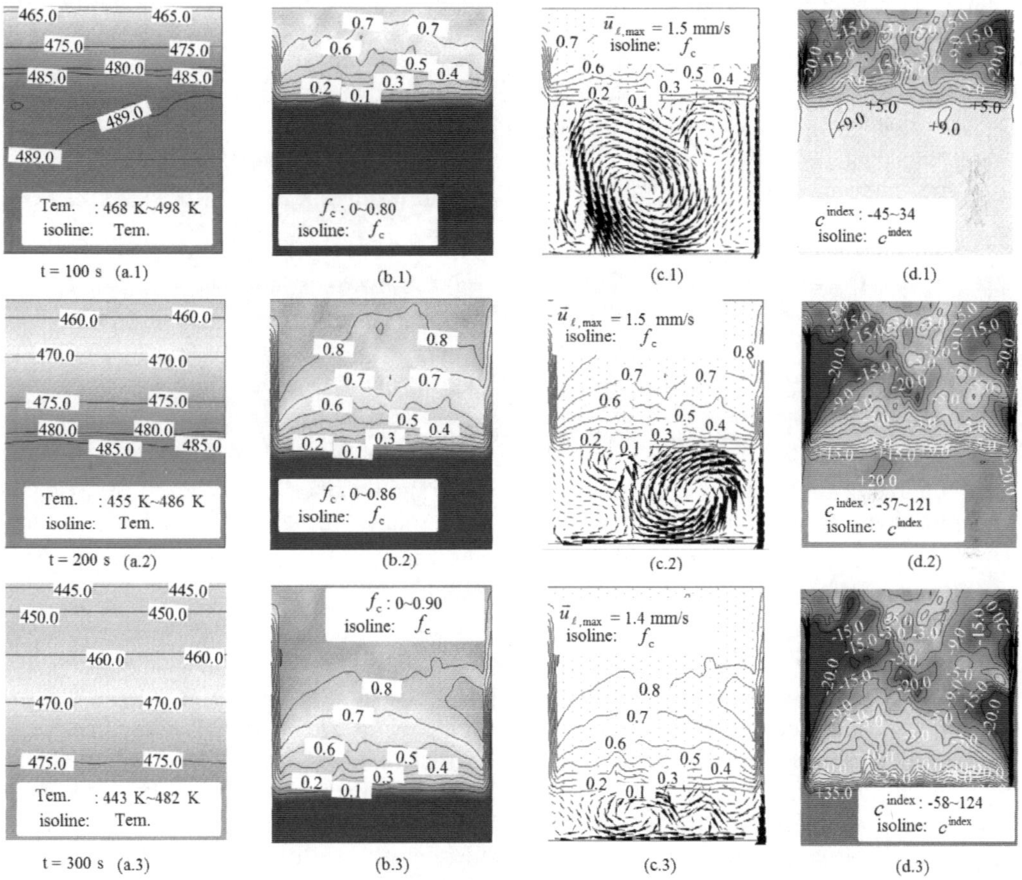


**Figure 3**. Simulation results of columnar solidification for Case 2: (a.1 - a.3) evolution of temperature contour; (b.1 - b.3) columnar phase fraction; (c.1 - c.3) liquid velocity; (d.1 - d.3) macrosegregation index. All of the qualities are shown in the color scales in addition to some isolines, with blue for the minimum (or negative extreme), and red for the maximum.

## 4. Discussion

### *4.1 Segregation by diffusion*

In order to understand the formation mechanism of macrosegregation in detail, here two additional simulations are conducted. Both simulations are based on Case 1. For the simulation 1, the liquid diffusion coefficient is set to a mini value $D_\ell = 1.0\times10^{-20}\mathrm{m}^2\cdot\mathrm{s}^{-1}$ instead of its physical value of $D_\ell = 4.5\times10^{-9}\mathrm{m}^2\cdot\mathrm{s}^{-1}$ to mimic the case without liquid diffusion, and no melt flow is included, i.e. the Navier-Stocks equation is 'switched off'. The simulation result of macrosegregation is presented in Figure 4(a). Since there is no flow and diffusion is also ignored, the mixture concentration $c_{\mathrm{mix}}$ does not change at all over the calculation domain, staying at the value of the initial condition. For the simulation 2, the liquid diffusion is introduced ($D_\ell = 4.5\times10^{-9} m^2\cdot s^{-1}$) while the other condition (no flow) is the same as the last simulation. The macrosegregation contour has obviously stratified, which is shown in Figure 4(b). Pb element is rejected into the interdendritic melt at the solid-liquid interface due to the solute partitioning ($k<1$), causing a concentration gradient in the interdendritic melt. Diffusion in the interdendritic melt leads to formation of the negative segregation in the casting surface at the bottom at the initial solidification stage. Obviously, this kind of macrosegregation is dominated by liquid diffusion. This process is very stable and smooth, resulting in a gradually change of solute concentration over the benchmark, with a negative segregation layer in the bottom and a positive segregation layer in the top. The macrosegregation index range is very narrow, only from -0.7% to 0.6%. Here we can conclude that if there is no liquid flow, or the liquid is merely with ignorable velocity, the final macrosegregation is not significant.

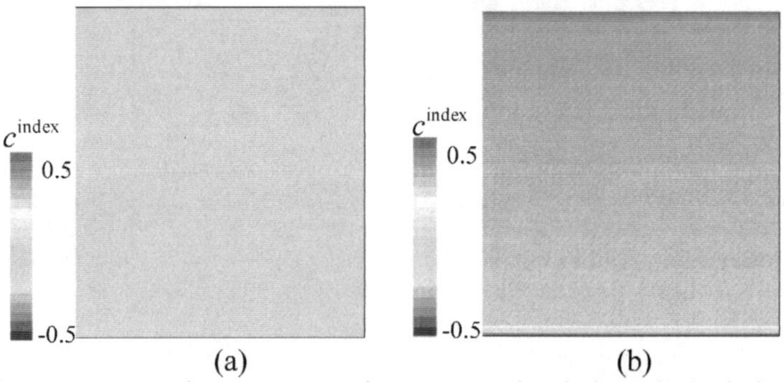


**Figure 4**. Final macrosegregation contours of two more simulations (calculations) based on Case 1: (a) Simulation 1: no flow is calculated and solute diffusion in liquid is ignored ($D_\ell = 1.0\times10^{-20}\mathrm{m}^2\cdot\mathrm{s}^{-1}$); (b) Simulation 2: no flow is calculated, but solute diffusion in liquid is included ($D_\ell = 4.5\times10^{-9}\mathrm{m}^2\cdot\mathrm{s}^{-1}$).

*4.2 Segregation by thermo-solutal convection*

Melt flow is not only sensitive to the thermal physical parameters, such as, thermal expansion coefficient, but also varied remarkably with solidification direction. For Case 2 the dominant factor for thermal solutal convection is the solutal expansion coefficient. For the Case 2, on the one hand, the solutal element Pb has significant potential to drive the liquid in the mushy zone move downwards. On the other hand, the solidification is from top to bottom, the 'heavy liquid' has sufficient space to move, thus presents the unstable flow in mushy zone, to generate different type of convections at different locations in the mushy zone. The interaction between those circulations can initiate some channel segregation.

The interdendritic melt, enriched with Pb, has a higher density than the bulk melt, and both the thermal and solute buoyancy forces lead to a downward flow in the solidification front, resulting in the global obvious convection in liquid region, which is indicated by white arrow in Figure 5(a). The magnitude of the velocity in the convection can reach $2.2\times10^{-2} m/s$. Despite the channels are originated from the unstable convention, complex convections are not all-sufficient conditions for the further growth, which needs some sensible species distribution in the liquid phase. Combining the species conservation equation and the mass conservation equation, and further considering the character of Pb-Sn phase diagram, the follow equitation can be derived:

$$\frac{\partial(c_\ell^* - c_\ell)}{\partial t} = (1-k)\frac{c_\ell}{f_\ell}\frac{\partial f_\ell}{\partial t} + \frac{1}{m}\frac{\partial T}{\partial t} + \vec{u}_\ell \cdot \nabla c_\ell \qquad (3)$$

The third RHS term in Equation 3, $\vec{u}_\ell \cdot \nabla c_\ell$, is the so called flow-solidification interaction term, which is the most critical for growing of the channels. In according to different interdendritic flow, the sign of this term can be positive or negative. Local solidification behavior depends closely on the sign of the flow solidification interaction term. In a region where the velocity direction is same as the liquid concentration gradient, flow-solidification interaction term is positive, vice versa.

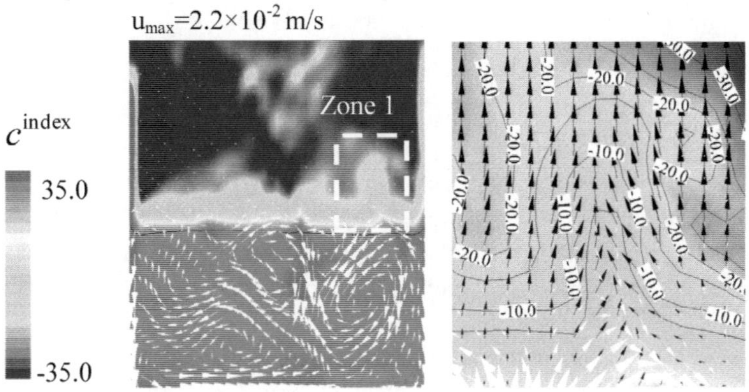


(a) (b)

**Figure 5**. Simulation result for Case 2 at t = 50 s: (a) macrosegregation overlaid by liquid velocity vectors (white); (b) macrosegregation of Zone 1 in (a) overlaid by liquid velocity vectors (white) and gradient of liquid solutal concentration (black).

Channel segregation can form in the mushy zone, which is shown in Zone 1 in Figure 5(a). An enlarged Zone 1 is presented in Figure 5(b). The white arrow indicates the flow direction of liquid, while the black one is the direction of liquid concentration gradient. It can be clearly seen that in the channel, which has higher Pb concentration comparing to the neighboring parts, the direction of liquid velocity and liquid concentration gradient are opposite, resulting in a negative flow-solidification interaction term and retarding the solidification. On the contrary, the above-mentioned two directions are same for the areas on both side of the channel; it means that in those parts of the benchmark solidification is accelerated.

*4.3 Macrosegregation and velocity magnitude*

Some important solidification results are summarized in Table 2 for the above-mentioned four solidification benchmark cases. The distribution of liquid fraction, cooperating the interaction of thermal and solutal convention, gives the possibility to the evolution of velocity during the solidification of benchmark. Moreover, from the investigation in the previous sections the result can be concluded that those cases with large velocity differences have the ability to bear severer macrosegregation. When the flow is not included, the benchmark has homogeneous solutal distribution as initial value. As the velocity magnitude increases, up to range of $10^{-5} \sim 10^{-3} m/s$, the final macrosegregation index is controlled by diffusion mechanism, resulting in mild species fluctuation (ca. 3%). Case 1 is one of the examples. However, when the velocity of the cases, such as Case 2, reached the magnitude about $10^{-2} m/s$, the dominant factor for macrosegregation transfers from diffusion to thermal-solutal flow, and it can result in about 150% of macrosegregation index fluctuation over the domain.

**Table 2.** Statistic of solidification results for the liquid-columnar solidification benchmark.

| Cases | Solidification time (s) | GMI at 600s (%) | Segregation range at 600s (%) | Range of $f_c$ at 600 s (%) | Range of liquid velocity during solidification (m/s) |
|---|---|---|---|---|---|
| Case 1 | 465 | 0.038 | -3.05~0.64 | 75.1~77.2 | 0.0~3.0×$10^{-5}$ |
| Case 2 | 465 | 38.22 | -58.2~156 | 13~91.3 | 0.0~3.0×$10^{-2}$ |
| Case 1without $D_\ell$ | 465 | 0.021 | -0.13~0.3 | 75.2~76.3 | 0.0~1.0×$10^{-5}$ |
| Case 1 without flow | 465 | 0.0 | 0.0 | 75.3~76.3 | 0.0~0.0 |

## 5. Summary

(1) For columnar solidification, thermal expansion and solute expansion coefficients of the melt are the key factors for the melt flow. The thermo-solutal convections and macrosegregation distribution are strongly influenced by the solidification direction.

(2) For columnar solidification model, macrosegregation is relative to convections velocity magnitude. For cases with flow convections velocity less than $10^{-3}$ m/s, the dominant factor for macrosegregation is the element diffusion in the interdendritic melt. The induced macrosegregation index fluctuation is about 3.0.

(3) When convections velocity is about $10^{-2}$ m/s, the dominant factor for macrosegregation is thermo-solutal convection, and it can result in about 150.0 fluctuation of $c^{\mathrm{index}}$ over the domain.

(4) Present study explored the effect of thermo-solutal convection on macrosegregation. It can give a comprehensive understanding to the mechanism of macrosegregation.

**Acknowledgements**

The authors acknowledge the financial support from Austrian Research Promotion Agency (FFG) through the project of Bridge Early Stage (No. 842441), as well as the technical support from the industrial partner Primetals Technologies (former Siemens VAI).

## Reference

[1] Beckermann C 2002 *Int. Mater. Rev.* **47** 243
[2] Flemings M C 2000 *ISIJ Int.* **40** 833
[3] Flemings M C and Nereo G E 1967 *Trans. Metall. Soc. Aime* **239** 1449
[4] Dantzig J A and Rappaz M.EPFL Press, 2009
[5] Li J, Wu M, Ludwig A and Kharicha A 2014 *Int. J. Heat Mass. Tran* **72** 668
[6] Wu M, Domitner J and Ludwig A 2012 *Metall. Mater. Trans. A* **43** 945
[7] Hachani L, Zaidat K and Fautrelle Y 2015 *Int. J. Heat Mass. Tran* **85** 438
[8] Ahmad N, Rappaz J, Rappaz M and Combeau H 1998 *Metall. Mater. Trans. A* **29** 617
[9] Li J, Wu M, Hao J and Ludwig A 2012 *Comp. Mater. Sci.* **55** 407
[10] Li J, Wu M, Hao J, Kharicha A and Ludwig A 2012 *Comp. Mater. Sci.* **55** 419
[11] Wu M, Li J, Ludwig A and Kharicha A 2013 *Comp. Mater. Sci.* **79** 830
[12] Ludwig A and Wu M 2005 *Mater. Sci. Eng. A* **413** 109
[13] Wu M and Ludwig A 2006 *Metall. Mater. Trans. A* **37** 1613
[14] Poirier D 1987 *Metall. Mater. Trans. B* **18** 245
[15] Wu M, Li J, Ludwig A and Kharicha A 2014 *Comp. Mater. Sci.* **92** 267
[16] Bellet M, Combeau H, Fautrelle Y, Gobin D, Rady M, Arquis E, Budenkova O, Dussoubs B, Duterrail Y and Kumar A 2009 *Int. J. Therm. Sci.* **48** 2013
[17] Bergman M I, Fearn D R, Bloxham J and Shannon M C 1997 *Metall. Mater. Trans. A* **28** 859
[18] Sarazin J and Hellawell A 1988 *Metall. Mater. Trans. A* **19** 1861
[19] Hebditch D and Hunt J 1974 *Metall. Mater. Trans.* **5** 1557
[20] Patankar S V.Hemisphere Pub. Corp., 1980

# Mechanical properties of low-alloy-steels with bainitic microstructures and varying carbon content

**A Weber[1,5], J Klarner[2], T Vogl[2], R Schöngrundner[2], G Sam[3] and B Buchmayr[4]**
[1] Materials Center Leoben Forschungs GmbH, Roseggerstraße 12, A-8700 Leoben
Email: andreas.weber@unileoben.ac.at, Österreich
[2] voestalpine Tubulars GmbH &Co KG, Kindberg, Österreich
[3] RAG Rohöl-Aufsuchungs Aktiengesellschaft, Wien, Österreich
[4] Chair of Metal Forming, Department of Product Engineering, Montanuniversitaet Leoben, Österreich

**Abstract.** Materials used in the oilfield industry are subjected to special conditions. These requirements for seamless steel tubes are between the priorities of strength, toughness and sour gas resistance. Steels with bainitic microstructure provide a great opportunity for those harsh environmental conditions. With different morphologies of bainite, like carbide free, upper or lower bainite, the interaction of high tensile strength and elongation is assumed to be better than with tempered martensite. To form carbide free bainite two ways of processing are proposed, isothermal holding with accurate time control or controlled continuous cooling. Both require knowledge of time-temperature transformation behaviour, which can be reached through a detailed alloying concept, focused on the influence of silicon to supress the carbide nucleation and chromium to stabilize the austenite fraction. The present work is based on three alloys with varying silicon and chromium contents. The carbide free microstructure is obtained by a continuous cooling path. Additionally different heat treatments were done to compare the inherent performance of the bainitic morphologies. The bainitic structures were characterized metallographically for their microstructure and the primary phase by means of transmission electron microscopy. The mechanical properties of carbide-free structures were analysed with quasi-static tensile tests and Charpy impact tests. Moreover, investigations about hydrogen embrittlement were done with focus on the effect of retained austenite. The results were ranked and compared qualitatively.

## 1. Introduction

Contemporary developments in the steel industry promise extraordinary balance of strength and toughness. It emphasizes the material as an essential part of modern applications [1]. Seamless steel tubes as an important part are used in structural, mechanical and engineering fields. The largest market share for seamless steel tubes contains oil country tubular good (OCTG) and line pipes [2]. The tense political situation in the Middle East and the changed economic market in summer 2015, led to downgraded forecasts for the demand of steel tubes [2], but investigation and exploitation of unconventional sources of oil will lead to a long-term increased demand [3]. Tools for the oil industry have to fulfil a variety of mechanical and chemical requirements [4], which are influenced by the natural conditions and the drilling techniques. Following requirements are postulated in the literature [5]:

- highest tensile strength,
- low yield-to-tensile ratio,
- high impact toughness at low temperatures (-20 / -40 °C),
- resistance against hydrogen induced cracking (HIC).

The aim of this contribution is to rank low-alloy carbon steels with high strength microstructures to check their practical usability for OCTG production.

### *1.1. Production route*
The typical production for seamless steel tubes includes a subseries of billet heating, cross rolling and piercing, push bench processing, reheating, stretch reduction, elongation and (accelerated) water cooling [6, 7]. Seamless pipes for applications in the oil and gas industry are typically made in the quenched and tempered martensitic condition as it is defined in the widely used API (American Petroleum Institute) standard [8]. For premium qualities, improvements are performed by thermomechanical processing [6].

### *1.2. Sour gas resistance*
Sour gas is a natural or any other gas containing significant amounts of hydrogen sulphide ($H_2S$). Steel is damaged by sour gas by the loss of ductility and formation of cracks. The free hydrogen $H^+$ is hindered to recombine at the material surface by the $HS^-$ from the dissociated $H_2S$. So, it can diffuse into the material and remains at dislocations, phase boundaries and cracks, where it recombines with other $H^+$ to $H_2$. This leads to local pressure and propagates cracks and embrittlement [9, 10].

## 2. Alloy design
The alloys of the present study are focused on martensitic and bainitic morphologies with varying silicon- and chromium contents. The chemical compositions of the investigated materials are shown in table 1. The carbon content in alloy 3 is slightly lowered in comparison with alloy 1 and 2. The silicon content from alloy 2 is by half the content of alloy 1 and 3, the chromium content of alloy 3 is a little bit higher than those of alloy 1 and 2.

**Table 1.** Chemical composition of the investigated steels with martensite-start-temperature ($M_s$).

| | C [wt%] | Si [wt%] | Mn [wt%] | Cr [wt%] | Mo [wt%] | B [ppm] | $M_s$ [°C] |
|---|---|---|---|---|---|---|---|
| **Alloy 1** | 0.25-0.3 | 0.5-1.0 | 1.0-1.5 | 0.5-1.0 | 0.3-0.7 | 5-10 | 360 |
| **Alloy 2** | 0.25-0.3 | 0.1-0.5 | 1.0-1.5 | 0.5-1.0 | 0.3-0.7 | 5-10 | 364 |
| **Alloy 3** | 0.2-0.25 | 0.5-1.0 | 1.0-1.5 | 1.0-1.5 | 0.3-0.7 | 5-10 | 380 |

The transformation points for martensite-start, bainite-start, lower-bainite-start and $A_{c3}$ were estimated with thermodynamical calculations after Buchmayr, 2002 [11]. Based on these calculations the austenitization temperature was set 50 °C higher than $A_{c3}$ (alloy 1 = 940 °C, alloy 2 = 910 °C, alloy 3 = 950 °C) and the holding time at this temperature was fixed with 5 minutes. The alloying concept was validated with the thermodynamic calculation software JMatPro. The melts in laboratory scale were produced (casted, rolled) at voestalpine Stahl GmbH in Linz. Jominy tests have proved a good hardenability of all three alloys. Continuous and isothermal time-temperature diagrams were carried out by a dilatometer Bähr DIL805A at the Chair of Physical Metallurgy and Metallic Materials of the Montanuniversitaet Leoben. The CCT and the TTT-Diagram of alloy 1 are shown in figure 1.

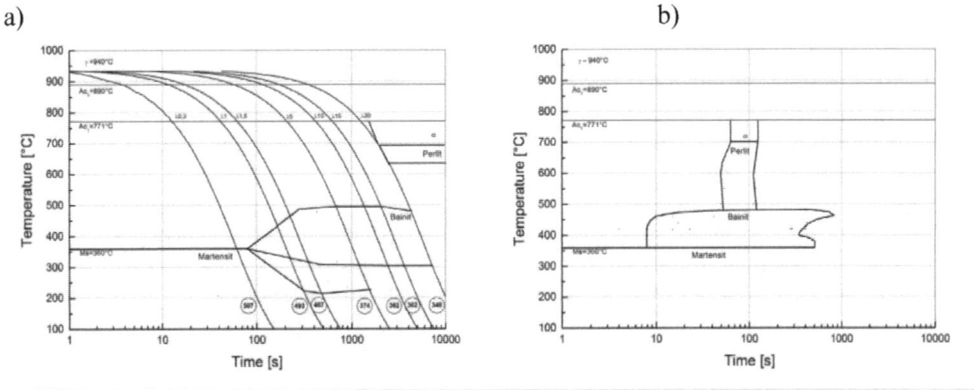


**Figure 1.** a) CCT diagram of alloy 1, b) TTT diagram of alloy 1.

## 3. Experimental

### *3.1. Mechanical properties*

The mechanical properties were determined using tensile tests according to DIN EN 10002-1 and Charpy V-notch impact tests according to DIN EN 10 045-1. The dimensions for the round tensile specimens are shown in figure 2 a). The experiments were carried out on a universal testing machine Zwick/Roell 250 at the Chair of Metal Forming at the Montanuniversitaet. The deformation analysis was done with the optical measurement system GOM-Aramis. With this method it is possible to track the whole deformation path and to identify local strains. Figure 2 b) shows an example of a sample with an overlaid analysis mask (coloured regions). The notch-impact tests were done at the Chair of Metal Forming with a pendulum impact tester from Zwick/Roell RKP 450 with a 300 J hammer.

### *3.2. Heat treatment*

The Charpy specimens were heat treated with a Gleeble 3800 at the Chair of Metal Forming. The heat treatments for the tensile specimens were realized with the inductive FTTU (Fast Thermal Treatment Unit) from the Thermo-Mechanical Treatment Simulator Servotest at the Chair of Metal Forming. All heat treatments are listed in table 2. Four heat treatments were performed, one in as quenched condition with air pressure cooling (approximately 60 seconds to room temperature), a continuously cooled condition following a cooling curve with a defined factor $\lambda 2$ (described in equation (1)), and two tempered conditions. For the quenched and tempered (Q&T) conditions two different cooling rates are performed (as quenched and $\lambda 2$). The tempering was done with reheating 15 $Ks^{-1}$ at 700 °C and holding for 10 minutes.

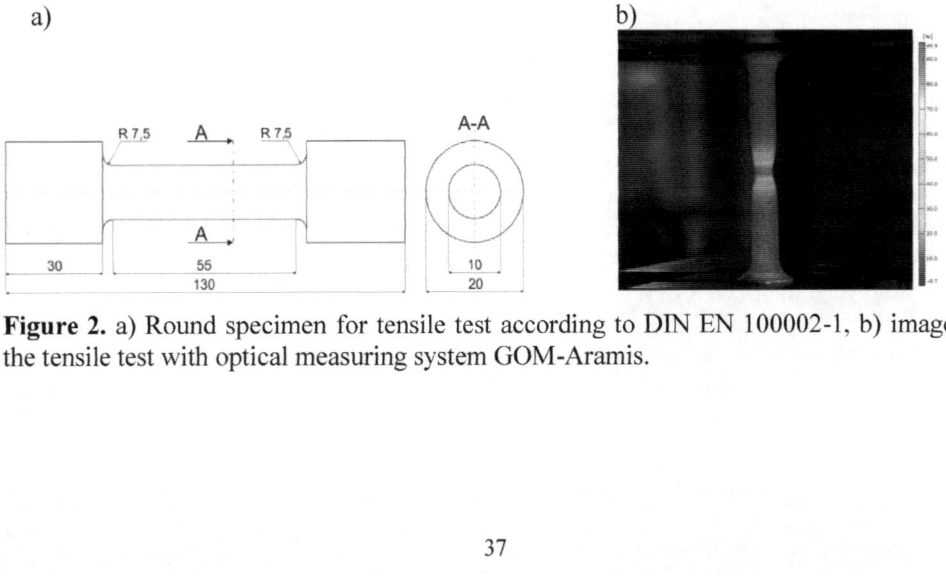


**Figure 2.** a) Round specimen for tensile test according to DIN EN 100002-1, b) image of the tensile test with optical measuring system GOM-Aramis.

**Table 2.** Matrix of heat treatments.

| Alloy 1 | Alloy 2 | Alloy 3 |
|---|---|---|
| as quenched | as quenched | as quenched |
| λ2 | λ2 | λ2 |
| λ2 tempered 700°C 10' | λ2 tempered 700°C 10' | λ2 tempered 700°C 10' |
| Q&T 700°C 10' | Q&T 700°C 10' | Q&T 700°C 10' |

The variable λ (equation (1)) describes the time for the degressive course of the cooling-rate:

$$\lambda = \frac{t_{800°C} - t_{500°C}}{100} \quad (1).$$

Finite element thermal calculations indicated for tube thicknesses between 8 and 24 mm, an average cooling rate in air after the hot rolling mill with a finish rolling temperature between 950–1000 °C of about 0,1–0,5 $Ks^{-1}$ [8]. This leads to a cooling time of approximately 30 minutes and a λ value of about 2, carried out by dilatometric experiments.

*3.3. Hydrogen Induced Cracking*

In addition to the mechanical properties a first estimation of sour service resistance against hydrogen induced cracking were done. Therefore the standard testing method from the NACE Standard TM0284-2011 [12] was modified for smaller samples. The selected samples were laid in a sour-gas environment with defined temperature, chemical composition of the medium and pH-value for 96 hours. After penetration the cross section of the sample is examined with a light microscope (100 x magnification). The identified cracks were measured as shown in figure 3 and related with the crack sensitive ratio (CSR), given in equation (2):

$$CSR = \frac{\sum(a*b)}{(W*T)}*100\,\% \quad (2).$$

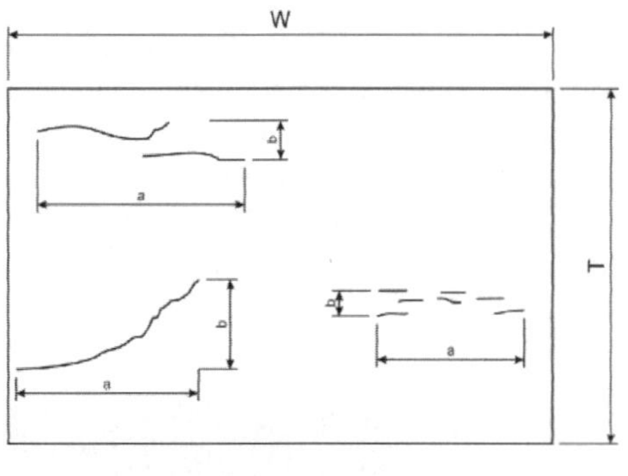


**Figure 3.** Schematic illustration of sour gas-cracks for HIC evaluation, a = crack length, b = crack thickness, W = section width, T = test specimen thickness.

*3.4. Investigations with transmission electron microscope*

To confirm the produced microstructures transmission electron microscope (TEM) investigations were made at the Alloy Development Group Seibersdorf (Dr. Krystina Spiradek-Hahn).

TEM samples were prepared by mechanical thinning and electrolytic polishing. Investigations were done with the transmission electron microscope Philips CM20 STEM and Tecnai F20 (high definition) with acceleration voltage of 200 kV.

## 4. Results and discussion

### *4.1. Microstructure*

It could be shown that for all alloys the as quenched conditions lead to a mainly martensitic morphology with different amounts of retained austenite and some carbides between the laths. Alloy 3 exhibits more coarse carbide-precipitations and partially self-tempered martensite. Figure 4 shows the morphology of alloy 1 and 3 for the as quenched conditions. The dark areas in a) are the martensite-laths, the thick bright dappled needle is a bainite lath. Figure 4 b) shows crossed precipitations, which are typical for tempered martensite. This phenomenon can be explained with the high martensite-start temperature of alloy 3 (listed in table 2) and the reduced carbon content. The first laths are formed at high temperatures, when the carbon has enough time to diffuse and to form precipitations [13]. The light grey film along the grain boundary is identified as retained austenite.

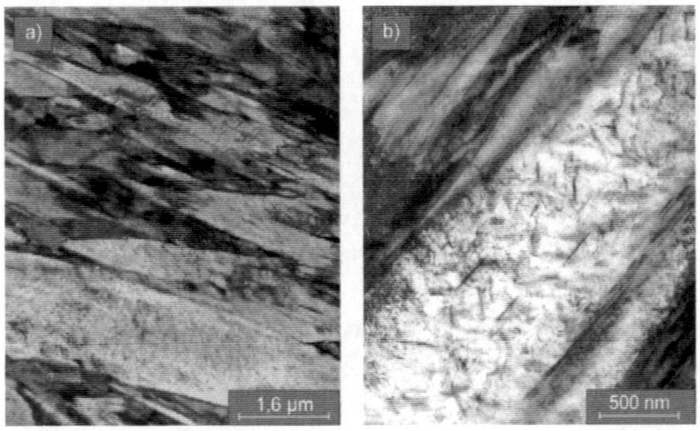


**Figure 4.** TEM micrographs, as quenched conditions, a) alloy 1, b) alloy 3.

For alloy 1 and 3 carbide free bainite could be identified at the continuously cooled conditions with λ2. Light microscope images show a similar structure for alloy 3, but the carbide free morphology is not investigated with TEM. Figure 5 shows the TEM micrograph from alloy 1 with the related diffraction image. Attributable to the lower austenitization temperature it could be seen, that the microstructure of alloy 1 is finer and more structured than alloy 3. It is worth to note, that the microstructures are primarily carbide free bainite, but related to the cooling conditions further bainitic morphologies can occur.

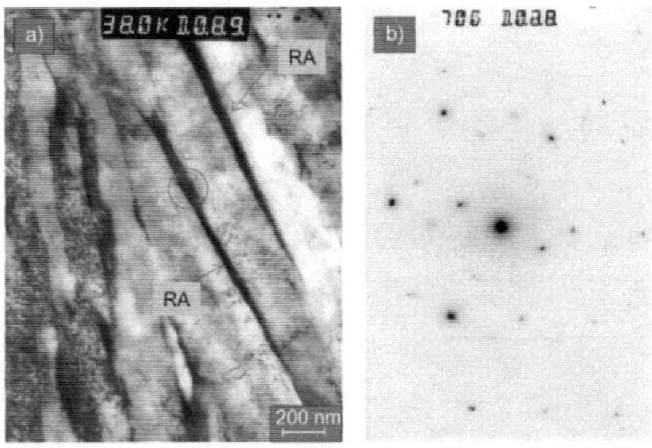


**Figure 5.** Alloy 1 continuously cooled at λ2 condition, a) TEM micrograph of alloy 1, RA=retained austenite, b) diffraction image to verify the carbide free morphology.

### *4.2. Mechanical properties and hydrogen resistance*

Mechanical properties of the alloys with their different morphologies were determined with tensile and Charpy V-notch tests. The results of the tensile tests are shown in table 3. The as quenched conditions show the highest values compared to the other conditions with reduction of area-values close to 40 %. The total elongation of alloy 3 as quenched is 6 % and lies significantly below the elongation of alloy 2 as quenched and alloy 1 as quenched which is 9.5 % and 11% respectively. Regarding to sour gas resistance the CSR values from the as quenched conditions are generally lower in comparison to the other morphologies. It is surprising, that alloy 2 with the highest tensile strength has a lower CSR than alloy 1. This is caused by the lower silicon- and the reduced boron-content in the alloying concept. Alloy 3 shows the lowest crack sensitive ratio. This may be caused by the higher chromium content with respect to the lowered carbon content. It is considered that chromium improves the hydrogen embrittlement resistance by contributing to improve the hydrogen trapping capability and to inhibit the crack propagation. Comparison of impact toughness to tensile strength shows the correlation between these factors. By decreasing strength the impact energy increases significantly. The values of the carbide free (λ2) and carbide free tempered morphologies (λ2&T 700°C 10 minutes) were surprising. Alloy 1, with the highest carbon and silicon content, shows the highest strength with 1413 MPa and 1043 MPa with the lowest total elongation of 9.6 % and 10.5 % respectively in both conditions. The reduction of area, with 52 % and 47 % is lower than for alloy 2 (55 % and 51 % respectively) and alloy 3 (57 % and 60 % respectively). Alloy 2 reveals a tensile strength of about 1255 MPa for λ2 condition and 1009 MPa after tempering. Alloy 3 has a strength about 1186 MPa for λ2 without tempering and 951 MPa for the tempered microstructure. The impact energy for the λ2 conditions is at about 48.4 J for alloy 3 and below 35 J for alloy 1 and 30 J for alloy 2. The total elongation of alloy 3 with cooling condition λ2 is by far highest at about 14.8 %. The values for the yield-to-tensile ratio from as quenched and continuously cooled condition are close to 0.7 and 0.75. Values for seamless steel pipe-applications with bainitic and martensitic microstructures are typically in the range between 0.7 and 0.8 [14].

For the carbide free bainitic microstructures, alloy 1 is most vulnerable to sour gas and hydrogen cracking with a CSR at 24 %. Alloy 2 has a CSR close to 10 % and alloy 3 close to 7 %. For sour gas resistance, the amount of the retained austenite is decisive. Austenite can be seen as a hindrance, it lowers the diffusion of hydrogen significantly. At lower temperatures, the austenite lattice sites behave like traps. There is too less thermal energy to overcome the energy barrier for hydrogen diffusion in austenite [15]. Investigations from Lee et.al. [16] and Choo et.al. [17] have shown, that the capability of austenite to hold hydrogen atoms is stronger than for dislocations. This is why they accumulate at the boundaries and recombines there [18]. As a consequence, morphologies with higher retained austenite content are less resistant against hydrogen induced cracking. Finally all tempered morphologies show a better HIC resistance than in the directly cooled conditions. This leads to the assumption, that the residual stresses may also influence the hydrogen behaviour of the material. It should be noted that the CSR value in this case is only to classify and rank the materials and morphologies. Seamless steel tubes for industrial applications need a HIC resistance of 0 % CSR for accreditation.

The observation for the transition temperature is listed in table 4. It should be noted, that the carbide free morphology (continuous cooling condition λ2) has a much higher transition temperature than with the as quenched and the tempered heat treatments. The other heat treatments fulfil the requirements as described in chapter 1 [5].

**Table 3.** Mechanical properties, tensile strength (TS), reduction of area (Z), total elongation ($A_{50}$), crack sensitive ratio (CSR), impact energy (KV) and yield-to-tensile ratio (YS/TS).

| Heat treatment conditions | TS [MPa] | Z [%] | $A_{50}$ [%] | CSR [%] | KV [J] | YS/TS [-] |
|---|---|---|---|---|---|---|
| **Alloy 1** | | | | | | |
| as quenched | 1675 | 39.8 | 11.0 | 2.4 | 28.4 ±1,4 | 0.74 |
| λ2 | 1413 | 51.9 | 9.6 | 24.1 | 34.4 ±4,0 | 0.69 |
| λ2&T 700°C 10’ | 1043 | 47.4 | 10.5 | 2.9 | 55.2 ±4,0 | 0.89 |
| Q&T 700°C 10’ | 1047 | 57.7 | 11.1 | 4.8 | 56.3 ±1,2 | 0.90 |
| **Alloy 2** | | | | | | |
| as quenched | 1785 | 42.5 | 9.6 | 1.4 | 28.2 ±1,2 | 0.76 |
| λ2 | 1255 | 54.7 | 11.7 | 9.9 | 29.1 ±3,2 | 0.75 |
| λ2&T 700°C 10’ | 1009 | 51.2 | 11.6 | 2.0 | 45.6 ±8,7 | 0.84 |
| Q&T 700°C 10’ | 1012 | 54.2 | 11.0 | 1.4 | 64.5 ±1,2 | 0.90 |
| **Alloy 3** | | | | | | |
| as quenched | 1480 | 38.7 | 6.1 | 0.6 | 39.7 ±1,4 | 0.75 |
| λ2 | 1186 | 56.9 | 14.8 | 7.2 | 48.4 ±0,9 | 0.74 |
| λ2&T 700°C 10’ | 951 | 59.6 | 12.6 | 3.2 | 41.1 ±8,9 | 0.87 |
| Q&T 700°C 10’ | 971 | 55.8 | 12.9 | 5.1 | 82.6 ±3,0 | 0.90 |

**Table 4.** Transition temperature.

| Heat treatment | Alloy 1 | Alloy 2 | Alloy 3 |
|---|---|---|---|
| as quenched | -16 °C | -27 °C | -35 °C |
| λ2 | -10 °C | 14 °C | -24 °C |
| λ2 tempered 700°C 10’ | -27 °C | -37 °C | -11 °C |
| Q&T 700°C 10’ | -36 °C | -60 °C | -48 °C |

## 5. Conclusion

Three low-alloy steels and their suitability for OCTG-applications were compared quantitatively. The alloy-concepts were strictly designed to ensure the desired microstructures. The laboratory heat treatment simulations follow commonly used industrial heat treatment routes in the field of seamless steel production. The as quenched heat treatments can be realized with an accelerated pipe cooling system. The alignment of a certain amount of pipes on the rake type cooling bed influences the continuous cooling time and further bainitic morphologies of a single pipe. A denser assignment leads to slower cooling rates of the pipes and further different morphologies. The comparison between the high strength morphologies have shown, that ultimate tensile strengths close to 1200 MPa with total elongations close to 15 % and reduction of area of about 55 % can be realized. Increasing the strength up to 1800 MPa with ductility values close to 40 % is easy to fulfil. Further, investigations to compare morphologies and classify them regarding the hydrogen resistance were done. It is shown, that the carbide free microstructures do not lead to a sufficient better result in case of hydrogen embrittlement as originally proposed. The mixed microstructures with martensitic/ bainitic-phases show however an excellent strength and toughness behaviour. It can be postulated that the retained austenite and the residual stresses have a major influence on the hydrogen induced cracking.

**Acknowledgements**
Financial support by the Austrian Federal Government (in particular from Bundesministerium für Verkehr, Innovation und Technologie and Bundesministerium für Wissenschaft, Forschung und Wirtschaft) represented by Österreichische Forschungsförderungsgesellschaft mbH and the Styrian and the Tyrolean Provincial Government, represented by Steirische Wirtschaftsförderungsgesellschaft mbH and Standortagentur Tirol, within the framework of the COMET Funding Programme is

gratefully acknowledged. Special thanks to Mrs. Dr. Spiradek-Hahn and Mr. Ing. Manfred Brabetz from the Montanuniversitaet Leoben for material characterizations with TEM. The authors gratefully acknowledge to the industrial-partners voestalpine tubulars GmbH & Co KG and RAG Rohöl-Aufsuchungs Aktiengesellschaft for support and discussion.

**References**

[1] Bleck W 2014 Advanced High Strength Steels for the automotive industry – from microstructures to nanostructures *stahl und eisen* 134 **7** 25-34, 2014

[2] Wirtschaftsvereinigung Stahlrohre e.V.: http://www.wv-stahlrohre.de/markt.html (22.01.2015)

[3] md/k 2014 Frackingboom und Schwellenländer treiben Stahlrohrnachfrage *stahl und eisen* 134 **3** 20-22

[4] Winter G, Klarner J, Keckes J and Buchmayr B 2015 *Proc. XXXIV. Verformungskundliches Kolloquium (Zauchensse)* (Chair of Metal Forming, Montanuniversitaet Leoben) p 203-209

[5] Kern A, Schriever U and Steinbeck G 2010 *Proc. of the Twentieth (2010) International Offshore and Polar Engineering Conference* (Beijing) ISBN 978-1-880653-77-7 (Set)

[6] Klarner J 2009 Thermomechanische Behandlung beim Walzen von nahtlosen Stahlrohren *PhD Thesis*

[7] Klarner J, Buchmayr B and Rainer W 2011 *BHM* 156 **5** 168–174

[8] Gomez G, Perez T, Bhadeshia H K D H 2009 *Mater. Sci. Technol.* **25** 1508-1512

[9] Zapffe C A 1941 Hydrogen embrittlement, internal stress and defects in steel *AIME 145* 225–271

[10] Song E J, Bhadeshia H K D H, Suh D 2013 Effect of hydrogen on the surface energy of ferrite and austenite *Corrosion Science* **77** 379–384

[11] Buchmayr B 2002 *Werkstoff – und Produktionstechnik mit Mathcad, Modellierung und Simulation* Springer (Berlin) 3-540-43014-8

[12] NACE Standard TM0284-2011 Item No. 21215 *Evaluation of Pipeline and Pressure Vessel Steels for Resistance to Hydrogen-Induced Cracking,* 2011.

[13] Sackl S, Clemens H and Primig S 2015 Investigation of the Self Tempering Effect of Martensite by Means of Atom Probe Tomography *PM 52* **7** 374-383

[14] Gomez R, Pérez T E, Bhadeshia H 2010 High strength bainitic steel for OCTG applications, Patent US 2010/0294401 A1

[15] Park Y D, Maroef I S, Landau A, Olson D L 2002 Retained austenite as a hydrogen trap in steel welds *Welding Journal* 27–35

[16] Lee H G and Lee J Y 1984 Hydrogen trapping by TiC particels in iron *Acta Metall.* **32** (1) 131-136

[17] Choo W Y and Lee J Y 1982 Thermal analysis of trapped hydrogen in pure iron *Metall. Trans* A **13** A 135-140

[18] Fielding L, Song E, Han D, Bhadeshia H K D H 2014 Hydrogen diffusion and the percolation of austenite in nanostructured bainitic steel *Proceedings of the Royal Society* A **470** 1–32

# idRHa+ProMod – Rail Hardening Control System

**L Ferro**[1]

Research&Development Engineer,
[1] Primetals Technologies Italy, Marnate, IT

E-mail: lorenzo.ferro@primetals.com

**Abstract**. idRHa+ProMod is the process control system developed by Primetals Technologies to foresee the thermo-mechanical evolution and micro-structural composition of rail steels subjected to slack quenching into idRHa+ Rail Hardening equipments in a simulation environment. This tool can be used both off-line or in-line, giving the user the chance to test and study the best cooling strategies or letting the automatic control system free to adjust the proper cooling recipe. Optimization criteria have been tailored in order to determine the best cooling conditions according to the metallurgical requirements imposed by the main rail standards and also taking into account the elastoplastic bending phenomena occurring during all stages of the head hardening process. The computational core of idRHa+ProMod is a thermal finite element procedure coupled with special algorithms developed to work out the main thermo-physical properties of steel, to predict the non-isothermal austenite decomposition into all the relevant phases and subsequently to evaluate the amount of latent heat of transformation released, the compound thermal expansion coefficient and the amount of plastic deformation in the material. Air mist and air blades boundary conditions have been carefully investigated by means of pilot plant tests aimed to study the jet impingement on rail surfaces and the cooling efficiency at all working conditions. Heat transfer coefficients have been further checked and adjusted directly on field during commissioning.
*idRHa+ is a trademark of Primetals Technologies Italy Srl*

## 1. Introduction

Rail market is increasingly asking for Premium rails, characterized by improved mechanical properties required to extend the useful lifecycle of high-speed and heavy-traffic railways. The answer of Primetals Technologies to the market call is idRHa+® an innovative and flexible rail hardening system, whose first application was installed at Baogang Iron and Steel plant in Baotou, China.

The system is able to in-line process both pearlitic and bainitic rails, with symmetric and asymmetric profiles, with a process velocity from 0.4 to 0.9 m/s and an inlet temperature range, from 700 to 850 °C. Depending on the requirements, several different cooling paths are to be followed to achieve the best metallurgical microstructure which requires a wide cooling regulation flexibility. This is made possible by special air-mist nozzles, which can modulate different water droplet sizes and velocities, as well as by high efficiency air blades which can decrease temperature in a very smooth way and complete the cooling process.

In order to satisfy market and producers' needs, Primetals Technologies has developed the software idRHa+ProMod which, basically, simulates the in-line rail hardening process. The software is a thermo-metallurgical model based on a 2D finite element simulator coupled with a routine which models the austenite decomposition into final product phases, and the subsequent release of latent heat occurring during the phase change. The software can work both off-line or in-line. The off-line mode allows the user to test different cooling strategies and to optimize all the process parameters. The in-line mode calculates the sensitivity of the system, compares the calculated results with the in-line pyrometric readings and automatically adjusts the pressure settings whether the temperature targets were not achieved due to changes in the nozzles behavior, due to wear or cloggings.

## 2. idRHA+ plant configuration with idRHa+ ProMod

Due to its modular design, idRHa+ProMod is very flexible and can be easily applied to any kind of plant configuration, having the possibility to input all the following components:

- Induction heating furnace units, position and length of each one.
- Plant sections, defined as the idRHa+ portions between two consecutive pyrometers and comprising some cooling modules.
- Control groups (CG), defined as ensembles of cooling modules sharing the same air and water pressure setting
- Cooling modules, both air mist or air blade ones
- Nozzles collectors, corresponding to rows of equal spaced or unequal spaced nozzles, with a proper orientation in space

Upper and lateral cooling section positions determine the way the rail head is impacted by the spray. In order to have a good spray coverage, idRHa+ProMod furnishes a view of the surfaces directly impacted by the jet (figure 1). This feature can be useful to decide a priori the upper collector height and then to optimize the jet impingement according to the simulated thermal paths.

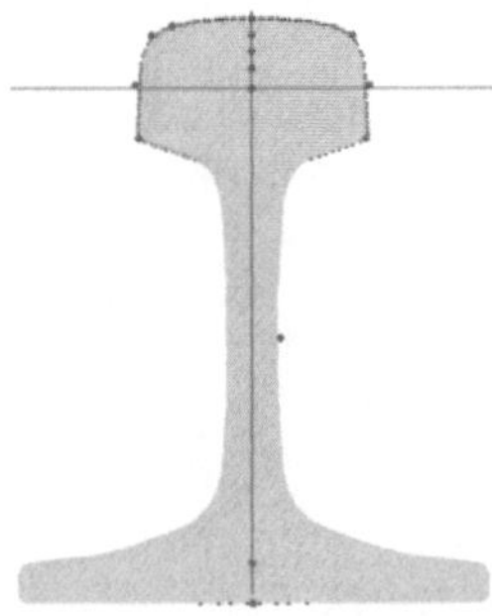

Figure 1. Spray impingement

## 3. Thermal model

The Fourier heat conduction equation has been integrated by a 2D finite element approach in order to achieve a very fast and accurate thermal model. The method permits to correctly represent every different rail shape, also making the application of the boundary conditions easier and accurate. The weak formulation of the differential problem was used.

4 nodes linear isoparametric elements with full integration [1] have been chosen in order to use the minimum number of elements (figure 2) keeping a very good representation of internal temperature gradients, particularly under the rail head surface.

The finite element algorithm has been implemented in ANSI C so as to achieve the maximum performance, particularly useful for in-line calculation purposes.

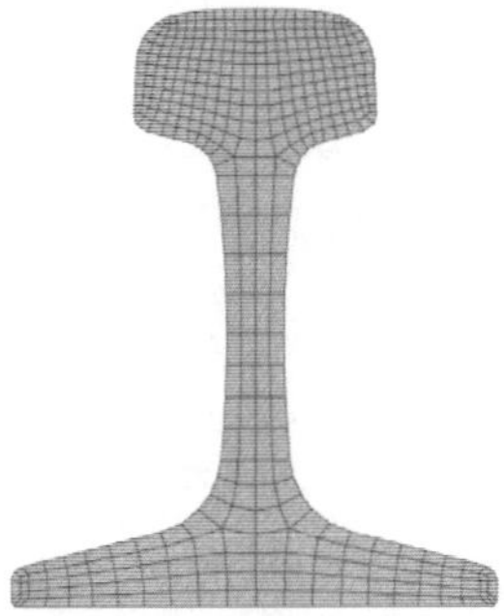

Figure 2. Mesh for a rail with 65 kg/m linear weight

## 4. Phase change model and thermo-mechanical material properties

For each rail material, the modeling of the metallurgical transformations by idRHa+ProMod requires the insertion of the proper Isothermal Transformation (IT) diagram. Nucleation and growth processes of diffusive, together with displacive transformations have been modelled (figure 3) according to the Scheil's Additivity Rule and the Johnson-Mehl-Avrami-Kolmogorov model (figure 4), and the amount of product phase can be evaluated as follows [2]:

$$F_i(T) = 1.0 - exp \cdot \left[-a(T) \cdot t(T)n^{(T)}\right]$$

where $F_i$ is the volume fraction of the product phase i, t is the time elapsed since the transformation beginning, i.e. after the nucleation process, a and n are material parameters dependent on the isothermal temperature and the forming phase. The diffusional coefficient a and the transformation exponent n can be evaluated by the IT diagram of the considered material [3].

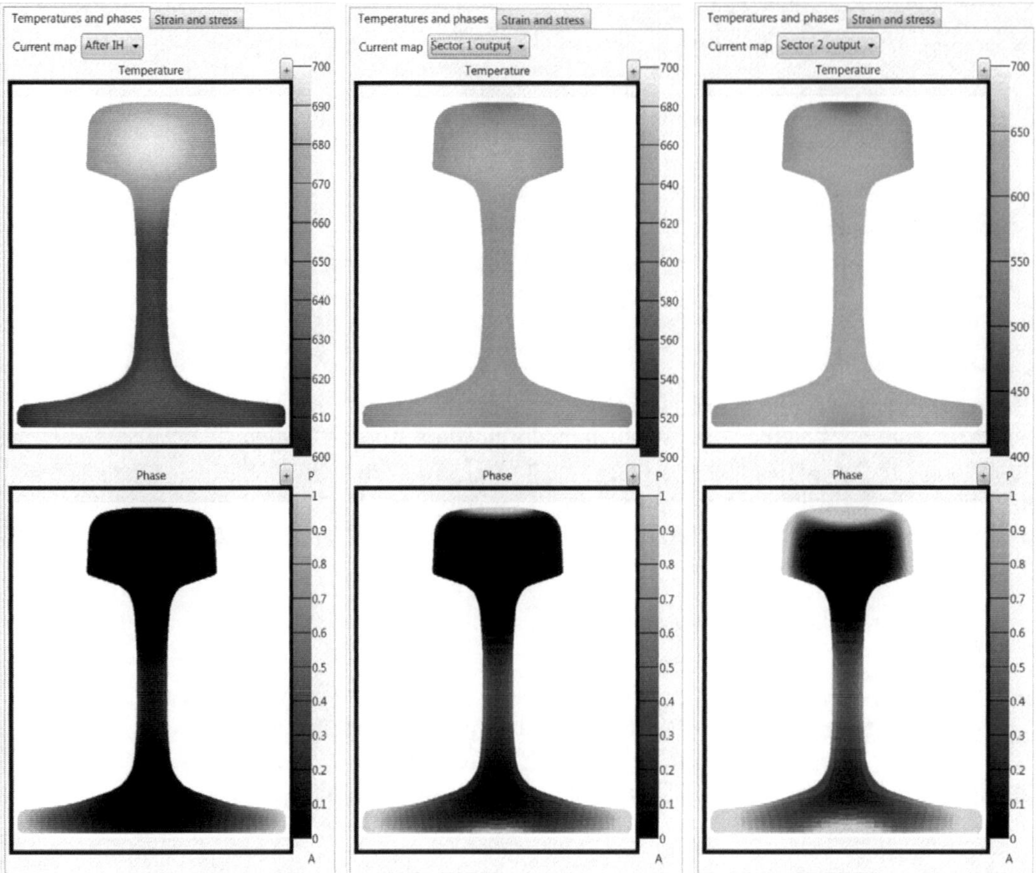


Figure 3. Rail internal field of temperature (top) and phase transformation (bottom) for the first two cooling sectors

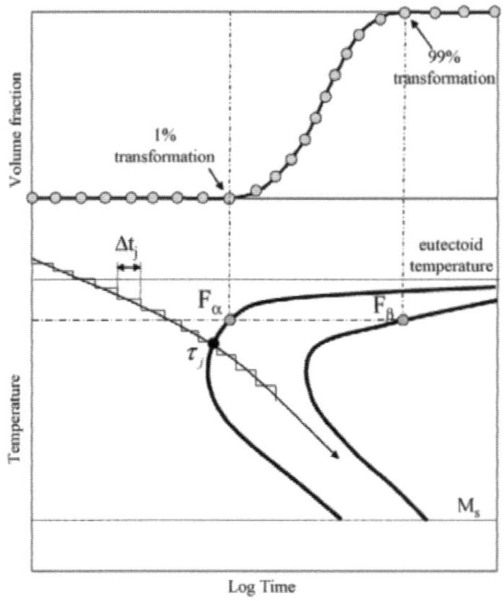


Figure 4. Scheil's additivity rule (from TTT to CCT)

The latent heat due to solid phase transformations has been taken into account and included into the energy balance equation, using the following heat generation rate form:

$$u_{gen} = \sum_i \Delta H_{i,j} \frac{\Delta F_{i,j}}{\Delta t_j}$$

where $\Delta H_{i,j}$ is the heat of transformation of the phase i formed during the time step j, at the temperature $T_j$; $\Delta F_{i,j}$ is the volume fraction of the transformed i phase, relatively to the same time step, and $\Delta t_{i,j}$ is duration of the time step j.

The amount of latent heat released during transformation from austenite to pearlite, bainite and martensite has been measured by calorimetric tests on several rail samples of different steel grades. Tests have been conducted at the Metallurgy laboratories of Politecnico di Milano (Polytechnic University of Milan).

On the one hand, in order to optimize the solution routine, saving computational time, the thermal and phase change models have been decoupled, and a different time step, for each model, has been adopted. On the other hand, in order to o reduce computational costs, a relatively larger time step has been used for the microstructural model, without any appreciable accuracy decreasing.

Temperature dependent material properties, taken from literature formulas [4] or tables [5], have been evaluated and updated at quite large time steps, so as to avoid frequent time consuming FEM matrix assemblies operations. Density, thermal conductivity, specific heat, yield stress and Young modulus have been estimated for each metallurgical phase available and have been further mixed together at each element integration point by means of a rule of mixture [4], taking into account the volume fractions of the formed phases.

## 5. Bending Estimation

Rail inflection deserves a particular attention, since a proper evaluation of the bending radius avoids clearance problems with the rail entering the cooling system. Rail curvature is affected by two opposite physical effects: the thermal contraction of metal due to the fast and inhomogeneous temperature reduction and the volume expansion due to the metallurgical transformations. A tradeoff is not easy to be evaluated a priori, because it strongly depends on the rail inlet conditions (steel grade, rail shape, temperature and phase content, in particular the amount of transformations already started on foot). Curvature evolution does not linearly depend on the cooling intensity, because the internal rail section stress and strain are affected by non-linear temperature dependent transformation rates and unfortunately also by the previous sectors cooling policy.

The internal stress and strain state, and consequently the integrated elastic-plastic bending curvature are calculated by the software along the length of the idRHa+ system (figure 5). A 2D approximation has been adopted and so rail section is supposed to remain plane during inflection. Rail section curvature (figure 6) and longitudinal contraction have been calculated by a fast iterative solver, making also considerations about the self weight contribution and running iterations till the achievement of the proper equilibrium of section forces and moments.

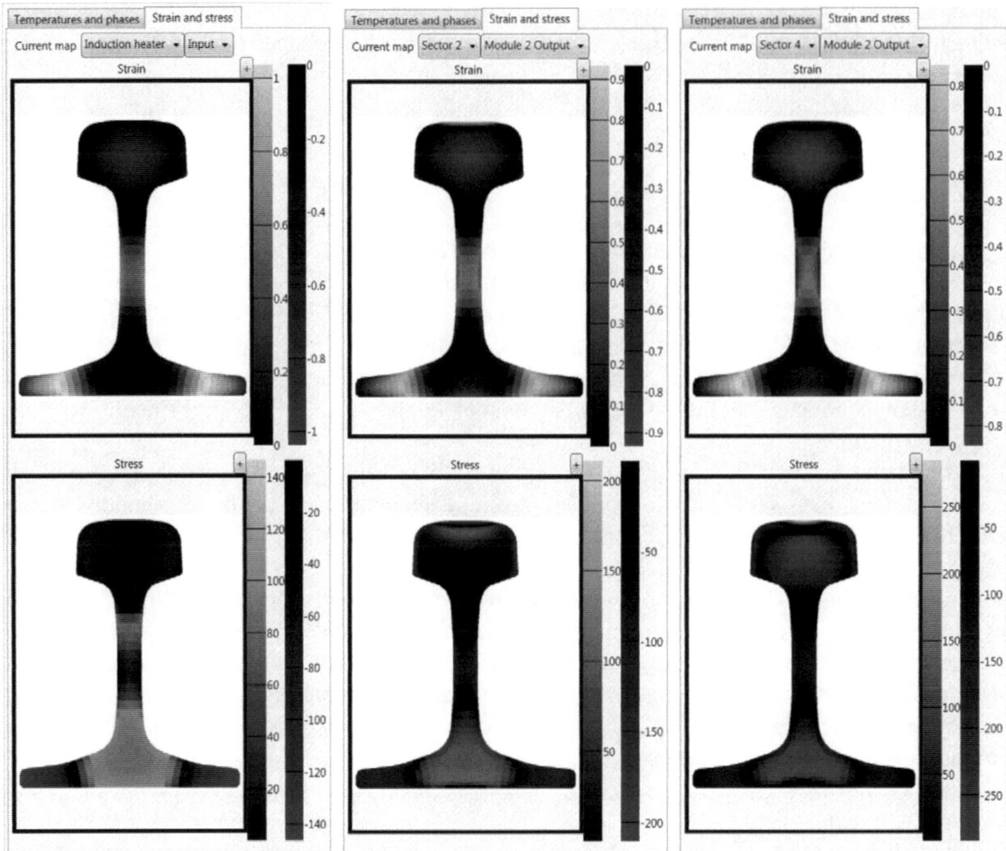


Figure 5. Evolution of rail internal field of plastic strain (top) and elastic stress (bottom) during heat treatment

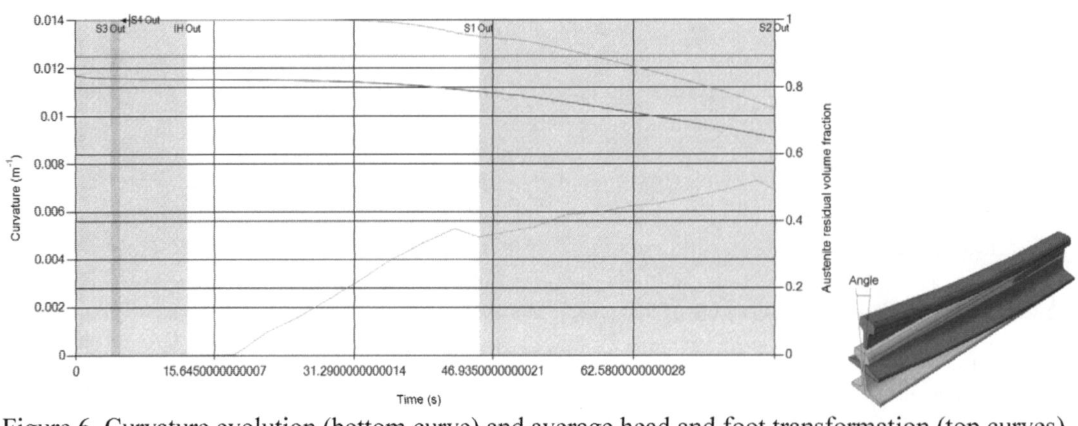


Figure 6. Curvature evolution (bottom curve) and average head and foot transformation (top curves)

The maximum curvature limit can be chosen by the user or, alternatively, default values, based on geometrical equipment limitations, can be kept. Curvature could be interpreted as the rotation angle (radians) formed by a 1 m long rail, as shown in figure 6. Automatically, idRHa+ProMod can simulate several scenarios in order to find the best pressure setting, taking into consideration both the optimal metallurgical structure and risks associated with excessive bending.

## 6. HTC measurement and modeling

Laboratory trials conducted by CSM (Centro Sviluppo Materiali) in Italy at the pilot plant in Dalmine (Italy) permitted the evaluation of the HTC (heat transfer coefficient) of both mist nozzles and air blades.

Cooling tests were performed on real rail specimens and on an austenitic plate, equipped with several thermocouples. The samples were heated up to 900 °C and then cooled down to ambient temperature using different nozzles pressures. All trials were modeled by a commercial finite element software and related HTCs were obtained by inverse modeling techniques (figure 7).

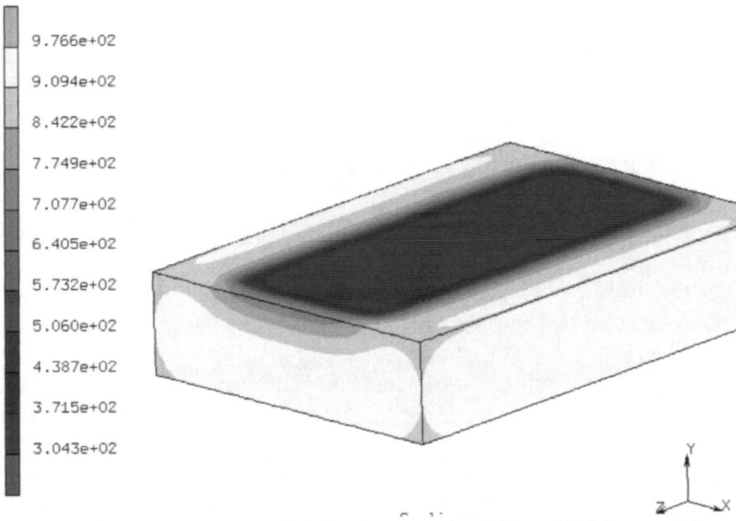


Figure 7. FEM model for inverse modeling (Celsius scale)

Air mist jet properties like water specific flows, droplet velocities and Sauter Mean Diameter ($D_{32}$) were measured (figure 8) and a comprehensive HTC correlation as function of air and water pressures was found (figure 9), based on the following literature equation [6]:

$$HTC \propto D_{32}^{k} \cdot v_{imp}^{l} \cdot N^{m}$$

where $v_{imp}$ and N are the impinging velocity of droplet and the number density of droplets, respectively.

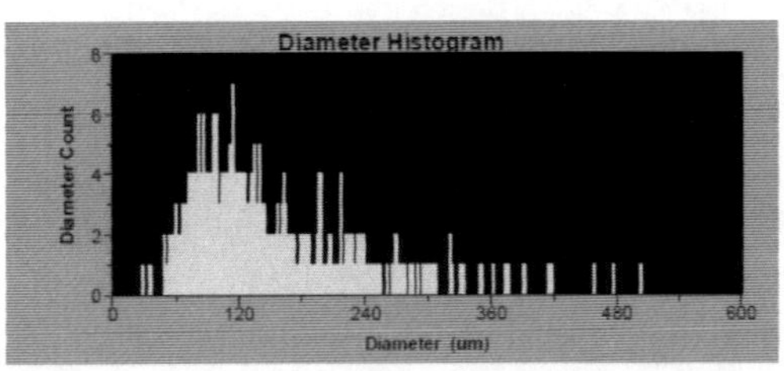


Figure 8. Sauter Mean Diameter measures

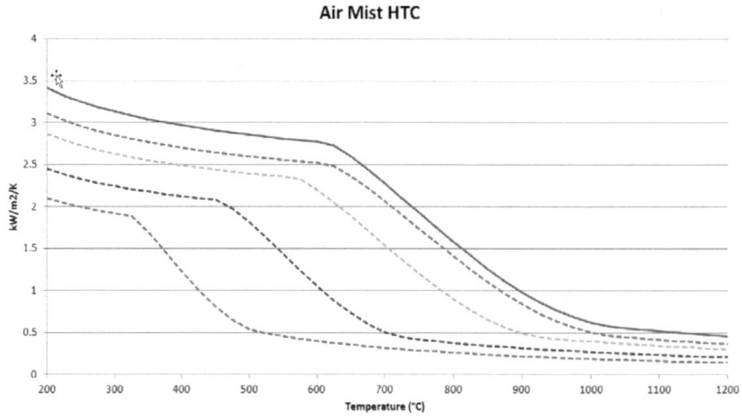


Figure 9. HTC values at different working pressures

Similarly, pressures and velocities of air blade jets were correlated to the experimental HTC values.

## 7. Induction heating modeling

Being the idRHa+ system equipped with induction heating furnaces, so as to homogenize the temperature along the rail length, a simplified and fast electromagnetic simulation model was tailored on the base of preliminary FEM analysis. This analysis was carried out by the induction heating system supplier with the use of the commercial finite element software ANSYS. Thus, it was possible to find geometrical correlations between the full FEM analysis and the well-known electromagnetic penetration depth formula,

$$\delta = \sqrt{\frac{\rho}{\pi \cdot \mu \cdot f}}$$

| | | |
|---|---|---|
| ρ | : | resistivity [Ω.m] |
| μ | : | magnetical permeability [H/m] (μ=μo.μr) |
| t | : | frequency [Hz] |

which is the solution of Maxwell's equations for semi-infinite bodies subjected to a uniform magnetic field but it's also suitable for cylindrical bodies with diameter bigger than δ. The specific heat produced inside the whole rail section was calculated at each element integration point and then summed up to the $u_{gen}$ term of the thermal model (figure 10). Results of the simplified electromagnetic model were in agreement with the full ANSYS FEM analysis.

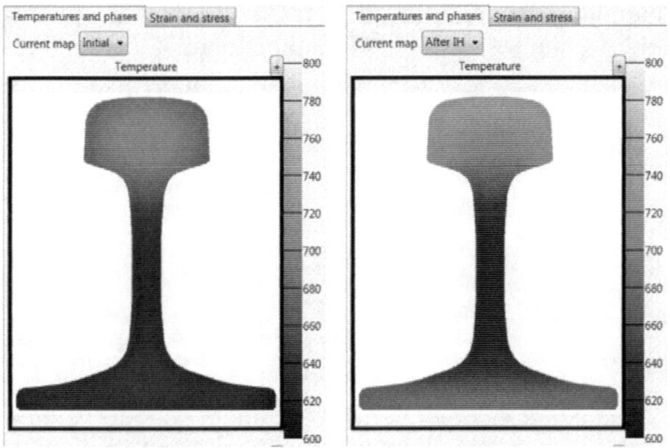


Figure 10. Induction heating simulation example

## 8. Industrial application

According to the temperatures measured by the in-line pyrometers positioned at the inlet of the cooling system, idRHa+ProMod can estimate the rail incoming thermal profile. Once the thermal map is established, idRHa+ProMod performs calculations at different pressure settings for each control group, then simulates the corresponding thermal path and finally, after a few iterations, finds out the optimized one. Cooling path optimization is based on the following criteria:

- Dropping down temperature as fast as possible at the early stages of the heat treatment. This choice permits to control metallurgical transformations as long as possible inside the idRHa+ system and to obtain hard structures deep inside the rail surface.
- Increasing the surface temperature over the pearlite limit (PL) so as to avoid bainite start.
- Keeping the surface temperatures very close to a predefined target temperature until micro-structural transformations have finished.
- Adjusting the pressures so as to keep the rail straight.

All the thermal profiles, cooling paths and calculation results are available for display on the user interface (figure 11).

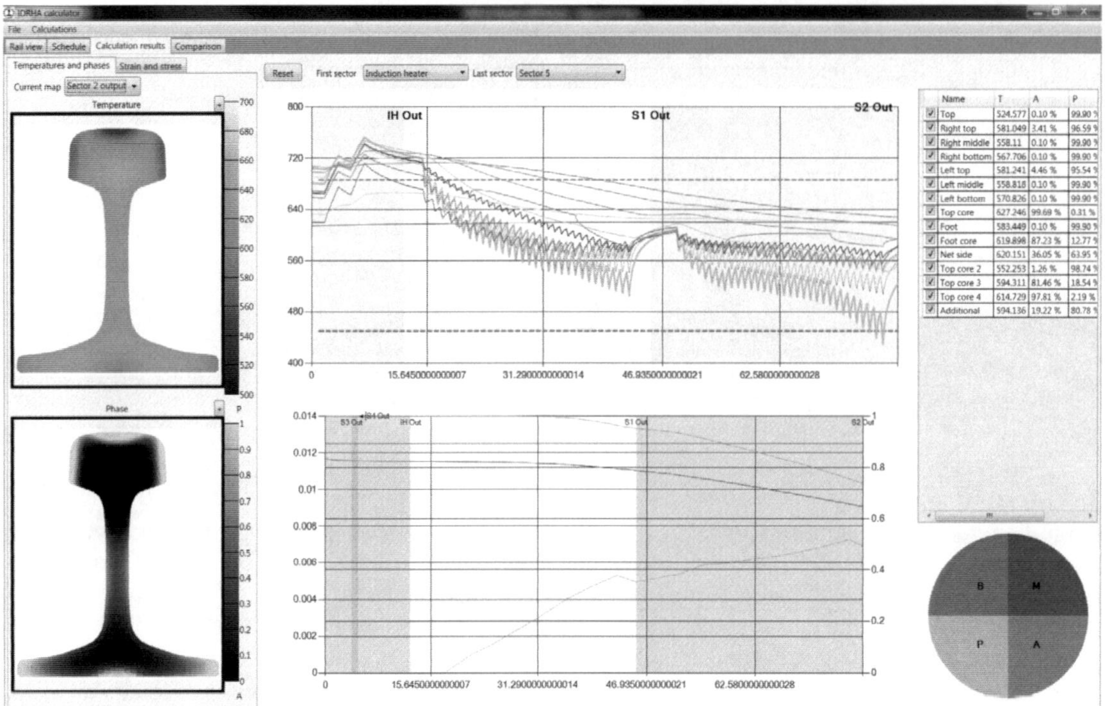


Figure 11. User interface and example of calculation output

Currently, idRHa+ProMod rail hardening control software is used by Baogang Iron and Steel in Baotou (China). This thermo-metallurgical software was first employed during the idRHa+ hot commissioning in April 2014, giving valuable information about the possible cooling strategies and suggesting the applicable pressure settings for different rail shapes, materials, temperatures and process velocities.

## References

[1] K.J.Bathe, Finite Element Procedures in Engineering Analysis, New Jersey: Prentice-Hall, Inc., 1982.

[2] Seong-Hoon Kang; Yong-Taek Im, "Three-dimensional thermo-elastic–plastic finite element modeling of quenching process of plain-carbon steel in couple with phase transformation," *International Journal of Mechanical Sciences,* 2007.

[3] C.Mapelli, "Modello curve CCT TTT," Politecnico di Milano, 2015.

[4] J., Miettinen, "Calculation of Solidification-Related Thermophysical Properties for Steels," *METALLURGICAL AND MATERIALS TRANSACTIONS B VOLUME 28B,* 1997.

[5] British Iron and Steel Research Association, "Physical Costants of some Commercial Steel at Elevated Temperatures," 1953.

[6] H.Fujimoto, N.Hatta, H.Asakawa,T.Hashimoto "Predictable Modelling of Heat Transfer Coefficient between Spraying Water and a Hot Surface above the Leidenfrost Temperature", ISIJ International, vol.37 (1997).

# Thermal Stability of Residual Stresses in Ti-6Al-4V components

**Aleksandar Stanojevic[1], Paul Angerer[2], Bernd Oberwinkler[3]**

[1]Montanuniversität Leoben, Franz-Josef Straße 18, 8700 Leoben, Austria
[2]Materials Center Leoben, Roseggerstraße 12, 8700 Leoben, Austria
[3]Bohler Forging, Mariazellerstraße 25, 8605 Kapfenberg, Austria

aleksandar.stanojevic@unileoben.ac.at

**Abstract**. The need for light weight design while maintaining a high safety is essential for many components, especially in the aircraft industry. Therefore, it's important to consider every aspect to reduce weight, improve fatigue life and maintain safety of crucial components. Residual stresses are a major factor which can positively influence components and fulfil all three requirements. However, due to the inconstancy of the behaviour of residual stresses during the life time of a component, residual stresses are often neglected. If the behaviour of residual stresses could be described reliably over the entire life time of a component, residual stresses could be taken into account and components could be optimized even further. Mechanical and thermal loads are the main reason for relaxation of residual stresses. This work covers the thermal stability of residual stresses in Ti-6Al-4V components. Therefore, exposure tests at raised temperatures were performed on specimens with different surface conditions. Residual stresses were measured by x-ray diffraction before and after testing. Creep tests were also carried out to describe the creep behaviour and thereby the ability for residual stress relaxation. A correlation between the creep rate and amount of relaxed stress was found. The creep behaviour of the material was described by using a combination of the Norton Power law and the Arrhenius equation. The Zener-Wert-Avrami model was used to describe the residual stress relaxation. With these models a satisfying correlation between measured and calculated data was found. Hence, the relaxation of residual stresses due to thermal load was described reliably.

## 1. Introduction

The importance of light weight design is constantly increasing and the ambition to optimize components forces materials to their limits. Optimal materials should offer low density while maintaining high static and cyclic mechanical behavior at elevated temperatures. To meet the demanding safety requirements and to achieve a high service life of components, certain finishing techniques become necessary.

The main reason for material fatigue can be described by crack nucleation and propagation. Cracks initiate mainly on the surface of components and continue to propagate inside the component. The process of crack nucleation and especially the subsequent short crack propagation can be greatly decelerated by a suitable surface treatment. Mechanical surface treatments (e.g., shot peening, rolling, hammering) or manufacturing processes (e.g., machining, low stress grinding) induce residual compressive stresses in near-surface areas of Ti-6Al-4V and cause hardening due to cold working

(Figure 1). While residual compressive stresses are responsible for crack closure effects, local hardening causes a shift of the local yield stress and therefore a modified local plastic deformation behavior. Both effects are responsible for higher fatigue life.

However, the positive effects of residual stresses can't be expected throughout the whole service life of components. There are numerous processes causing a relaxation of residual stresses. For example, high temperatures cause a relaxation of residual stresses through creep, while high stress or strain amplitudes cause a relaxation through local plastic deformation.

Residual stresses are often neglected in the design process of components, mostly because of the insufficient description of their stability over time, mechanical load and temperature. To improve the comprehension of residual stresses in fatigue lifetime estimation, the effects causing a relaxation of residual stresses have to be studied and precisely described.

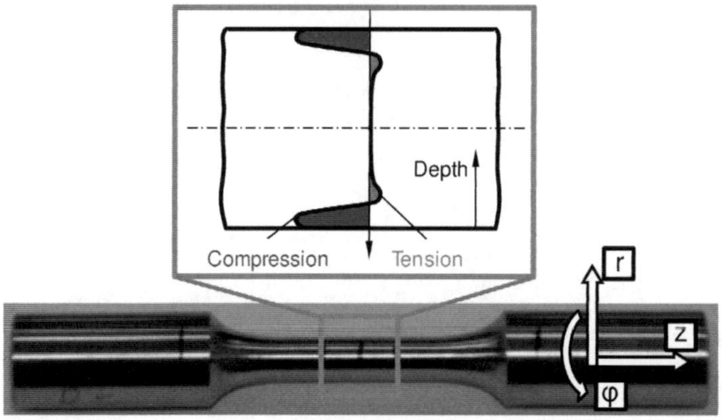


**Figure 1.** Used specimen with a qualitative residual stress distribution (blue: compression; red: tension)

This work focuses on the residual stress relaxation due to thermal loads. In order to describe the relaxation behavior reliably, a continuum mechanical approach was necessary. The main driving force behind thermal residual stress (RS) relaxation was identified by occurring creep in the specimen. Therefore it was important to specify the creep behavior of the material through creep tests at all necessary temperature and stress levels. Furthermore temperature exposure tests were performed and the resulting residual stress was measured by x-ray diffraction. The measured residual stress was used to verify the thermal stability of residual stress through creep. Also the Zener-Wert-Avrami parameters were determined and the results were compared.

## 2. Experimental procedure

In order to obtain all necessary data for the model, creep tests and time exposure tests with residual stress measurement through x-ray diffraction were carried out. The creep tests were performed at 100°C, 250°C, 370°C as well as at room temperature at loads between 550MPa and 800MPa.

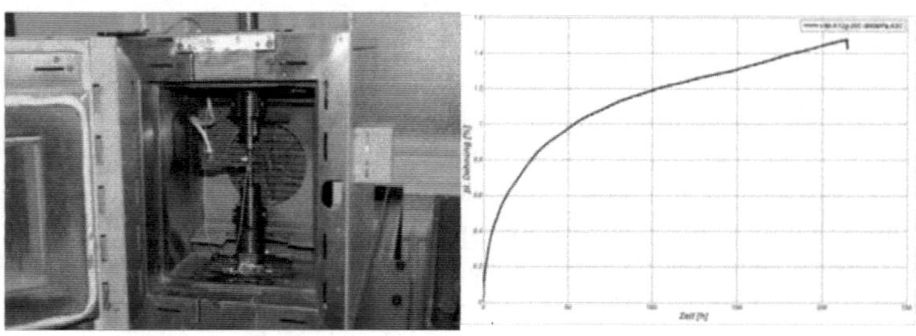

**Figure 2.** Creep test rig and a measured creep curve

Temperature exposure tests at 100°C, 250°C and 370°C were also performed. At defined time increments the specimens were measured through x-ray diffraction (XRD).

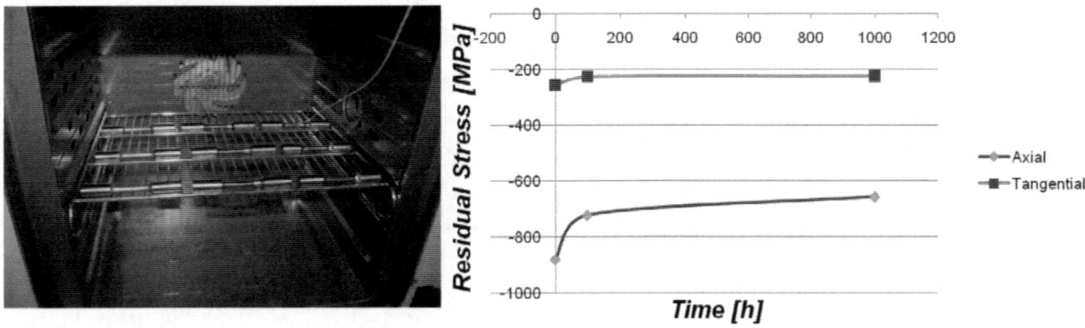


**Figure 3.** Temperature exposure chamber and measured RS after a certain time period

After XRD measurements, residual stresses and the Full Width at Half Maximum (FWHM) are obtained. The FWHM is proportional to the hardening. The hardening of the specimen can be observed through increased FWHM in the surface layer (Figure 4).

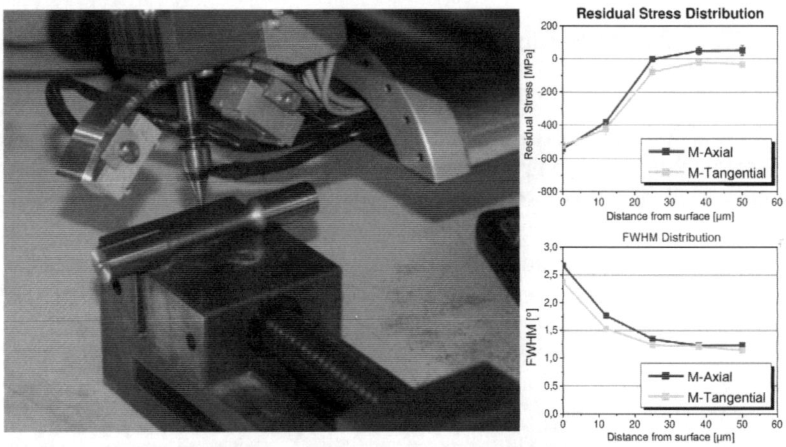


**Figure 4.** X-ray diffraction with measured Residual Stress and FWHM depth profiles through incremental surface layer erosion

## 3. Analysis and Simulation

To describe the residual stress relaxation behavior through creep it is necessary to identify all present load cases depending on the distance from the surface. A compressive residual stress with initial hardening can be found in the surface layer of the specimen. The subsurface layer beneath that contains tensile residual stresses without any initial hardening. The crucial part for the stability of residual stress due to thermal loads is the compressive residual stress, which is higher than the tensile part. The increased level of dislocation in the surface layer through cold working leads to a shift of the starting point along the creep curve [1]. This means that creep won't occur until the initial amount of dislocations due to cold working is reached (Figure 5). Only after that the creep will continue along the measured creep curve. Therefore as an approximation primary creep was neglected and thermal residual stress relaxation due to thermal loads was calculated based on only secondary creep.

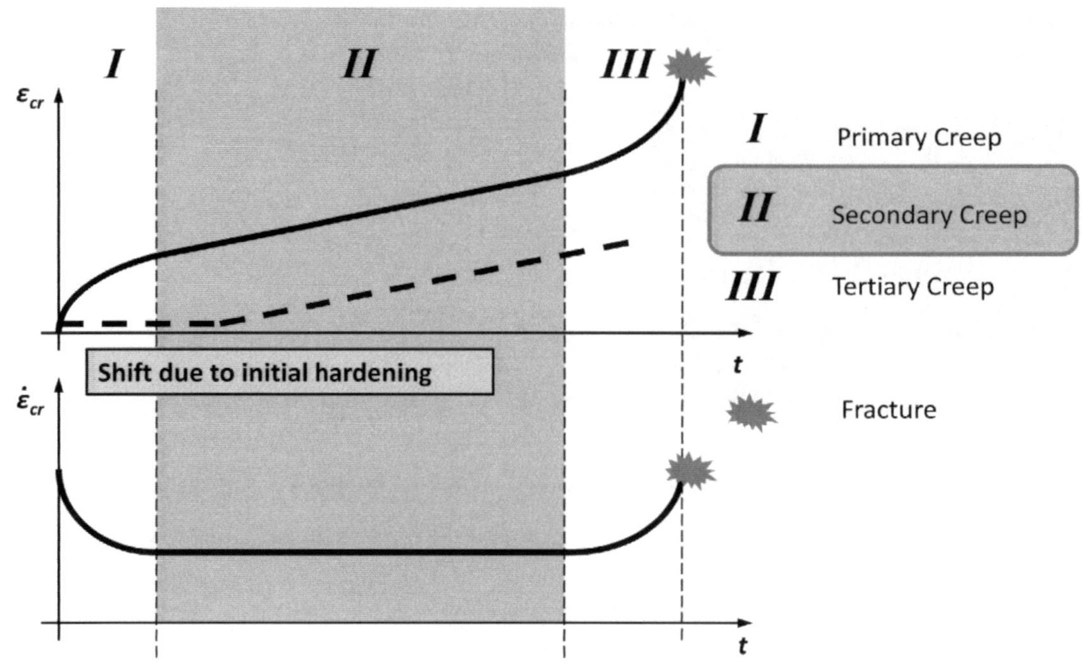


**Figure 5.** Creep curve with a shift of the initiation of creep due to initial hardening in the surface layer

The amount of relaxed residual stress is calculated by the local occurring creep strain in the surface layer of the specimen. With increasing temperatures and the same surface condition (in this case only machined specimen were examined) the creep strain also increases. Every portion of creep strain causes a reduction of the elastic strain leading to a reduction of the remaining residual stress (Figure 6). The steadily decreasing residual stress will also cause lower strain rates. This iterative process of interaction between strain rate and residual stress will end at a saturated level of residual stress where the creep strain is low enough to not cause any further relaxation.

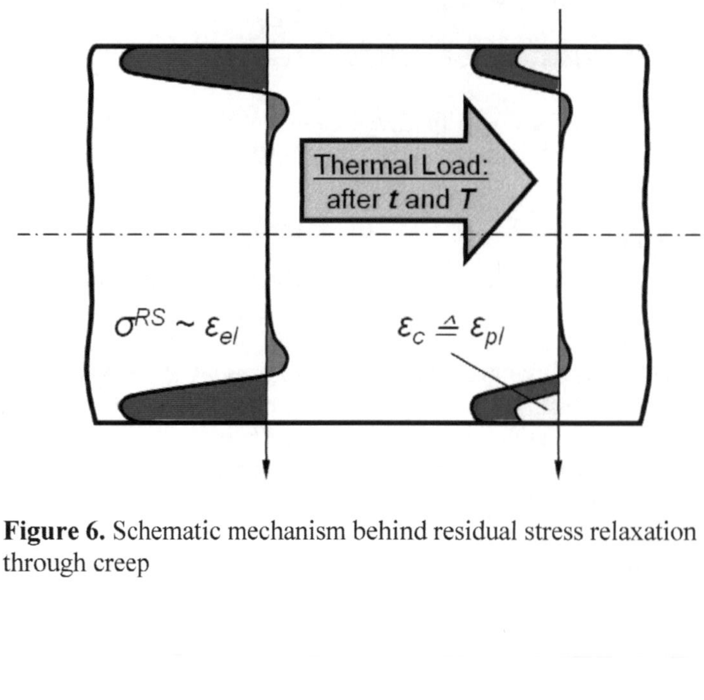


**Figure 6.** Schematic mechanism behind residual stress relaxation through creep

### *3.1. Determination of the Creep Parameters*

The description of creep was achieved through the definition of the stationary (secondary) creep rate due to temperature and stress. A combination of the Norton Power law and the Arrhenius equation was used [2]:

$$\dot{\varepsilon}_c = C\left(\frac{\sigma}{E(T)}\right)^n e^{-\frac{Q_c}{RT}} \tag{1}$$

Equation (1) shows the dependence of the stationary creep rate due to temperature $T$ and stress $\sigma$. The necessary parameters for this equation are the velocity constant $C$, the power law exponent $n$ and the activation energy $Q_c$. They can be determined using the secondary creep rates from creep tests. It was observed that the power law exponent is almost independent from temperature (Figure 7) but the activation energy is strongly affected by the temperature. For low temperatures the activation energy is nearly zero up to 240°C. Beyond that the activation energy increases with increasing temperature (Figure 8).

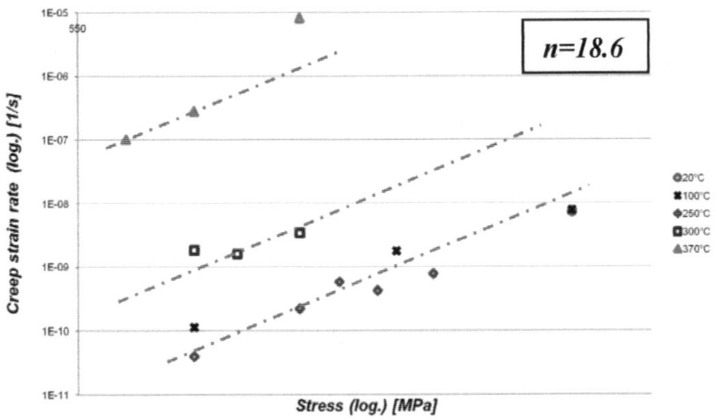


**Figure 7.** Dependence of the stationary creep rate on stress at different temperature levels (Assumption: constant Power Law Exponent (red dashed line))

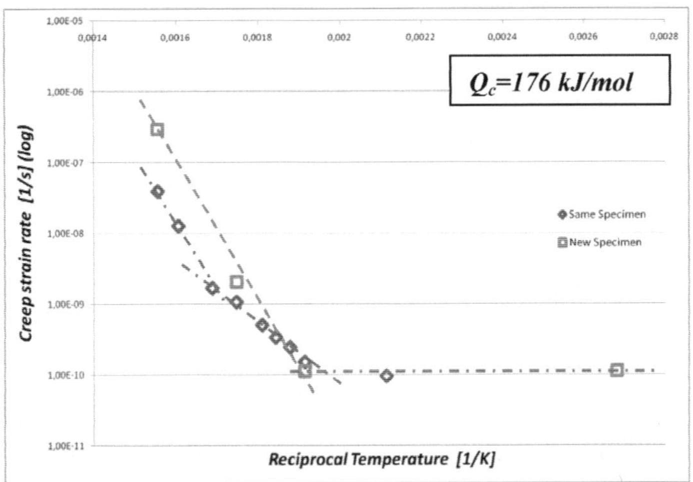


**Figure 8.** Dependence of the stationary creep rate on temperature at 600MPa (Possible change of activation energies due to varying active creep mechanisms (red dot and dashed line);Simplification: constant activation energy (green dashed line))

Figure 8 shows a comparison between two different investigations of the creep rate. On the one hand the creep curves were measured on new unloaded specimens (green squares) and on the other hand one specimen was used increasing the temperature after every creep test on the same specimen consecutively (purple diamonds). Experiments prove that the secondary creep rate is not affected by the temperature until 240°C but also show that the secondary creep rate at room temperature is just as high. Beyond 240°C a slight increase of the temperature sensitivity of the secondary creep rate was observed, increasing again after 320°C. This matter can be observed as an increased slope in Figure 8 (red dot and dashed line).

However, due to the fact that literature mentions only one slope (constant activation energy) at higher temperatures (above 600°C) and the insufficient existence of detailed creep investigations at the required temperature range (between 20°C and 370°C) the mentioned behavior has to be proved at more stress levels. Therefore as a simplification, constant activation energy for creep was chosen between 250°C and 370°C (green dashed line). For lower temperatures the temperature sensibility was set to zero.

### *3.2. Verification of creep rates in a FE-simulation model*

The determined creep parameters were implemented into Abaqus (Dassault Systems). When a unified cube (one element) with precisely set material parameters is loaded with a constant force adequately to the performed creep tests, the simulated results match the measured data.

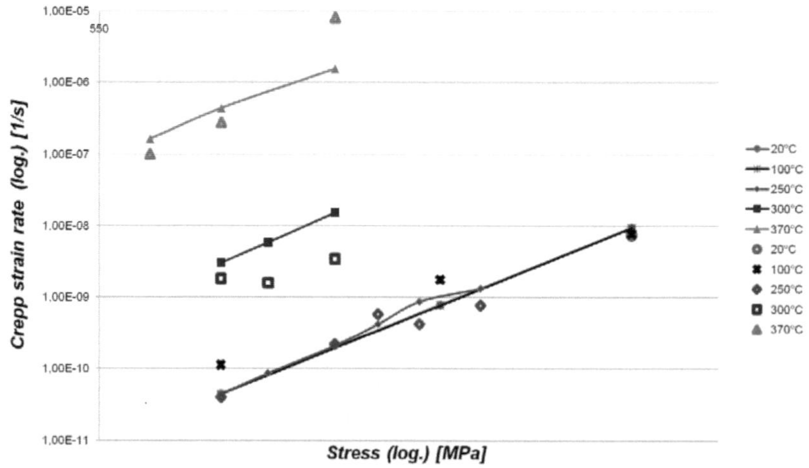


**Figure 9.** Comparison of measured data (points) and FE-simulated stationary creep strain data (lines) vs. stress at different temperature levels

### *3.3. Consideration of effects during the heating process*

During the heating process of the specimen different effects may occur, which can cause a residual stress relaxation. If the temperature gradient during heating is too high and causes plastic deformation, the residual stress will decrease permanently. The deviation of mechanical properties with increasing temperature has to be considered as well. On the one hand the Young's Modulus decreases causing an elastic, temporary relaxation of the residual stress (initial residual stress recurs after cooling) on the other hand the yield stress decreases and can cause a local plastic deformation due to high compressive residual stresses in the surface layer which leads to a permanent residual stress relaxation. Every other permanent diminishment of the residual stress is caused by creep (in the investigated temperature range).

**Table 1.** Effective residual stresses at a certain temperature: The decrease is caused by temporary, elastic relaxation

| Temperature (°C) | Effective residual stress (MPa) |
|---|---|
| 20 | 540 |
| 100 | 518 |
| 250 | 473 |
| 370 | 450 |

*3.4. Calculation of the residual stress relaxation*

The residual stress relaxation was calculated using a FE-model with a mapped residual stress distribution. The specimen was implemented as an axial symmetric model containing one quarter of the round specimen. The residual stress was mapped on each integration point of every element according to the measured residual stress distribution. For more detailed information about the application of residual stresses see [3].

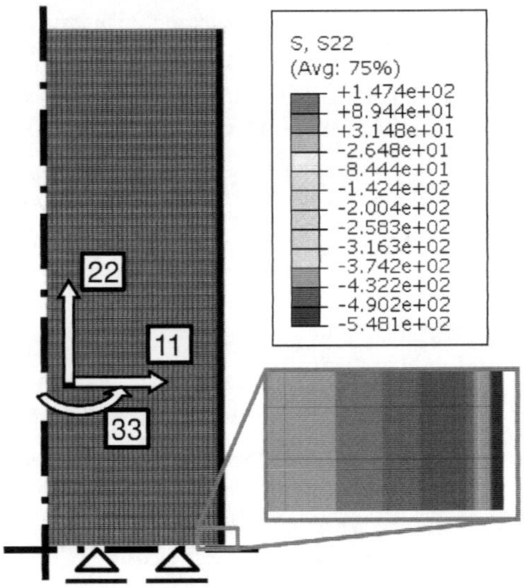


**Figure 10.** Model of the FE-supported calculation of the residual stress distribution.

The model assigned with a desired residual stress distribution was heated at a defined temperature and remained at that elevated temperature for a defined amount of time. Afterwards the calculated remaining residual stress distribution was compared with the measured data from the iterative exposure test measured by XRD.

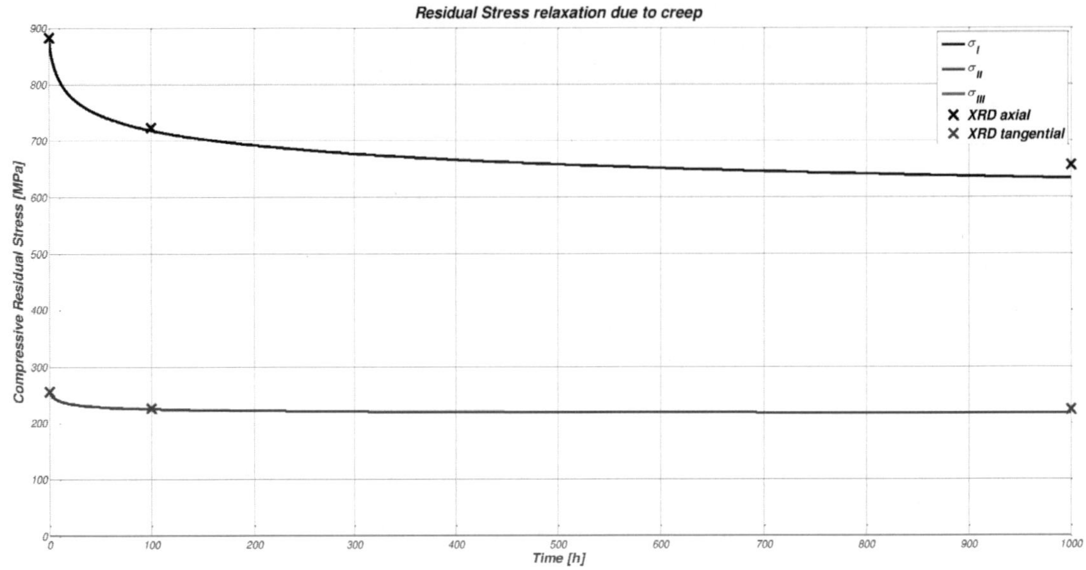


**Figure 11.** Comparison of the measured and calculated remaining residual stresses over temperature and time

Figure 11 shows an example of the evolution of the residual stress on the surface of a machined specimen exposed to a defined temperature for a certain amount of time. It can be observed, that the higher axial residual stress shows a high relaxation, especially in the beginning. The tangential residual stress doesn't show any relaxation due to creep. The residual stress decreases foremost due to the transversal strain of the axial direction.

Furthermore the parameters for the Zener-Wert-Avrami model were determined [4]. Therefore equation 2 was approximated to measured values of residual stress in axial and tangential direction respectively (Figure 12).

$$\frac{\sigma^{RS}(t,T)}{\sigma^{RS}(t=0)} = e^{-\left(C\,t\,e^{-\frac{\Delta H_A}{k_B T}}\right)^m} \qquad (2)$$

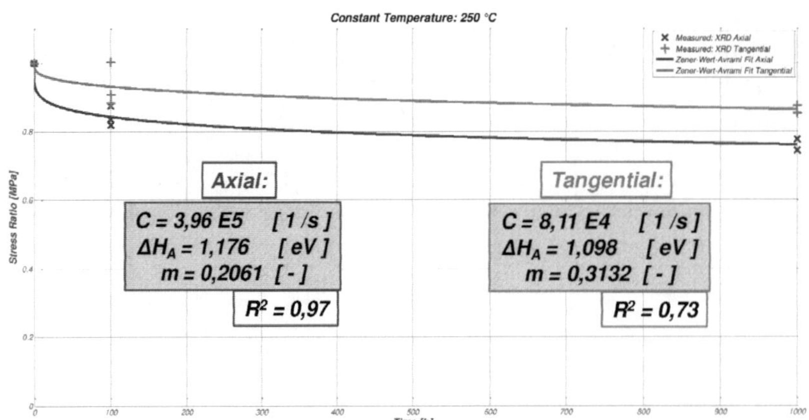


**Figure 12.** Fit of the Zener-Wert-Avrami parameters for axial (blue) and tangential (red) direction

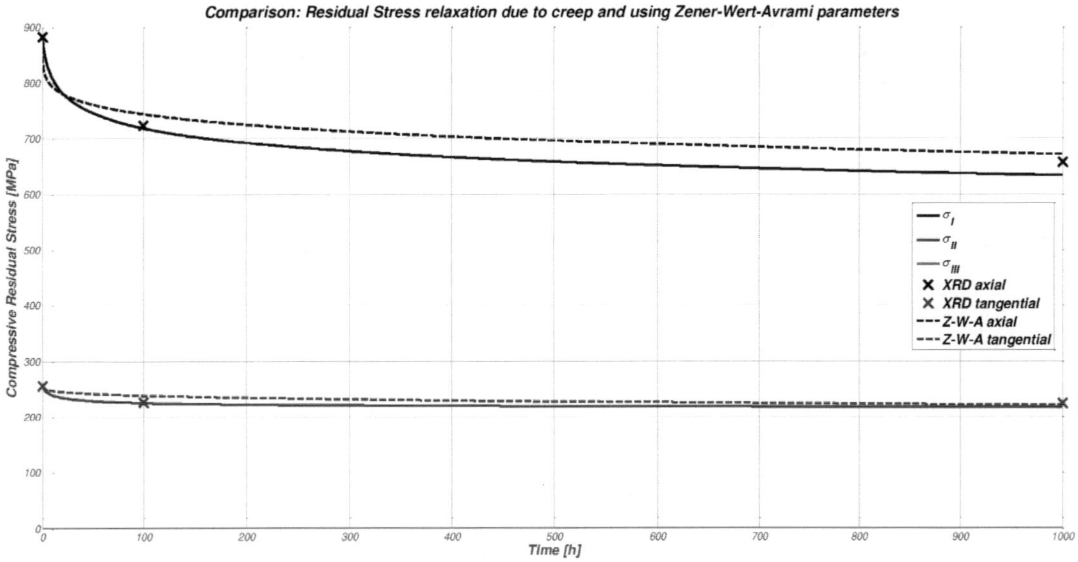


**Figure 13.** Comparison of the measured and FE-simulated remaining residual stresses over temperature and time with the Zener-Wert-Avrami Fit (dashed Lines)

Figure 13 shows a comparison between the calculated residual stress relaxation through creep and the characterization of the residual stress relaxation by the Zener-Wert-Avrami equation. It can be observed that both approaches lead to satisfying and accurate results. Considering the scatter of the XRD measurement of 50MPa, both approaches for calculation of the residual stress relaxation fit the scatter band adequately.

## 4. Summary and Outlook

In this work, a basic overview about the description of thermal stability of residual stresses through creep was given. The determination of creep material properties was explained as well as the influence of the heating process on the residual stress distribution. Also the amount of relaxed stress was matched using the approach based on creep. Finally the results were compared with the Zener-Wert-Avrami approach and discussed.

Further studies and investigations will cover following topics:

- Accurate evaluation of the effect of hardening due to cold work on the creep behavior [1]
- Description of the stability of residual stresses due to the combined thermal and cyclically mechanical load
- Calculation of different surface conditions (Shot peened, Low Stress Grinded)
- Identification of local creep behavior including primary creep

## 5. Acknowledgement

Financial support by the Austrian Federal Government (in particular from Bundesministerium für Verkehr, Innovation und Technologie and Bundesministerium für Wissenschaft, Forschung und Wirtschaft) represented by Österreichische Forschungsförderungsgesellschaft mbH and the Styrian and the Tyrolean Provincial Government, represented by Steirische Wirtschaftsförderungsgesellschaft mbH and Standortagentur Tirol, within the framework of the COMET Funding Programme is gratefully acknowledged.

## References

[1] Ilschner, "Hochtemperatur-Plastizität", Springer, 1973
[2] Frost, Ashby, "Deformationmachanism maps", Pergamon press, 1983

[3] Stanojevic, Maderbacher, Angerer, Oberwinkler, "Stability of Residual Stresses in Ti-6Al-4V Components due to Mechanical Loads"
[4] Jürgen Hoffmeister, "Characterization of residual stress relaxation in shot peened Inconel 718 due to thermal, quasi-static and cyclic loads" (Ph.D. thesis, KIT Karlsruhe, 2009), 136

IOP Conf. Series: Materials Science and Engineering **119** (2016) 012008 doi:10.1088/1757-899X/119/1/012008

# Influence of Surface Roughness in Electron Beam Welding

**C. Wiednig[1], F. Stiefler[2], N. Enzinger[1]**
[1]Institute of Materials Science and Welding, Graz University of Technology
[2]Kremsmüller Industrieanlagenbau KG, Schwechat, Austria

E-Mail: christopher.wiednig@tugraz.at

**Abstract.** The requirements of welded components are rising continuously through increasing demands in engineering. But in engineering not only the quality of welds is important also an economic and timesaving production is crucial. Especially in welding of large cross sections economization potential is existing and significant. Beside the welding technique itself the joint preparation is a major part of work. Electron beam welding has some major advantages in this area. Due the high energy density a very short welding time as well as a small heat affected zone can be achieved. Furthermore the joint preparation can be held simple. Nevertheless, a careful machining and cleaning of the joint surfaces is recommended in literature. In addition to geometric tolerances a specific surface roughness should be kept. These statements are quite general and unspecific. In this contribution a systematic investigation on the influence of joint preparation on the joint properties is presented. By performing several welding experiments with different surface roughness this study provides empirical conclusions. Beside the microscopic investigation of different cross sections and mechanical tests of the welded samples also the process stability during welding was reviewed.

## 1. Introduction

In many industrial fields, welding is a major part of manufacturing. Welded joints not only have to offer a consistent high quality but they also have to be economic in production. When it comes to large cross sections this fact is becoming even more crucial. Here, welding effort and especially joint preparation can become a major part of production time and cost.

Electron beam welding (EBW) has some mature advantages in welding of large cross section and in joint preparation. Welding is done with a single pass and the joint preparation is, independent of the wall thickness, a simple butt preparation. Hence the surface roughness of the welding faces has to be well considered in view of weld quality and joint preparation time. So far little attention has been paid on to the influence of the SR on the final weld and on the welding process. In literature the statements according the surface texture of the welding surface are quite general or differ widely. Often the surface texture is not even mentioned [1], [2] or the recommendations differ widely. In literature surface preparing suggestions differ from the need of a smooth machined surface [3] up to the statement that a too smooth surface may hinder degassing of the weld [4].

The aim of this paper is to analyse the influence of the surface roughness on the welding process and the final weld quality. For two different wall thicknesses, welding specimens with four different surface roughness values were prepared and welded. The welding process was monitored and microscopic investigations, as well as tensile tests were carried out to examine the welds.

### *1.1. Electron Beam Welding*

Electron beam welding is a fusion welding process where the kinetic energy of electrons is used as heat source. Electrons are accelerated and impact on a work piece with high velocity; when hitting the metal surface, the electrons are decelerated in the material. Thereby the incoming electrons transfer their kinetic energy to the lattice atoms in the workpiece which increases the local temperature immediately [3], [5].

By focusing the EB the energy density of the beam spot can be controlled. At a power density above $10^5$W/cm², the deep welding effect occurs. Conduction of heat now is secondary; locally the material evaporates instantly and a steam capillary (keyhole), surrounded by molten metal is established [4], leading to a very small heat affected zone (HAZ).

By applying relative movement between the workpiece and the electron beam the weld seam is formed. EBW offers a very high seam depth to width ratio, which is advantageous especially in thick walled welding [6]. Compared to laser beam welding the EB remains almost uninfluenced by metal vapour or plasma, [7].

However an interaction with the metal vapour, especially when metals with high vapour pressure are welded can happen in the EB gun. This so called arcing is a spark discharging in the electron gun. If too much metal vapour reaches the EB gun a flashover between the anode and cathode can occur. The susceptibility to arcing is given especially at high acceleration voltages and thicker walls. Arcing is also forced when gas pores and impurities in the metal, or dirt on the welding faces lead to small explosions. These explosions send a higher amount of metal vapour to the EB gun.

Depending on the stability of the established arcs, flashovers are distinguished in soft- and hard flashovers. Hard flashovers are characterized by the formation of a non-self-extinguishing arc between parts at high voltage potential and ground. Hard flashovers lead, by activation the overcurrent protection, to an immediate interruption of the welding process, and therefore mostly irreparable welding defects. Soft flashovers have a self-erasing arc and have less intensity. They are mostly established between the beam source and the Wehnelt (bias) cylinder. They cause a short term current change in the electron beam and may cause welding defects [3], [8], [9]. The higher the acceleration voltage, and the more pores and contaminations are in the material and at the welding faces, the more likely flashovers can happen.

### *1.2. Surface roughness*

Surface roughness is a component of surface texture. It is quantified by the deviations in the direction normal of a surface from its ideal form. Surface roughness is a sort of shape deviation and according to DIN 4760 the macroscopic surface roughness can be subdivided to 3$^{rd}$ order roughness and 4$^{th}$ order roughness.

Roughness can be measured as a surface profile with a profilometer. For the detection and classification of the surface roughness various parameters are available. For technical usage typically the mean roughness index Ra, expressed in microns, is used. Ra is the mean deviation of the roughness profile and is defined as "the arithmetic average of the absolute values ($y_i$) of the profile height deviations from the mean line, recorded within the evaluation length" [11].The Ra index is represent the combined 3$^{rd}$ and 4$^{th}$ order surface deviation.

$$Ra=\int_0^1 |y|_i \times d_x \tag{1}$$

| **Shape deviation** (as profile section with vertical exaggregations) | **Examples for the type of deviation** | **Examples for the cause** |
|---|---|---|
| 1st order: form deviation  | straightness, flatness, roundness deviation, etc/ | incorrect clamping of workpiece, machine or workpiece deflection |
| 2nd order: waviness  | waves (according to DIN 4761) | machine or tool vibrations |
| 3rd order: roughness  | grooves (according to DIN 4761) | cutting edge shape, feed or depth of cut of the tool |
| 4th order: roughness  | grooves, scales, lumps (according to DIN 4761) | process of chip formation, material deformation |

**Figure 1.** Classification of shape deviation (DIN 4760) [10]

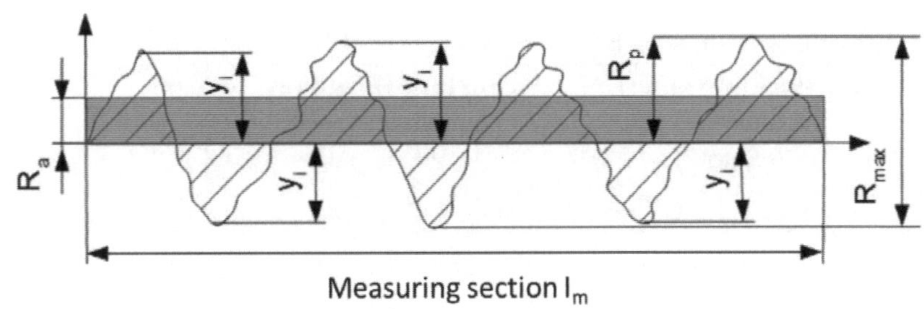


**Figure 2.** Geometrical illustration of Ra [11]

Every machining operation leaves a characteristic surface texture depending on the machining process, the type and shape of the cutting tool and the chip formation. In the following table the surface roughness values are connected to the milling processes. A milling operation with a higher cutting efficiency usually produces a higher surface roughness.

**Table 1.** Surface roughness values of different milling processes

| Ra (μm) | Milling operation |
|---|---|
| 12.5 | Pre roughing |
| 6.3 | Roughing |
| 1.6 | Finishing |
| 0.8 | Superfinishing |

## 2. Experiments

In the following subchapters the used material, welding equipment, specimen setup and investigation methods are described.

### *2.1. Material*

For the experiments a stainless soft martensitic steel X3CrNiMo13-4 (1.4313) was used. Its application is wide spread in hydraulic engines, turbines and naval engineering. The chemical composition is given in Table 2. The tensile strength in tempered conditions (tempering temperature 550 – 600°C) is between 780 and 980 MPa [12].

**Table 2.** Composition X3CrNiMo13-4

| C | Si | Mn | Cr | Mo | Ni | N |
|---|---|---|---|---|---|---|
| < 0.029 | 0.40 | 0.85 | 13.00 | 0.55 | 4.20 | 0.04 |

### *2.2. Welding setup*

For investigation, weld samples with a plate thickness of 20mm and 100mm were used. The samples were prepared as butt welds and welded in PA position (1G flat position, according to ISO 6947) without preheating. For welding the 100mm samples a metal backing was used to prevent the weld pool of fall trough. Welding was carried out with a probeam universal chamber machine, equipped with a 45kW EB-gun. Optimal welding parameters were determined in prior experiments. The welding surfaces of the specimen were face milled to the defined surface roughness. For every mean roughness index and plate thickness three samples were welded. The experimental matrix is shown in Table 3. Before welding the welding surfaces were brushed with a stainless steel brush and subsequently cleaned with an isopropanol soaked industrial wiping paper.

**Table 3.** Experiment Matrix

| Plate thickness (mm) | Surface Roughness Ra (µm) |
|---|---|
| 20mm | 12.5 / 6.3 / 1.6 / 0.8 |
| 100mm | 12.5 / 6.3 / 1.6 / 0.8 |

### *2.3. Investigation Methods*

The welding process was recorded with an inbuilt CCD camera to determine the process stability.
Several micro sections were taken and investigated by means of light microscopy to check the weld on defects. The micro sections taken can be seen in Figure 3. The arrows mark the examination surface. Additionally for every thickness and surface roughness, one longitudinal section of the whole weld was produced. Therefore the welded samples were cut alongside the weld and the longitudinal section was grinded till the weld centreline. Cross tensile test where carried out for basic characterisation. Tensile test were performed according to DIN50125, specimen geometry B with 8mm diameter. The specimen removal positions are marked in Figure 3.

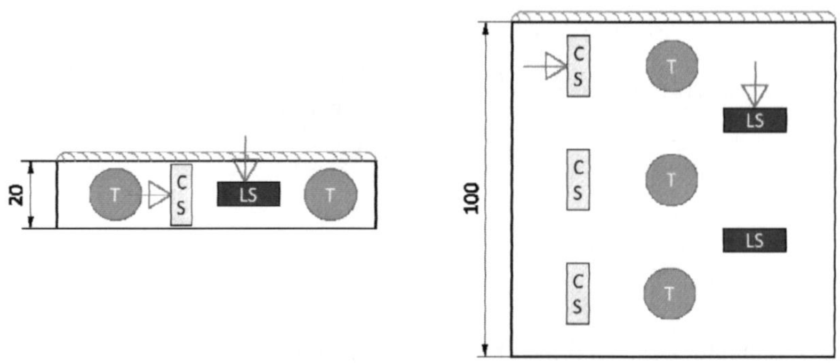


**Figure 3.** Position and amount of micro sections an tensile specimen; T- tensile specimen, CS-cross section, LS- longitudinal section

## 3. Results

This chapter is sub divided in macroscopic and microscopic investigation of the weld seam. Followed by the subchapter presenting the results of the tensile test and the observations of the welding process.

### *3.1. Macroscopic Pictures*

For every surface roughness and thickness a longitudinal and two cross sections were made (Figure 4 to Figure 7). No macroscopic visible defects were found at any specimen. The microstructure in the weld centre line shows dendrites in front view with no irregularities. The cracks which can be seen at the end in the 100mm specimen (Figure 5) occurred due to a too short slope out procedure. The cross-sections offer no macroscopic defects, when an arcing free process was carried out.

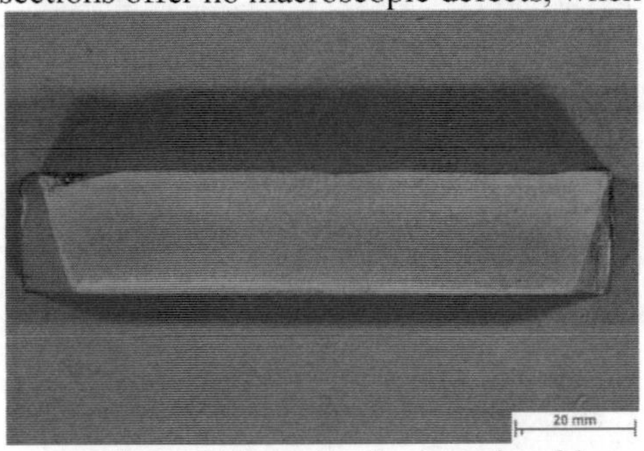


**Figure 4.** Longitudinal-section 20mm

**Figure 5.** Longitudinal-section 100mm

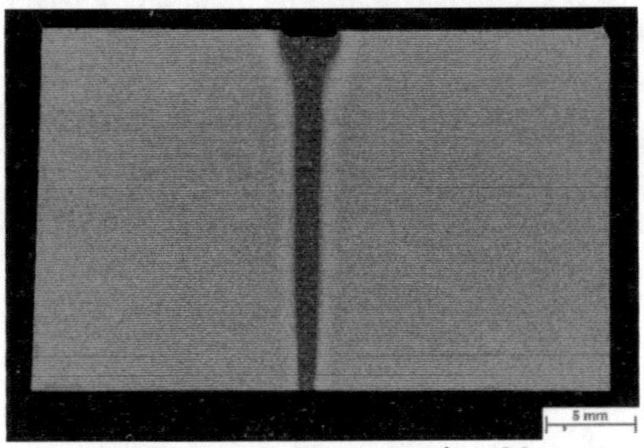


**Figure 6.** Cross-section 20mm

**Figure 7.** Cross-section 100mm

The 20mm weld was made without backing, so a root formation was necessary. The specimens with a surface roughness of Ra 12.5 show a distinctive undercut at the weld face. Figure 8 shows two half cross sections to compare the undercut of the Ra12.5 to the other SR values, represented by Ra1.6. The SR values Ra6.3 and Ra0.8 provide the same shape as the Ra 1.6 (right side in Figure 8).

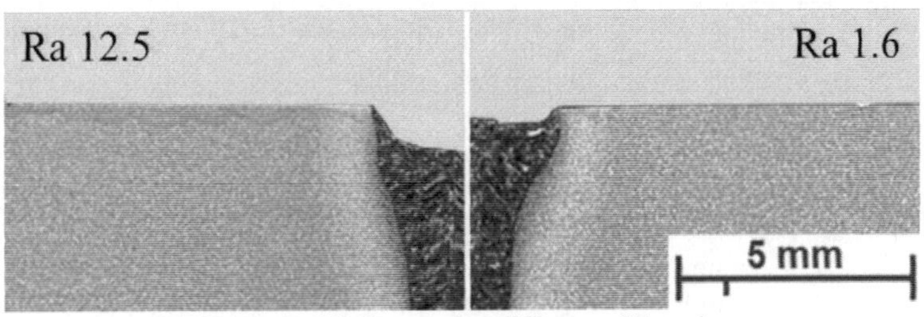


**Figure 8.** Undercut for different SR values

### *3.2. Microscopic Pictures*

28 microscopic sections were investigated by means of light microscopy. For no specimen a remarkable increased porosity was found. Some welds showed pores whereas other welds did not. A correlation of porosity to surface roughness could not be discovered. The following two figures show an example for a seam with pores (Figure 9) and a seam without pores (Figure 10). Pores found in the weld seams did not exceed a size of 20µm.

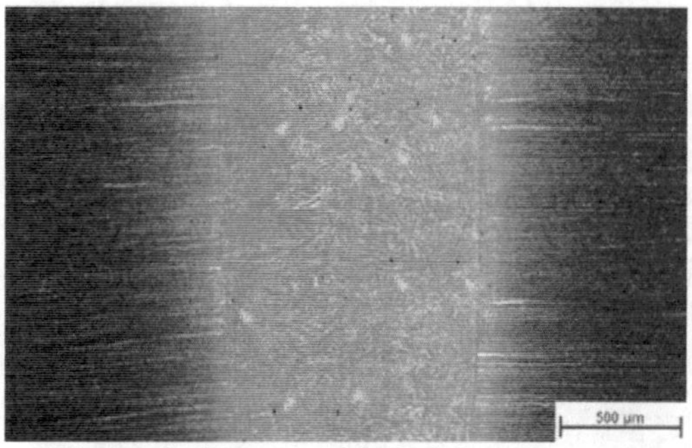


**Figure 9.** 20mm weld, Ra0.8, pores visible

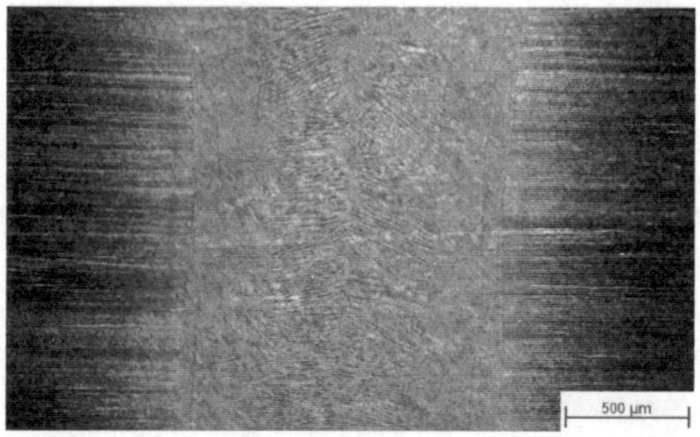


**Figure 10.** 20mm weld, Ra6.3, no pores visible

### *3.3. Tensile Test*

All tensile specimen ruptured in the base material. The tensile strength is conform the material data sheet and identical for all surface roughness values with the same wall thickness. No distinctive variation in tensile strength for the 100mm top, middle and bottom tensile specimen occurred. The 20mm specimen have an approximately 30MPa higher tensile strength compared to the 100mm specimens. The two Ra12.5 20mm specimen have about 2mm less nominal strain. In Figure 11 and Figure 12 the stress-strain diagrams are shown.

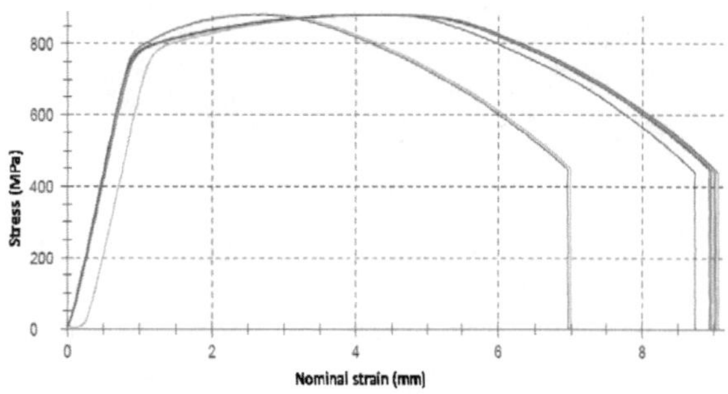


**Figure 11.** Stress strain curves 20mm specimen

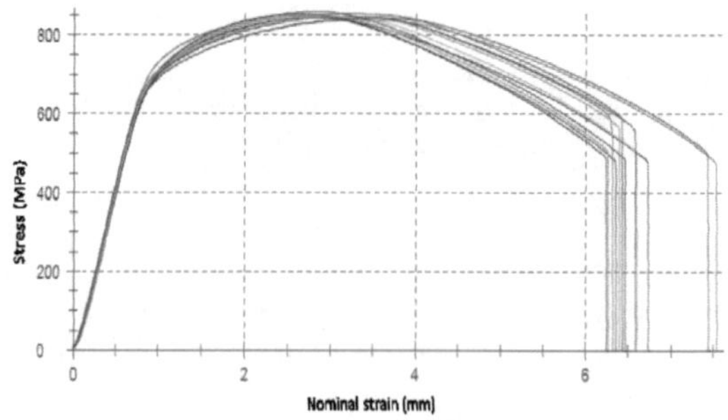


**Figure 12.** Stress strain curves 100mm specimen

### *3.4. Welding Process*

The welding process, when welding the 20mm samples, was stable. No influenced by the different surface roughness values was given. When welding the 100mm thick parts the process become more prone to arcing due to the higher amount of metal vapour. Welding the 100mm specimen with Ra0.8, and Ra1.6 surface roughness, no flashovers occur. While welding the Ra6.3 specimen, one soft flashover occurred but no welding defect was found in the weld. For the Ra12.5 specimens several soft flashovers happened during the welding procedure. One high intense flashover forced a small explosion in the keyhole and led to a major ejection of liquid material. In Figure 13 the stable keyhole can be seen just before the spark over. Figure 14 shows a short term beam breakdown, caused by this high intense flashover, and the resulting spattering. This specimen, revealed a cavity in the weld centre at the position of this flashover (Figure 15). No hard flashovers occurred at any specimen.

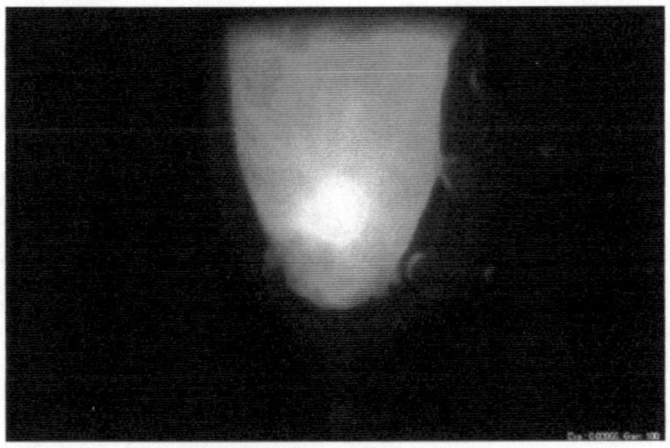

**Figure 13.** Stable Keyhole

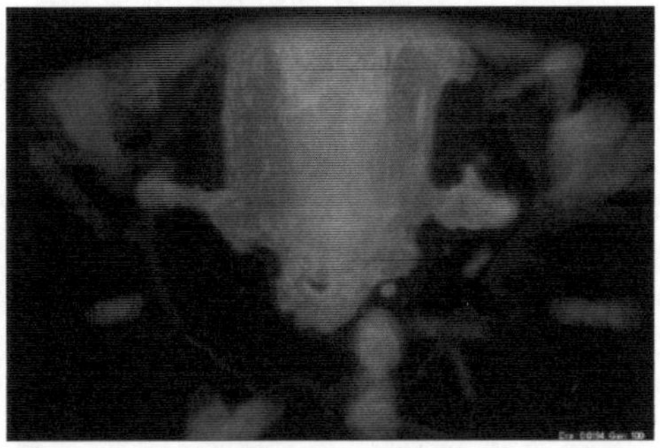

**Figure 14.** Flashover, with broken down keyhole and spatter formation

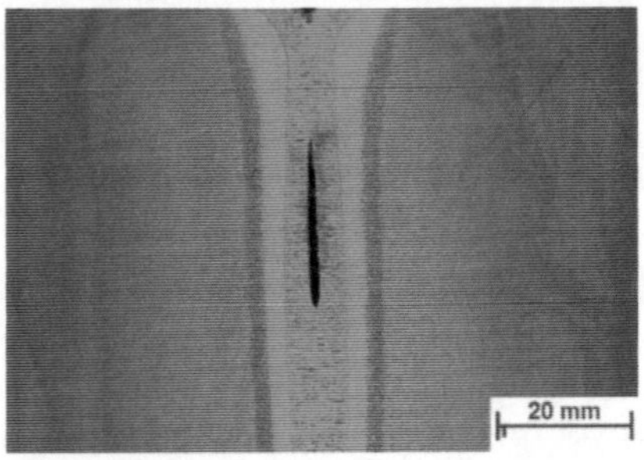


**Figure 15.** Welding defect based on the soft flashover shown in Figure 14.

## 4. Summary and Conclusion

The influence of roughness on the joint quality for 20 and 100mm electron beam welds of steel 1.4313 was systematically investigated.

It was observed, that the surface roughness directly has only a minor influence on the weld seam quality of electron beam welds. A direct influence was just observed for welded 20mm plates. For this wall thickness a distinctive undercut at the weld face was observed. This may happen due to the wider gap (missing volume), which exists when surfaces with high roughness are placed together. A difference in porosity or in tensile strength could not be observed. However, a major influence, is that surfaces with a higher roughness are harder to clean accurately and a not properly cleaned surface may lead to arcing during the process. This is especially the case for welding thick walled parts.

The results of this work may be summarised as follows:

- Surface roughness has no effect on the tensile strength of the weld
- For the investigated surface roughness values, no difference in the porosity of the weld was recognised
- When welding 20mm plates in full penetration technique, a higher surface roughness caused a distinctive undercut
- A higher surface roughness lead to arcing during welding the 100mm specimens. No arcing occurred during welding the 20mm specimen

The following conclusions can be drawn from this investigation

- Cleaning of the welding surface is a crucial point. Surfaces with a rougher surface texture are harder to clean. Due to an insufficient cleaning a contaminated surfaces can cause arcing.
- Producing an "easy to clean" surface is more important than the roughness itself.

- A suitable milling surface roughness for electron beam welding is between Ra1.8 and Ra6.3. For a wall thickness where full penetration welding is carried out a low surface roughness (Ra1.8) is recommended to avoid severe undercut.

## 5. Outlook

This paper provides a recommendation for the surface preparation for electron beam welding. However, some question remain unanswered and will be investigated in ongoing work.
It has to be considered that these investigations are limited to face milled surfaces, future research will be done on surfaces produced with different processes (grinding, sawing). Additional research will be required to cover the influence of the $1^{st}$ and $2^{nd}$ order shape deviation. To analyse the influence on toughness, Charpy V-notch tests should be carried out.
Concerning the process, the acceleration voltage will be added as a subject of investigation. By using a lower accelerating voltage arcing may be reduced for specimens with high surface roughness.

## 6. References

[1] EN, "EN 1011-7 - Empfehlung zum Schweißen von Metallischer Werkstoffe, Elektronenstrahlschweißen," Wien, 2004.
[2] DVS, "DVS MB 3213 - Empfehlungen zum Reinigen der Fügezone für das Elektronenstrahlschweißen," Düsseldorf, 2010.
[3] H. Schultz, *Elektron beam welding*. Cambridge: Woodhead Publishing Ltd, 2004.
[4] D. Dobeneck v., T. Löwer, and V. Adam, *Elektronenstrahlschweißen - Das Verfahren und seine industrielle Anwendung für höchste Produktivität*. Burg, Germany: Verlag moderne Industrie, 2001.
[5] V. Adam, U. Clauß, D. Dobeneck, T. Krüssel, and T. Löwer, *Electron Beam Welding - The Fundamentals of a fascinating Technology*. Burg, Germany: pro-beam AG & Co. KGaA, 2011.
[6] Y. Luo, J. Liu, C. Du, H. Xu, and C. Li, "The evaporation effect of front keyhole wall in penetration welding with an electron beam," *Mater. Sci. Forum*, vol. 686, pp. 355–360, 2011.
[7] U. Reisgen, S. Olschok, and S. Jakobs, "International Electron Beam Welding Conference," in *A comparison of electron beam welding with laser beam welding in vacuum*, 2012, pp. 119–127.
[8] Y. S. Panzer, "Vermeidung technologisch- und werkstoffbedingter Hochspannungsüberschläge in der Elektronenkanone beim Elektronenstrahlschweißen," 1978.
[9] D. Dobeneck, *An International History of Electron Beam Welding*, 1st ed. Pro-beam AG,, 2007.
[10] DIN, "DIN 4760 - Gestaltabweichungen; Begriffe, Ordnungssystem," 1982.
[11] K.-H. Grote and J. Feldhusen, *Dubbel - Taschenbuch für den Mschinenbau*, 21st ed. Berlin-Heidelberg: Springer Verlag, 2005.
[12] Böhler, "Werkstoffdatenblatt 1.4313 - Böhler," Kapfenberg, N400 DE- 03.99 - EM-WS, 2013.

IOP Conf. Series: Materials Science and Engineering **119** (2016) 012009 doi:10.1088/1757-899X/119/1/012009

# Single-asperity contributions to multi-asperity wear simulated with molecular dynamics

**S J Eder, U Cihak-Bayr, and D Bianchi**
AC2T research GmbH,
Viktor-Kaplan-Straße 2, 2700 Wiener Neustadt, Austria
E-mail: stefan.eder@ac2t.at

**Abstract.** We use a molecular dynamics approach to simulate the wear of a rough ferrite surface due to multiple hard, abrasive particles under variation of normal pressure, grinding direction, and particle geometry. By employing a clustering algorithm that incorporates some knowledge about the grinding process such as the main grinding direction, we can break down the total wear volume into contributions from the individual abrasive particles in a time-resolved fashion. The resulting analysis of the simulated grinding process allows statements on wear particle generation, distribution, and stability depending on the initial topography, the grinding angle, the normal pressure, as well as the abrasive shape and orientation with respect to the surface.

## 1. Introduction

Most molecular dynamics (MD) simulations treat two-body wear as a nanometric cutting process and focus on the interaction between a single abrasive particle and a usually flat substrate [1, 2, 3]. However, wear processes may involve circumstances such as changing rough topographies, random abrasive orientation distributions or wear debris build-up between neighbouring abrasives, which cannot be studied using the above approach. The implementation of a statistically based surface roughness in an MD model combined with multiple abrasive particles acting on the surface is therefore useful for more realistic modelling of wear phenomena [4, 5, 6, 7]. In this way, the interactions between abrasives/asperities are explicitly included in the model, while the results, if properly evaluated, retain most of the information from the single-asperity level.

In this work, we simulate the abrasion process of an atomically rough ferrite surface with 16 hard abrasive particles implicitly bonded into a matrix using MD. By dynamically identifying the atoms which are worn away from the substrate, the wear volume can be quantified and monitored in a time-resolved fashion. Moreover, every nanoscopic wear particle is affiliated with the abrasive particle by which it is produced using an iterative clustering algorithm. This allows a thorough analysis of the individual contributions of the abrasive particles to the wear volume. We describe our MD model as well as the simulation procedure in section 2, where we also give a short summary of how the clustering algorithm identifies the wear particles. In section 3 we first briefly present some exemplary worn topographies, followed by a more in-depth discussion of the time dependence as well as the distribution of the individual abrasive contributions to the wear volume, based on two selected examples. We also bring up two special cases, eventually showing the limits of the clustering algorithm. We conclude with a listing of five major influences on wear particle stability in section 4.

## 2. Computational details

### *2.1. Model*

Our single-crystal ferrite substrate has lateral dimensions of $28.5 \times 28.5$ nm$^2$, an initial average height of 28 monolayers, and exposes a (001) surface to the abrasives. Its periodically replicated topography features a near-Gaussian height distribution with a root mean squared roughness of 0.78 nm, an arithmetic mean height of 3.94 nm, a nominal asperity width of 2 nm, and a maximum random lateral asperity site offset from the regular grid of 1 nm, see figure 1. More information on the topography generation procedure can be found in [4]. At the substrate base, three rigid monolayers of Fe atoms form support, covered by two monolayers that are kept at a temperature of 300 K. This is done via a Langevin thermostat acting only in $y$ direction so that it does not overly interfere with the movement of the abrasive particles and the applied pressure. The abrasive particles are modelled as rigid bodies arranged in a staggered $4 \times 4$ grid as shown in figure 1, rotated randomly between 0 and 90 degrees about all three Cartesian axes, and they are initially placed several nanometres above the surface. Two abrasive particle geometries are considered: spherical with a radius of 3.25 nm and cubic with a side length of 4.2 nm, chosen to provide a high degree of surface coverage without allowing overlap between neighbouring abrasives. A state-of-the-art Finnis-Sinclair potential is used for the Fe–Fe interactions [8], while the interactions between abrasive particles and the substrate is modelled using a Lennard-Jones potential with the parameters $\varepsilon = 0.125$ eV and $\sigma = 0.2203$ nm. These values are comparable to [2, 9, 10, 11] and yield an interaction roughly one order of magnitude weaker than the one for Fe–Fe, leading to mild adhesion.

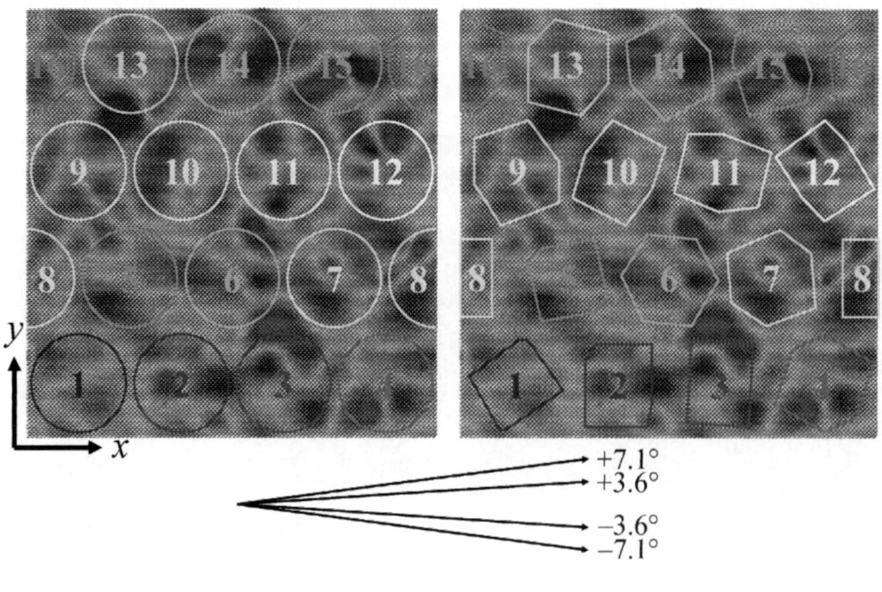


**Figure 1.** Top view of the initial substrate topography coloured according to local height (red is high, blue is low). The numbered outlines of the 16 spherical (left) and cubic (right) abrasives in their initial positions are superimposed with individual rainbow-style colouring for future reference. Note that abrasives #8 and #16 straddle a simulation box boundary. Below, the four grinding angles $\alpha$ are schematically shown.

### *2.2. Simulation procedure*

All wear simulations were performed using the open-source MD code LAMMPS [12]. We apply periodic boundary conditions in the $x$ and $y$ directions, meaning that if an atom leaves the simulation box through one boundary, it re-enters at the opposing one. The abrasive particles are dragged across the surface at a sliding velocity in $x$ direction of $v^{(\mathrm{max})} = 16$ m/s at four different angles $\alpha$ of $\pm 7.1°$ and $\pm 3.6°$ with the $x$-axis, so that the abrasives re-enter the simulation box at different $y$ positions every time they pass the periodic box boundaries and therefore do not immediately follow in their own grinding marks, see the lower part of figure 1. Since the $y$ component thus introduced increases the total sliding velocity by less than 1%, only the $x$ component is discussed henceforth. The particle positions relative to each other are locked throughout the simulations, and particle rotation is not allowed as would be the case for a grindstone. Furthermore, the particles cannot change their $z$ positions individually depending on the topography, but only collectively. This leads to occasional loss of surface contact for individual abrasive particles for some time intervals, while the remaining particles share the increased load. The abrasive particles themselves are not subject to any wear. The normal pressure $\sigma_z$ on the abrasives is kept constant at values of 0.1, 0.5, and 1.0 GPa for the entire simulation time of 15 ns, corresponding to a total abrasive translation of $\Delta x = 240$ nm. It therefore takes a given abrasive 1.8 ns to perform one sweep of the entire simulation box.

The $x$ component of the Fe atom velocities, $v \equiv v_x$, is Gauss-filtered with a total filtering window of 0.4 ns and a standard deviation of 0.2 ns to yield the pure atomic drift velocities without any thermal contribution. Atoms with a momentary drift velocity of $v > 0.9\, v^{(\mathrm{max})}$ are considered wear particles, those essentially at rest with $|v| < 0.1\, v^{(\mathrm{max})}$ are part of the substrate, while all remaining atoms that do not belong to the abrasive particles are currently under shear [4]. The number of atoms in the wear particles can then be multiplied with a per-atom volume of 11.6 Å$^3$, leading to a wear volume $V_\mathrm{w}(t)$ equivalent to the wear depth $h_\mathrm{w}(t)$ defined in [6].

### *2.3. Identification of individual wear particles*

As discussed in further detail with respect to contact zones in [4], it is possible to clusterize the contributions of the individual abrasives to the total wear volume $V_\mathrm{w}(t)$ by iteratively searching for wear particle atoms which lie closer than $d_\mathrm{clust} = 0.4$ nm to abrasive particles or other wear particle atoms whose abrasive affiliation is already known. The number of affiliated wear particle atoms may stop growing due to problems arising from the periodic boundary conditions, namely unaffiliated atoms with their affiliated neighbors at the other side of the simulation cell. This can be solved by once mapping all atoms in question which lie closer than $d_\mathrm{clust}$ to the lower cell boundary across the upper cell boundary. Thus for the next iteration, a sufficient number of affiliated atoms is available on the correct side of the cell boundary so that further affiliations can be made. Finally, single outliers and small groups of outlying atoms join the closest clusters with known affiliation. This last step introduces a small error to the break-down of the total wear volume, as isolated wear particles without a single initial particle affiliation may be falsely affiliated with close yet unrelated abrasives in this way. The above method works very well for non-coalescent wear particles. As the build-up of wear debris between two abrasives at higher normal pressures becomes sufficient to form a single uninterrupted wear particle, it becomes a matter of definition which part of that particle should be identified with a given abrasive. Our algorithm initially searches for affiliated atoms only in sliding direction, which largely hinders the iterative "backwards growth" of abrasive affiliation observed in coalescing wear particles.

## 3. Results and Discussion

In figure 2, we give the reader an idea of what the final surfaces, including the wear particles, look like after grinding at different normal pressures. While some features from the original topography remain visible as blue areas for the lowest normal pressures (left), the other topographies are dominated by the geometries of the abrasive particles. It is also obvious from the comparison of the images in the right column that cubic abrasives produce larger wear particles by way of cutting, which also have a tendency to coalesce, while spherical abrasives produce smaller, more evenly distributed wear particles that result mainly from ploughing.

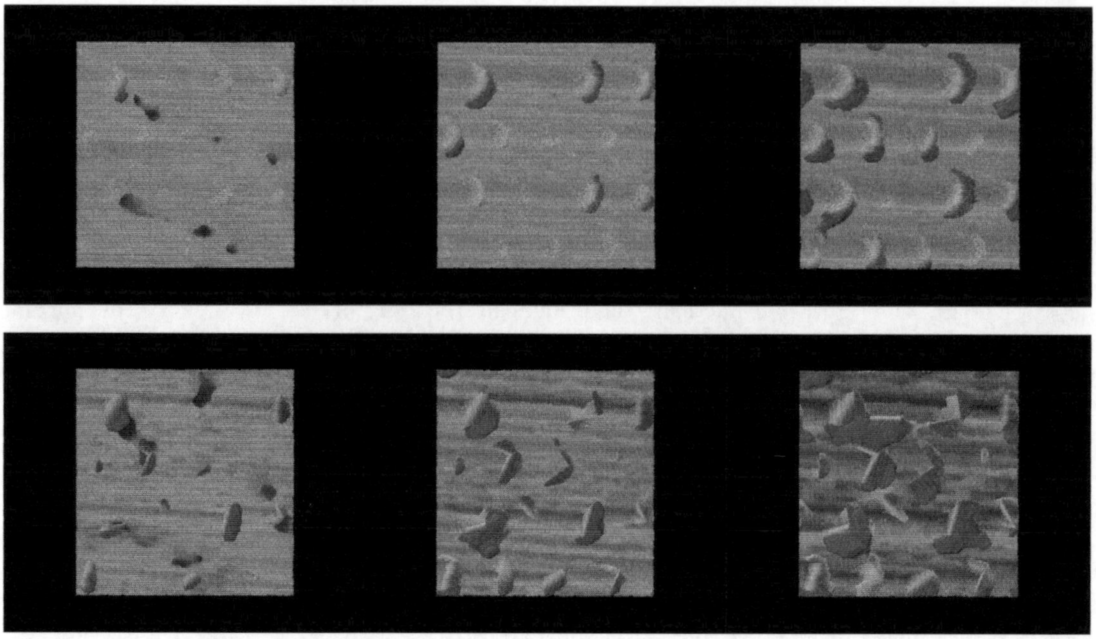

**Figure 2.** Exemplary final topographies after grinding at normal pressures of 0.1, 0.5, and 1.0 GPa (from left to right) with spherical abrasives (top) or cubic abrasives (bottom). The surface atoms are coloured according to their local height, where blue denotes valleys, yellow is the original mean height, elevated parts are shown in green and the extreme tips in red. The grinding angle was $\alpha = -3.6°$ in both examples.

We discuss the individual abrasive contributions to the total wear volume by means of some examples shown in figure 3. There we selected two examples of what we consider (based on the large body of available load- and grinding-angle-resolved data) two typical cases as well as two special cases that require separate discussion. In the left panel of each row, we show the time development of the total wear volume broken down into the individual abrasive contributions, coloured according to the abrasive key colour scheme introduced in figure 1 and used throughout this contribution. This representation allows the tracking of the emergence, stability, and possible destruction of wear particles as well as the occurrence of periodic phenomena. The center column in figure 3 shows a side-by-side comparison of time-averaged wear volume contributions with error bars representing the time variance over the entire grinding duration. This bar chart layout lends itself well to study emerging patterns which might be explained with the initial topography, the normal pressure, the grinding angle, the asperity shape or orientation. Finally, on the right, we have included a top-view snapshot of the wear particles only, coloured according to their abrasive affiliation. This allows us to check how well the clustering algorithm determined abrasive affiliations and how it handled wear particle coalescence. It also allows us to estimate the order of magnitude of the error made when a wear particle is in physical contact with more than the one abrasive that produced it, thus leading to a number of false affiliations.

Rows 1 and 2 in figure 3 represent typical results at a normal pressure of 0.5 GPa for spherical and cubic abrasives, respectively. It is obvious that the cubic abrasives produce more than

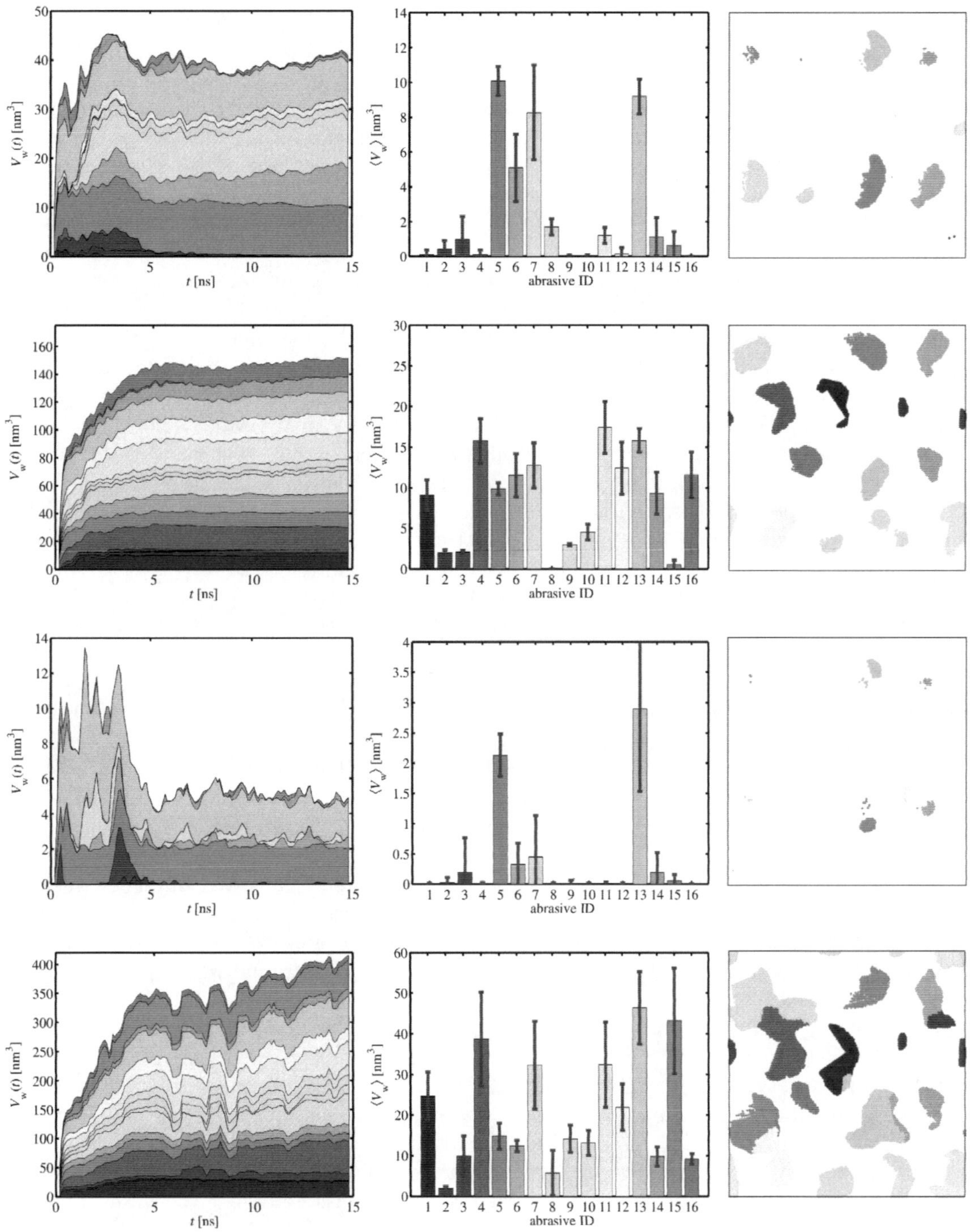


**Figure 3.** Left column: time dependent break-down of wear volume with respect to abrasive particle; centre column: comparison of time-averaged per-abrasive wear volumes; right column: top view of distribution and affiliation of wear particles at $t = 15$ ns. Rows from top to bottom: (1) spherical abrasives at 0.5 GPa, $\alpha = -7.1°$, (2) cubic abrasives at 0.5 GPa, $\alpha = +3.6°$, (3) spherical abrasives at 0.1 GPa, $\alpha = -7.1°$, (4) cubic abrasives at 1.0 GPa, $\alpha = -3.6°$.

three times as much wear as the spherical abrasives. However, when breaking the wear volume down into individual abrasive contributions, it becomes clear that while spherical abrasives can produce wear particles with an average volume of $\approx$ 10 nm$^3$, only three out of 16 do so. A comparison of row 1 with figure 1 explains why abrasives #5 and #8 immediately produce sizeable wear particles at a grinding angle of $\alpha = -7.1°$, as two large asperities of the initial topography lie directly on their respective trajectories. When looking at the time-dependent representation at the left of row 1, we see that wear particles #6 and #7 are formed after $\approx$ 1.5 ns and then remain stable, while the smaller wear particle #3 is re-incorporated into the substrate after 5 ns.

By contrast, ten out of the 16 cubic abrasives produce similarly sized or even larger wear particles, see row 2 in figure 3. Of the remaining six, abrasives #2, #3, and #8 have disadvantageous orientations for producing large wear particles (see figure 1), as they do not present a cutting edge to the surface, and they also hardly come into substrate contact at this intermediate normal pressure. The two neighbouring abrasives #15/#16 have somewhat "complementary" orientations so that their ability to produce wear particles depends greatly on the grinding angle, with abrasive #16 favouring positive grinding angles and #15 negative ones. The same applies, to a lesser extent, to abrasives #9/#10. A comparison of the wear volume time development between rows 1 and 2 shows that the wear particles produced by cubic abrasives, once formed, are more stable over time than those formed by spherical abrasives.

Row 3 in figure 3 represents a special case of grinding with spherical abrasives at low normal pressure, a process characterized by small wear volumes, unstable wear particles, as well as an initial peak/plateau of the total wear volume that subsides after 2.5–5 ns (when the roughness has reached its minimum), followed by a small, constant residual wear volume. The only two (stable) wear particles worth mentioning in the example shown in row 3 of figure 3 are #5 and #13, both of which are formed at the very beginning of the process due to the positions of their abrasives with respect to the initial asperities.

The second special case is presented in row 4 of figure 3. Here the top-view snapshot of the wear particles coloured according to their abrasive affiliation exhibits several examples of possibly false affiliation. At high normal pressure, cubic abrasives can form such large wear particles that some of them may touch the abrasive preceding them, which leads to false affiliation of several atoms. However, the error introduced to the break-down of the total wear volume in this way is small compared to the size of the wear particles. By contrast, wear particle coalescence, as seen between wear particles #12 and #15, as well as #4, #7, and #8, may present a greater challenge. Here, the general question arises how to correctly determine which part of a large wear particle was caused by which abrasive.

Taking into account the grinding angle of $\alpha = -3.6°$, it seems obvious for a human observer that a sizeable portion of what the clustering algorithm considers wear particle #12 should actually be considered a part of wear particle #15. Likewise, much of wear particle #4 is produced by abrasives #7 and #8. While the clustering algorithm employs some of the information (such as the main grinding direction) available to a knowledge-based analysis, it is currently unable to decide that, e.g., a wear particle is unlikely to be produced by the trailing edge of an abrasive. The implementation of such complex considerations into the algorithm is the focus of ongoing work.

The onset of false affiliations cannot be linked to specific theshold values such as the total wear volumes or the normal pressure. In fact it is a combined effect of the load, the inital topography, the size correlation between abrasives and asperity shapes and well as grinding direction relative to topographic features. Therefore the illustrations of the wear particle affiliations in the right column of figure 3 prove to be the best indicator for the limitations of a chosen simulation parameter set. Note, however, that a slightly incorrect particle affiliation does not influence the total wear volume. In contrast to our "dry" simulations under vacuum conditions, even in presence of a medium (e.g., $O_2$ or a lubricant), the same atoms would form wear particles, though possibly with some interface layers or free surfaces.

The false affiliations, however, are not the cause for the wear volume oscillations that can be seen in the left panel of row 4. These are rather the result of a large wear particle breaking off to the side, being re-incorporated into the substrate, and then being picked up by one of the following abrasives. Such oscillations only occurred for cubic abrasives at high normal pressures and negative grinding angles and seem to be connected with coalescing wear particles, as these have a higher likelihood of re-incorporation into the substrate between two abrasives.

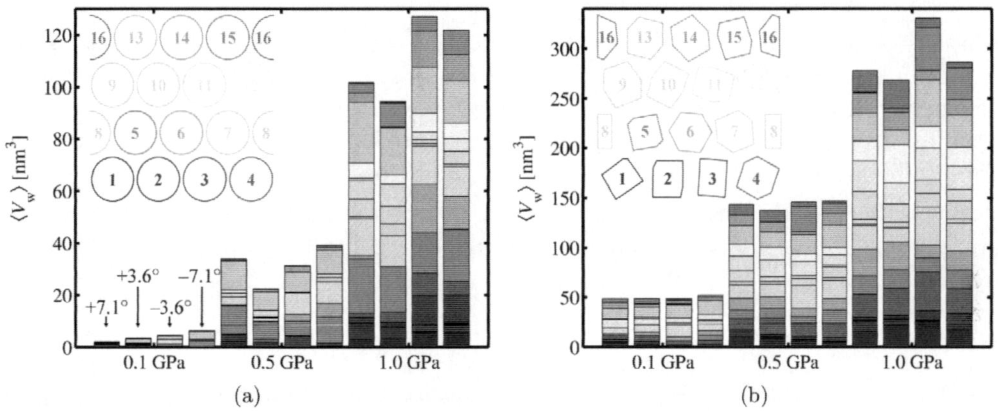


**Figure 4.** Break-down of the mean wear volume into individual abrasive contributions for spherical abrasives (a) and cubic abrasives (b), grouped by normal pressure. Grinding angle $\alpha$ within each group (from left to right): $+7.1°$, $+3.6°$, $-3.6°$, $-7.1°$.

In figure 4 we summarize the individual abrasive contributions to the total wear volume for all our simulations. The time-averaged particle-resolved wear volumes are displayed as stacked bars coloured according to abrasive affiliation and arranged in a way that allows the analysis of the dependence on the normal pressure as well as the grinding angle. The left panel shows the results for spherical abrasives and the right one those for cubic abrasives. We observe that the total wear volume is almost grinding-angle independent for cubic abrasives at low to intermediate normal pressures (varying only by approximately 3%), while the relative spread of the data produced by spherical abrasives is roughly ten times as high. At 1.0 GPa, the total wear volume varies by about 10%, depending on the grinding direction, for both abrasive geometries, and this dependence seems somewhat correlated between the two shapes. This serves as an indicator that at this normal pressure the orientation of the individual abrasives has little overall influence on the wear volume. Finally, the representation of the data in figure 4 also shows that, independent of normal pressure, the distribution of wear particles among the abrasives is much more stable with respect to varying grinding angles for cubic abrasives.

## 4. Summary and Conclusion

We have employed a recently proposed and partly knowledge-based clustering algorithm to break down the total wear volume occurring in nanoscopic grinding simulations into contributions of the individual abrasive particles. Based on the above, we can conclude that the stability of individual wear particles is influenced by:

(i) The initial position of the abrasive with respect to the original topography as well as the grinding angle. If the first topographic feature an abrasive encounters on its path upon striking the surface is a pronounced asperity, it will quickly produce a relatively large wear particle.

(ii) Existence of deep pits along the trajectory of the wear-particle-producing abrasive. Especially if these pits are encountered soon after the first wear particle has formed, there is a high chance that a large portion of the wear particle will be re-incorporated into the substrate there.

(iii) Abrasive shape. Spherical abrasives do not exhibit any large surfaces consisting of crystallographic planes, so wear debris attached to them can be more easily sheared off than from the cubic abrasives, which exhibit bcc (100) planes.

(iv) Abrasive orientation. While almost irrelevant for spherical abrasives, orientation is of high importance for the cubic abrasives as it and the grinding angle determine the rake angle, depending on which a given abrasive will tend more towards cutting or ploughing. Certain orientations (e.g., crystallographic planes nearly parallel to the surface) may even rule out appreciable wear particle formation except at the highest normal pressures.

(v) Wear particle detachment. Fortuitous detachment of a wear particle from the surface may prevent subsequent re-incorporation into the substrate and thus leads to a perfectly stable wear particle.

Our multi-asperity MD simulations have the advantage over single-asperity simulations that they can reflect the interactions between wear particles produced by individual abrasives in real grinding processes (and partly also wear processes). This could be shown by the wear particle affiliations and the sensitivity study with respect to the grinding angle. Compared to other simulation techniques (e.g., FEM), we believe that MD combined with proper surface topography analysis techniques is a suitable tool for describing real grinding processes at the nano-scale.

## Acknowledgments

This work was partly funded by the Austrian COMET-Program (Project K2, XTribology, no. 849109) and carried out at the "Excellence Centre of Tribology". The authors also acknowledge the support of the Province of Niederösterreich (Project "SaPPS", WST3-T-8/028-2014).

## References

[1] Komanduri R, Chandrasekaran N and Raff L M 2000 *Wear* **240** 113–143
[2] Pei Q, Lu C, Lee H and Zhang Y 2009 *Nanoscale research letters* **4** 444–451
[3] Si L, Guo D, Luo J and Lu X 2010 *Journal of Applied Physics* **107** 064310
[4] Eder S J, Bianchi D, Cihak-Bayr U, Vernes A and Betz G 2014 *Comput. Phys. Commun.* **185** 2456–2466
[5] Eder S J, Cihak-Bayr U and Pauschitz A 2015 *Wear* **http://dx.doi.org/10.1016/j.wear.2015.03.006**
[6] Eder S J, Feldbauer G, Bianchi D, Cihak-Bayr U, Betz G and Vernes A 2015 *Phys. Rev. Lett.* **115**(2) 025502
[7] Eder S J, Cihak-Bayr U, Betz G and Vernes A 2015 *Journal of Physics D: Applied Physics* **JPhysD-106042.R1** awaiting decision
[8] Mendelev M I, Han S, Srolovitz D J, Ackland G J, Sun D Y and Asta M 2003 *Philosophical Magazine* **83** 3977–3994
[9] Maekawa K and Itoh A 1995 *Wear* **188** 115–122
[10] Zhang L and Tanaka H 1997 *Wear* **211** 44–53
[11] Fang T H and Weng C I 2000 *Nanotechnology* **11** 148–153
[12] Plimpton S J 1995 *J. Comput. Phys.* **117** 1–19

International Conference on Materials, Processing and Product Engineering 2015 (MPPE 2015) IOP Publishing
IOP Conf. Series: Materials Science and Engineering **119** (2016) 012010 doi:10.1088/1757-899X/119/1/012010

# Refinement of digital image correlation technique to investigate the fracture behaviour of refractory materials

**Y.Belrhiti[1], O. Pop[2], A.Germaneau[3], P.Doumalin[3], J.C.Dupré[3], M.Huger[1], T.Chotard[1]**
[1] Science des Procédés Céramiques et Traitements de Surface
(SPCTS UMR CNRS 7315), CEC, 12 rue Atlantis, 87068 Limoges Cedex, France
[2] Groupe d'Etude des Matériaux Hétérogènes (GEMH), Université de Limoges, 12 rue Atlantis, 87068 Limoges Cedex, France
[3] Institut Pprime, CNRS, Université de Poitiers, 11 Boulevard Marie et Pierre Curie, 86962 Futuroscope Chasseneuil Cedex, France

E-mail: pascal.doumalin@univ-poitiers.fr

**Abstract.** Refractory materials exhibit a heterogeneous microstructure consisting in coarse aggregates surrounded by fine grains that form an aggregate/matrix composite. This heterogeneous microstructure often leads to a complex mechanical behaviour during loading. This paper is devoted to the study, thanks to an optical method, Digital Image Correlation (DIC), of the fracture behaviour of two industrial refractory materials in relation with their microstructure resulting from both the chosen constituents and the sintering process. The aim is here, specifically, to highlight and to characterize the evolution of kinematic fields (displacement and strain) observed at the surface of sample during a wedge splitting test typically used to quantify the work of fracture. DIC is indeed a helpful and effective tool, in the topic of experimental mechanics, for the measurement of deformation in a planar sample surface. This non-contact optical method directly provides full-field displacements by comparing the digital images of the sample surface obtained before and during loading. In the present study, DIC has been improved to take into account the occurrence of cracks and performed so as to better identify the early stage of the cracking behaviour. The material transformation, usually assumed homogeneous inside each DIC subset, is thus more complex and a discontinuity of displacement should be taken into account. Then each subset which crosses a crack can be cut in two parts with different kinematics. By this way, it is possible to automatically find the fracture paths and follow the crack geometries (length, opening).

## 1. Introduction

Refractories, widely used in many industrial areas such as steel, cement, lime, non-ferrous metals and glass processing, are known for their ability to be used at very high temperature and to resist, in the same time, to significant levels of mechanical stress and strain. In addition, their ability to sustain thermal shocks is also, in many cases, of prime importance to avoid premature failure. This property of thermal shock resistance is known to be closely related to the crack growth behaviour of these materials ([1]-[3]). In this aim to promote crack growth resistance, an initial network of microcracks can be voluntary introduced within the microstructure which moves the mechanical behaviour, from a classical pure elastic to a material with pronounced R-curve behaviour, thus enhancing work of fracture as well as strain to rupture. The tensile mechanical behaviour (Figure 1) is then typically characterized by a rather low stress-to-rupture and an important compliance (flexibility) of the material.

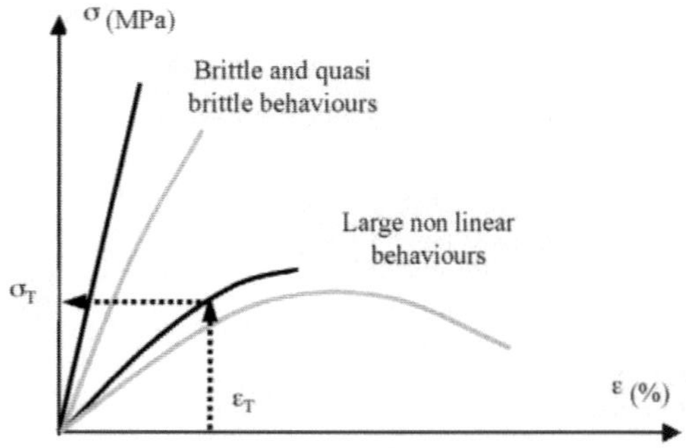


**Figure 1:** Typical stress/strain behaviours observed in tension on refractory materials [3]

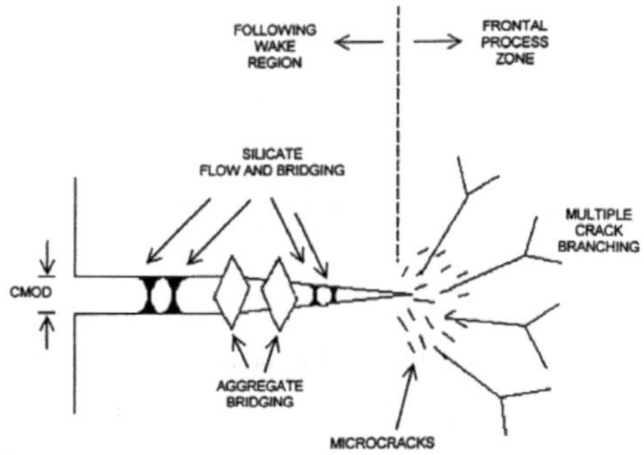


**Figure 2:** Schematic representation of the crack propagation zone for refractories [4]

During macrocrack opening, due to the initial network of microcracks intentionaly introduced within the microstructure of the material, different complex inelastic phenomena will occur and consume a large amount of energy which will slow down the macrocrack propagation, leading to an increase in work of fracture [4]. The crack region can be divided to two areas: the frontal process zone and the following wake region as represented in Figure 2. The frontal process zone is localized ahead of the main crack and corresponds mainly to micro-cracking. It may lead to large crack branching phenomena depending on the initial microstructure. Thus, the size of this frontal process region depends on the type of refractory. The following wake region is located behind the crack front and across the fracture surface. The most important mechanisms are here the bridging across the created crack surfaces and the frictional effects due to the pull-out of the aggregates between the two sides of the passing crack.

The aim of the present paper is specifically to illustrate those different mechanisms by their direct visualisation at the surface of sample with the help of specific numerical treatments of images recorded during mechanical test. Non-contact optical methods are commonly used as for example Digital Image Correlation technique (DIC). This technique has been developed since several years to study 2D surface motions and it is a powerful method to study the behaviour of materials [5]-[8]. DIC is based on the analysis of successive images of the surface of the studied specimen during a mechanical test. At each state of load, a displacement field is calculated by measuring the degree of similarity (assuming a material transformation) of series of subsets between the image corresponding to an initial and the one of the deformed studied state. The performance of this technique is closely related to local gradients of grey levels inside images [8]. These grey level values come from a speckle pattern at the surface which can be the natural texture of the specimen surface or an artificial pattern (deposed by spraying white and/or black paints).

Using DIC gives available results as shown in [6] but a bias exists [9]. Indeed the material transformation which is assumed homogeneous on a DIC subset cannot take into account the real discontinuous displacement field near the crack tip. For subsets crossing cracks, the measurement of displacement is disturbed by an unpredictable bias and the calculus of strains near these subsets is wrong. To avoid this problem, a manual inspection can be useful to detect the discontinuity path and to keep only the subsets on each side of the crack. This approach has been used a lot in fracture mechanics for straight cracks [7] and also with local analysis around the crack to determine opening displacements [10], [11]. The detection can be improved by using a mask given the location of cracks [9], [12] and the definition of subsets on the sides of cracks can be automated. Another solution consists in changing the material transformation usually assuming homogeneous: a subset can be split in two parts with different kinematics [9], [13]. In this study, we propose a new DIC technique based on this last principle and especially developed for the study of the behaviour of flexible ceramics. The fracture characterization is here conducted through wedge splitting test (WST) which is well known for its ability to produce reliable data on stable crack propagation.

## 2. Materials and methods

### *2.1. Material*

Magnesia-Spinel materials (MSp) have been investigated in this study. Thanks to a very high degree of refractoriness and to their resistance to basic environments, these refractory materials are typically used in steel making vessels or rotary cement kilns. Different MSp products exist, depending on the type of the spinel aggregates. Industrial bricks are usually obtained by uniaxial pressing of the raw materials and a sintering thermal cycle up to about 1600°C in order to obtain a rather dense refractory material (porosity about 15-16%). Due to thermal expansion mismatch between magnesia matrix (~13.5 $10^{-6}$ °$C^{-1}$) and spinel inclusions (less than 10.3 $10^{-6}$ °$C^{-1}$), microcracks are usually developed in the matrix around the spinel inclusions during the cooling stage of the process. Two types of industrial materials were studied here: MSp1 for which inclusions were aggregates of magnesium aluminate spinel ($MgAl_2O_4$) and MSp2 for which inclusions were aggregates of iron aluminate spinel (Hercynite-$FeAl_2O_4$). Density and porosity of these composite materials are given in Table 1 as well as elastic properties measured by ultrasonic means. These measurements were evaluated on 5 samples of each material type. Since these two materials MSp1 and MSp2 don't contain the same kind of spinel exhibiting different coefficients of thermal expansion (~7.6 x $10^{-6}$ °$C^{-1}$ for $MgAl_2O_4$ in comparison to ~10.3 $10^{-6}$ °$C^{-1}$ for $FeAl_2O_4$), the microcrack network developed during cooling is not the same leading to some differences in elastic properties. From each type of material, two cubic-shaped specimens with cross-section of 100x100x70 $mm^3$ were tested, although, in this paper only one will be presented since the results are quite similar. For image acquisition during mechanical loading and further numerical treatments, a great attention must be paid to sample surface preparation in order to avoid artefacts of correlation results between deformed and reference states. A black (or white) paint layer has been spread on specimen surface, and after drying, fine white (or black) droplets have been projected in order to constitute a speckle pattern useful for DIC. For this study, white droplets on a black surface were used.

**Table 1.** Properties of the studied MSp1 and MSp2 materials

| | MSp1 | MSp2 |
|---|---|---|
| Density (g.$cm^{-3}$) | 2.94 (+/-0.01) | 3.01 (+/- 0.02) |
| Porosity (%) | 16.07 (+/- 0.50) | 15.11 (+/- 0.29) |
| Young's modulus E (GPa) | 31.43 (+/- 3.43) | 39.26 (+/- 3.56) |
| Shear modulus G (GPa) | 13.48 (+/- 1.79) | 16.89 (+/- 1.61) |
| Poisson ratio ν | 0.17 (+/- 0.03) | 0.16 (+/- 0.02) |

### *2.2. Wedge Splitting Test*

As other test methods like the notched beam test and the compact tension method, the wedge splitting test according to Tschegg [14], [15] allows to specify the mechanical fracturing properties of refractory materials under uniaxial load (mode I loading). This test, patented in 1986, is nowadays often applied on refractory materials to determine fracture mechanical properties (e.g. the specific fracture energy or the notch-tensile strength). The action of a wedge and the rather high rigidity of the equipment ensure a low energy accumulation in the specimen and the testing machine. Stable crack propagation can thus be easily achieved by displacement-controlled loading with a simple mechanical or hydraulic testing device. The test configuration is represented in Figure 3. The specific sample geometry, with a groove in order to apply the splitting load and a starter notch, is manufactured by machining. The vertical action of a metallic wedge, placed on the upper part of the sample and in contact with movable rolls, entails horizontal forces thank to load transmission pieces. Finally, the evolution of the horizontal force $F_H$ versus the horizontal displacement δ is registered. $F_H$ is evaluated from the applied vertical load Fv taking into account the angle of the wedge and omitting the presence of friction.

From the curve horizontal load vs. horizontal displacements, elastic energy is estimated by assuming that an elastic linear unloading of the material reaches the initial state before mechanical solicitation. It corresponds to the area enclosed in a triangle whose vertexes are the origin of the curve, the displacement and the applied load. The total energy is calculated by integrating the load-displacement curve.

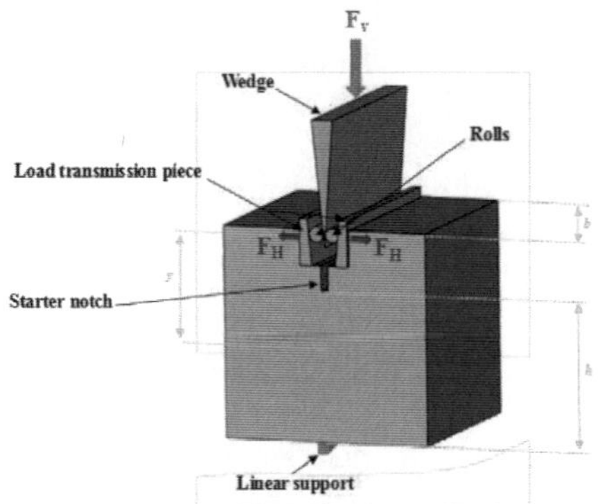


(a)

(b)

**Figure 3:** Principle of wedge splitting test with sample's geometry (a) and image acquisition setup (b)

### *2.3. 2P-DIC*

#### *2.3.1. Principle*

The new DIC technique, named two-parts DIC (2P-DIC) has especially been developed for the type of fracture observed in wedge splitting tests on refractory materials *i.e.* with a preferential direction of crack (here vertical) and a quasi-brittle mechanical behaviour. This technique is designed to detect the presence, the position and the opening of a crack. It uses the same principle of usual DIC for a homogeneous material transformation $\phi$ which is described by the vector $\underline{q}=\left(u, v, \frac{\partial u}{\partial x}, \frac{\partial u}{\partial y}, \frac{\partial v}{\partial x}, \frac{\partial v}{\partial y}\right)$ gathering two translation parameters (*u*,*v*) and four components of local gradient [7]. The expression of correlation coefficient used to this study is given by the following formula: $C(\underline{q})=\sum_D c^2(q)$ with

$$c(q)=\frac{f(\underline{x})}{\overline{f(\underline{x})}}-\frac{g(\phi(\underline{X}))}{\overline{g(\phi(\underline{X}))}} \tag{1}$$

where *D* is a subset, *f* and *g* represent the grey levels of pixels, $\overline{f}$ and $\overline{g}$ are their averages in *D* and *ϕ(D)*, $\underline{X}$ and $\underline{x}$ are the coordinates of grey levels for the reference and the deformed state respectively.

In the new proposed version of DIC, the subset *D* is divided vertically at the position XC in two parts $D_1$ (x < XC) and $D_2$ (x > XC) as shown Figure 4. The pixels at x=XC are not taken into account in the matching process.

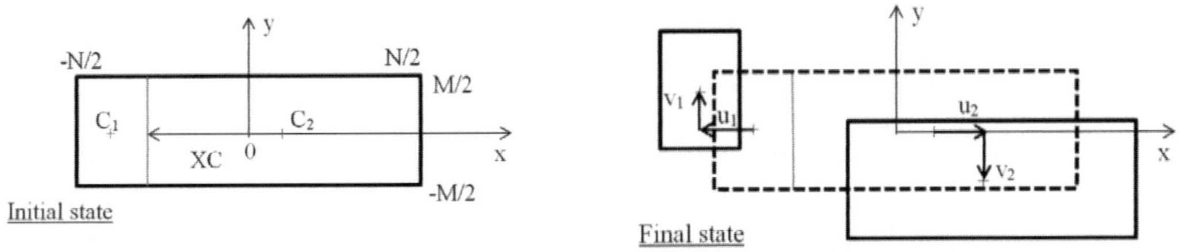


**Figure 4.** Principle of the 2P-DIC, decomposition of a subset.

For a given value of XC, the expression of correlation coefficient becomes:

$$C_{XC}(\underline{Q}) = \sum_{D_1} c^2(\underline{q_1}) + \sum_{D_2} c^2(\underline{q_2}) \tag{2}$$

With $\underline{q}_i = \left( u_i, v_i, \frac{\partial u_i}{\partial x}, \frac{\partial u_i}{\partial y}, \frac{\partial v_i}{\partial x}, \frac{\partial v_i}{\partial y} \right)$ the vector of material transformation parameters in the part $D_i$.(i=1,2). The vector $\underline{Q}$ represents all kinematic unknowns of $D$ and is the gathering of two vectors $\underline{q}_1$ and $\underline{q}_2$. The matching consists in searching the best values of XC and $\underline{Q}$ which minimize $C_{XC}(\underline{Q})$. For each value of XC varying by one-pixel step from -N/4 to N/4, a mean-square minimization process is used to achieve the best estimation of $\underline{Q}$. The values of XC smaller than -N/4 and larger than N/4 are not considered because they lead to a too small part $D_i$ giving a displacement with an insufficient accuracy. We also have to note that considering the quasi-brittle behaviour of our material for this study we have neglected the local gradient parameters, and only the four components of rigid-body displacements are calculated. As crack propagation is almost vertical, we can here choose a rectangular subset with a height (M) smaller than the width (N) as shown Figure 4.

### *2.3.2. Crack detection*

In previous studies [10], standard DIC was used to detect cracks according to the following method. Strain fields were calculated by finite differences [7] from displacement fields and a strain threshold $\varepsilon_S$ was defined to highlight the crack zone. Indeed crack presence corresponds to local higher strain values. This threshold $\varepsilon_S$ was set in order to extract the signal corresponding to reliable mechanical strains from fluctuations only due to DIC uncertainty. Thus cracks were detect with a rather large gauge length imposed by the initial choice of the subset size and of the grid step used to calculate strains by finite differences.

The proposed 2P-DIC gives a field of vectors $\underline{Q}$ for each subset of the zone of interest of the sample and we can deduce a crack opening field from the difference of the displacements of each part of the subset $D$. To separate physical opening values to fluctuations due to 2P-DIC uncertainty, a thresholding is necessary. To have the same analysis way than the one for standard DIC, a pseudo strain field is calculated from the next expression $\varepsilon=(u_2-u_1)/(XC_2-XC_1)$ (for a vertical crack) and the threshold $\varepsilon_S$ is applied on this field. In this case, we have to note that the detection of horizontal crack position is determined with a resolution equal to one pixel in x direction.

## 3. Results

### *3.1. Load-displacement curves obtained by WST*

Figure 5 represents the typical horizontal load-displacement curves of the two studied materials obtained by WST. The mechanical behaviour is characterized by three stages: (1) the first stage corresponds to linear elastic behaviour, (2) the beginning of non-linearity is related to the occurrence of very first microcracks within the microstructure close to the notch, (3) the maximum of the curve is linked to the macro-crack onset (the following part of the curve representing the macrocrack propagation). The horizontal load-displacement curve of MSp1 exhibits a lower maximum load in comparison to MSp2, thus the replacement of $MgAl_2O_4$ aggregates by hercynite aggregates seems here to increase the fracture energy which moves from 172 $J.m^{-2}$ for MSp1 to 227 $J.m^{-2}$ for MSp2.

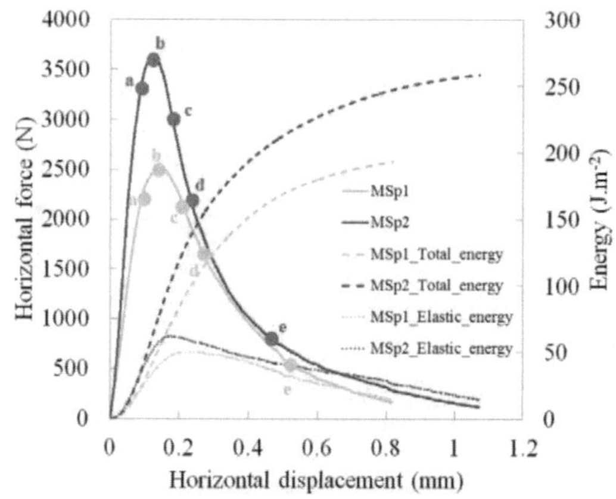


**Figure 5.** Horizontal load and energies versus horizontal load displacement for MSp1 and MSp2

In order to visualise the microcrack onset (stage 2) and macrocrack propagation (stage 3), different states, from (a) to (e) have been chosen (Figure 5) to compare results obtained by 2P-DIC and standard DIC. Taking into account fluctuations in $\varepsilon$ only due uncertainty for each treatment, related to subset size and step between subsequent subsets used to calculate $\varepsilon$, the thresholds $\varepsilon_S$ have been adjusted to 0.002 for DIC and to 0.003 for 2P-DIC.

### *3.2. Visualisation of microcracks onset and macrocrack propagation in the case of MSp1*

From the same images acquired during the same wedge splitting test on MSp1 material, Figures 6 and 7 present the results of standard DIC and 2P-DIC respectively for states (a) to (e) previously indicated in Figure 5. At first glance, one could observe that both approaches lead to a rather unique crack for states (a) to (c) and then, that large crack branching can be detected in states (d) and (e). It seems here that DIC and 2P-DIC reveal the same cracks but with a much better spatial resolution for the latter. Looking more carefully to the different thumbnails on Figure 7, particularly usefully in the case of 2P-DIC treatment for which each individual crack thickness is typically drawn at least with one pixel, one could notice that even in the early stages (states (a) and (b)), not only one single microcrack is observed (in contrast to standard DIC for these same first states). In same trend, one could also observe that, for state (d), 2P-DIC is able to reveal crack branching in the upper part of the global picture which is not detected by standard DIC. At this state, in some zones, the thickness of the cracks is larger than one pixel and one can observe their opening.

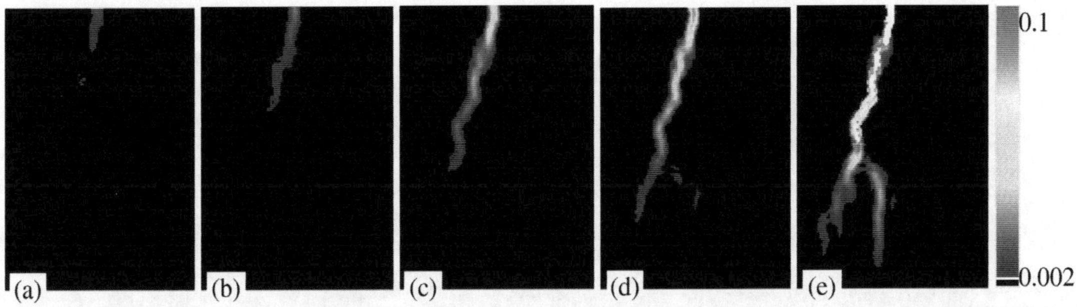


**Figure 6.** Standard DIC horizontal strain field with $\varepsilon_S$=0.002, subset size 32x32 pixels, subset shift 8 pixels, strain gauge length 32 pixels

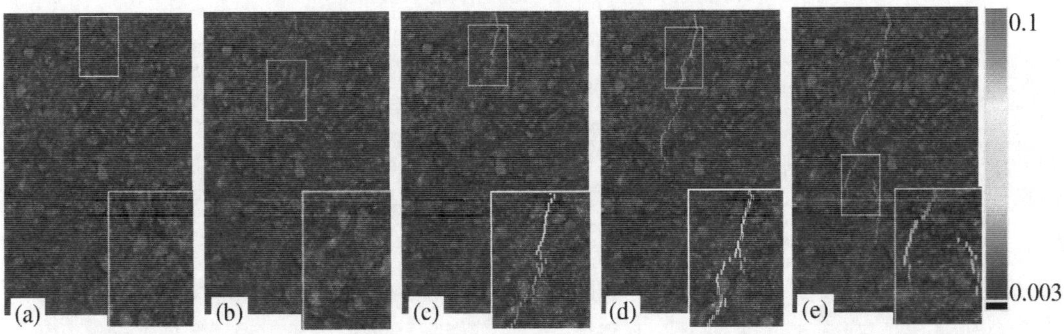


**Figure 7.** 2P-DIC crack opening by plotting $\varepsilon$ for $\varepsilon_S$=0.003, subset size 64x8 pixels.

*3.3. Visualisation of microcracks onset and macrocrack propagation in the case of MSp2*

In a similar way to previous results, from the same images acquired during the same wedge splitting test on MSp2 material, Figures 8 and 9 present the results of standard DIC and 2P-DIC respectively for states (a) to (e) previously indicated in Figure 5. Same general comments can be done here: DIC and 2P-DIC detects the same global crack path but with a much better spatial resolution for 2P-DIC. Nevertheless, one could notice that the crack branching phenomenon takes place earlier (state (c) here in comparison to state (d) for MSp1) and is much more developed in further states (states (d) and (e)). This important microstructural feature leads to the larger specific fracture energy measured on this second material. Again, looking more carefully to the different thumbnails on Figure 9, it seems that even in the very early stages (stage (a) and (b)), the different very small microcracks are possibly more dispersed (in comparison to same stages on material MSp1), which leads further to more numerous branches when actual crack branching phenomenon will take place (especially in stage (d)). This very interesting point has to be related to voluntary introduced network of microcracks in the initial microstructure of these materials.

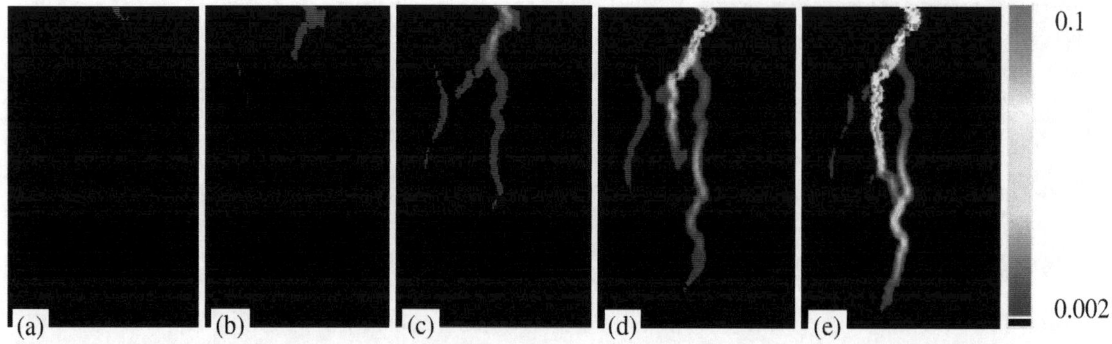


**Figure 8.** Standard DIC horizontal strain field with $\varepsilon_S$=0.002, subset size 32x32 pixels, subset shift 8 pixels, strain gauge length 32 pixels

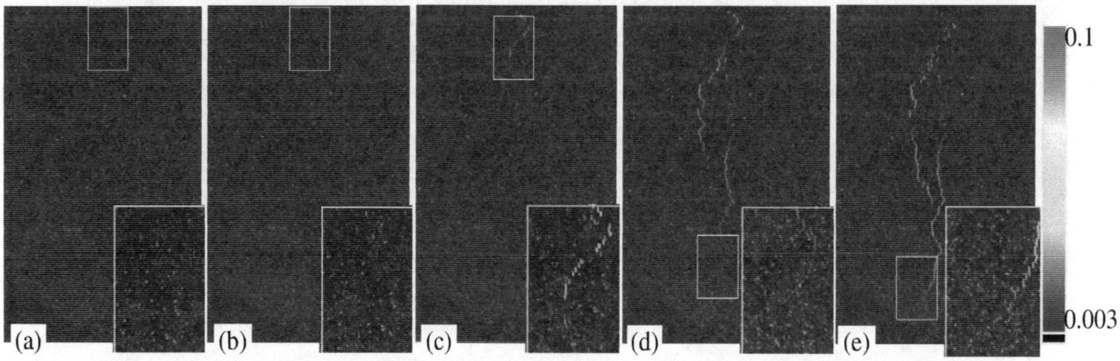


**Figure 9.** 2P-DIC crack opening by plotting ε for $\varepsilon_S$=0.003, subset size 64x8 pixels.

## 4. Conclusion

Thermal shock resistance of refractories is known to be closely related to the crack growth behaviour. This property can be optimized in industrial materials by a voluntary introduced micro-cracked network within the initial microstructure moving the mechanical stress-strain law from a elastic behaviour to a nonlinear one, thus enhancing surface fracture energy as well as strain to rupture. The recent development of optical methods for kinematic field measurement and associated dedicated numerical treatments is a promising route for improvement in the understanding of fracture mechanisms in these heterogeneous materials. Recent refinement of Digital Image Correlation technique with an adapted 2P-DIC treatment applied in this study on two magnesia-spinel materials illustrates the advantage which can be obtained in the case of wedge splitting test. The obtained results highlighted in particular the efficiency of this new technique to early detect crack path with an optimal spatial resolution. For the studied materials, it allows to underline the occurrence of the crack branching phenomenon, induced by the thermal expansion mismatch between the different phases with a significant differentiation between materials containing $MgAl_2O_4$ aggregates or hercynite aggregates. Besides, this optimized 2P-DIC method that allows to automatically determine the crack path with an interesting spatial accuracy, opens new way to correlated actual crack(s) length with the so-called specific fracture energy which plays a key role in the thermal shock resistance of refractories.

## Acknowledgements

This work is a part of a global research program engaged through collaboration within the F.I.R.E network. The authors are thankful to RHI AG for providing the materials.

## References

[1] G. A. Gogotsi, Y. L. Groushevsky, K. K. Strelov, The significance of non-elastic deformation in the fracture of heterogeneous ceramic materials, *Ceramurg. Int.*, 4, 1978, 113-118.

[2] H. Harmuth, E. K. Tschegg, A fracture mechanics approach for the development of refractory materials with reduced brittleness, *Fatigue Fract. Eng. Mater. Struct.*, 20, 1997, 1585-1603.

[3] M. Huger, N. Tessier-Doyen, T. Chotard, C. Gault, Microstructural effects associated to CTE mismatch for enhancing the thermal shock resistance of refractories: Investigation by high temperature ultrasounds, *CFI Ceram. Forum Int. 84*, 2007, E93-E102.

[4] R. C. Bradt, Problems and possibilities with cracks in ceramics, *Science of Ceramics*, 1981, 349-360.

[5] H. A. Bruck, S. R. McNeill, M. A. Sutton and W. H. Peters, Digital image correlation using newton-raphson method of partial differential correction, *Experimental Mechanics,* vol. 29, no. 3, pp. 261-267, 1989

[6] S. Choi and S. P. Shah, Measurement of Deformations on Concrete Subjected to Compression Using Image Correlation, *Experimental Mechanics,* vol. 37, no. 3, pp. 307-313, 1997.

[7] Y. Barranger, P. Doumalin, J. Dupré and A. Germaneau, Strain Measurement by Digital Image Correlation: Influence of Two Types of Speckle Patterns Made from Rigid or Deformable Marks, *Strain,* 2012.

[8] M. Bornert, F. Brémand, P. Doumalin, J.-C. Dupré, M. Fazzini, M. Grédiac, F. Hild, S. Mistou, J. Molimard, J.-J. Orteu, L. Robert, Y. Surrel, P. Vacher and B. Wattrisse, Assessment of digital image correlation measurement errors: Methodology and results, *Experimental Mechanics,* vol. 49, no. 3, pp. 353-370, 2009.

[9] J. C. Dupré, P. Doumalin, H. A. Husseini, A. Germaneau and F. Brémand, Displacement Discontinuity or Complex Shape of Sample: Assessment of Accuracy and Adaptation of Local DIC Approach, *Strain, Article first published online*: 29 JUL 2015, DOI: 10.1111/str.12150.

[10] Y. Belrhiti, O. Pop, A. Germaneau, P. Doumalin, J.C. Dupré, H. Harmuth, M. Huger, T. Chotard; Investigation of the impact of micro-cracks on fracture behavior of magnesia products using wedge splitting test and digital image correlation , *Journal of European Ceramic Society*, 35(2), pp823-829, 2015

[11] J. Réthoré and R. Estevez, Identification of a cohesive zone model from digital images at the micron-scale, *Journal of the Mechanics and Physics of Solids,* vol. 61, p. 1407–1420, 2013

[12] Q. Zhang and J. Zhao, Determination of mechanical properties and full-field strain measurements of rock material under dynamic loads, *International Journal of Rock Mechanics & Mining Sciences,* vol. 60, p. 423–439, 2013

[13] L. Nunes and J. Reis, Estimation of crack-tip-opening displacement and crack extension of glass fiber reinforced polymer mortars using digital image correlation method, Materials and Design, vol. 33, p. 248–253, 2012

[14] E.K. Tschegg, Equipment and appropriate specimen shapes for tests to measure fracture values, 31.1.1986, AT no. 390328, Austrian Patent Office, Austria (Prüfeinrichtung zur Ermittlung von bruchmechanischen Kennwerten sowie hiefür geeignete Prüfkörper, 31.1.1986, Patentschrift Nr. 390328).

[15] E.K. Tschegg, New equipment for fracture tests on concrete, *Mater. Test. (Mater)*, 33, 1991, 338-342.

IOP Conf. Series: Materials Science and Engineering **119** (2016) 012011 doi:10.1088/1757-899X/119/1/012011

# Progress in development of coated indexable cemented carbide inserts for machining of iron based work piece materials

**C Czettl**[1] **and M Pohler**[1]

[1] CERATIZIT Austria GmbH, Metallwerk-Plansee-Straße 71, A-6600 Reutte, Austria

E-mail: christoph.czettl@ceratizit.com

**Abstract.** Increasing demands on material properties of iron based work piece materials, e.g. for the turbine industry, complicate the machining process and reduce the lifetime of the cutting tools. Therefore, improved tool solutions, adapted to the requirements of the desired application have to be developed. Especially, the interplay of macro- and micro geometry, substrate material, coating and post treatment processes is crucial for the durability of modern high performance tool solutions. Improved and novel analytical methods allow a detailed understanding of material properties responsible for the wear behaviour of the tools. Those support the knowledge based development of tailored cutting materials for selected applications. One important factor for such a solution is the proper choice of coating material, which can be synthesized by physical or chemical vapor deposition techniques. Within this work an overview of state-of-the-art coated carbide grades is presented and application examples are shown to demonstrate their high efficiency. Machining processes for a material range from cast iron, low carbon steels to high alloyed steels are covered.

## 1. Introduction

The introduction of cemented carbide as tool material changed the chip forming machining industry significant in the beginning of the last century. With the introduction of exchangeable tips in the mid of the last century, this progress moved further on. A further milestone was reached around 1969 when the first coated cemented carbide inserts came to the market. These first TiC coatings synthesized via chemical vapor deposition (CVD) extended the lifetime of the tools by factors of 6 to 10. Rapidly the progress moved on by using multilayer coatings of CVD TiCN-TiC to meet the required material properties for metal cutting, which are mainly wear resistivity and toughness. In the 70ies of the last century CVD $Al_2O_3$ as an additional coating material could be synthesized in production scale. Especially the high temperature properties like hardness at elevated temperatures and low thermal conductivity showed big improvements in steel and cast iron turning operations. In the further decades, the CVD TiCN-$Al_2O_3$ multilayer systems where upgraded in detail, which means deposition technology and microstructure. To increase the toughness of the tools in cutting operation and avoid the brittle eta-phase in the cemented carbide-coating interface, the deposition process was changed to a medium temperature process using $CH_3CN$ instead of $CH_4$ as carbon feed. In parallel, the post treatment of such coatings was improved to decrease the surface roughness as well as to adjust the residual stresses. [1-6]

Beside the synthesis of hard and wear resistant coatings using CVD, the physical vapor deposition (PVD) raised significantly for cemented carbide cutting tools with the introduction of TiAlN coatings at the end of the last century. Despite the huge variety of PVD coating systems used for solid carbide tools, TiAlN based coatings are still dominant when indexable inserts are used. [7-9]

The major proportion of cutting applications using indexable inserts can be covered using the CVD TiCN-$Al_2O_3$ and PVD TiAlN based coatings, in tailored coating thickness. For some special applications like milling of stainless austenitic steels and titanium alloys an outstanding coating system is CVD $TiB_2$ which is covered within this work. [10, 11]

The further development of steels in high demanding applications like gas turbines regarding high temperature strength and creep resistivity, lead to difficult du machine materials with high demands on the cutting tool. Beside the machining strategy, macro and micro geometry of the tool as well as machining conditions, the proper choice of insert grade is crucial. An all-in-one solution suitable for every purpose is usually not possible, so the aim of this paper is to point out the strengths of the different systems in the high demanding machining of turbine blades.

## 2. Experimental

### *2.1. Cutting test*

To compare the machining performance of cemented carbide inserts a helirough 5-axis machining process was chosen. Therefore a martensitic stainless steel X22CrMoV12-1 (1.4923) and an austenitic stainless steel X12CrNiWTi16-13 (1.4962) in the dimensions 100×60×400 mm were selected. The cutting parameters for both materials were: cutting speed $v_c$ = 280 m/min, feed per tooth $f_z$ = 0.3 mm, engagement ae = 25.88 mm and maximum depth of cut $a_{p,max}$ = 3 mm. All cutting tests were carried out without coolant. A Ceratizit MaxiMill A251.40.R.04-12-RS tool holder with four bottom type cutting inserts in RPHX1204MOSN in F50 and M50 geometry, shown in Figure 1, was used. One "entry" describes the material removal around the turbine blade at one longitudinal position. The lifetime criteria is a maximum flank wear $v_{B\,max}$ = 0.5 mm or insert breakage.

**Figure 1.** Ceratizit MaxiMill A251.40.R.04-12-RS tool holder for turbine blade machining and the helirough machined turbine blade in the background.

*2.2. Cemented carbide substrate materials*

To evaluate the cutting performance within a wide range of material properties, cemented carbide grades with varying composition from ISO P20 to P45 were used and combined with CVD TiCN-$Al_2O_3$, PVD TiAlN and CVD $TiB_2$ coatings. An overview of the used combinations is given in Table 1.

**Table 1.** Carbide grades and coatings used in the tooling tests

| ISO | Cobalt content | HV30 | Coating | Ceratizit Grade |
|---|---|---|---|---|
| P20 | 8 % | 1500 | CVD TiCN-$Al_2O_3$ | CTCP220 |
| P20 | 8 % | 1500 | PVD TiAlN | CTPP225 |
| P30 | 10.5 % | 1400 | CVD TiCN-$Al_2O_3$ | CTCP230 |
| P30 | 10.5 % | 1400 | PVD TiAlN | CTCP235 |
| P40 | 12.5 % | 1380 | CVD TiCN-$Al_2O_3$ | CTCM235 |
| P40 | 12.5 % | 1380 | PVD TiAlN | CTPM240 |
| P45 | 10 % | 1330 | CVD TiCN-$Al_2O_3$ | experimental |
| P45 | 10 % | 1330 | PVD TiAlN | experimental |
| P45 | 10 % | 1330 | CVD $TiB_2$ | experimental |

*2.3. Investigated coating systems*

The coatings where synthesized in industrial scale ARC-PVD and thermal CVD devices. The PVD TiAlN coatings were alloyed with Ta as described elsewhere [12, 13]. The CVD TiCN coatings consists of a TiN interlayer (0.7 μm), a medium temperature TiCN layer (5.8 μm) [14], a bonding layer and corundum type α-$Al_2O_3$ layer in 0001 texture (3.5 μm). The CVD $TiB_2$ coating is deposited on a 1 μm TiN interlayer and shows despite most CVD coatings residual compressive stresses [10]. The microstructure of the coatings is shown in Figure 2.

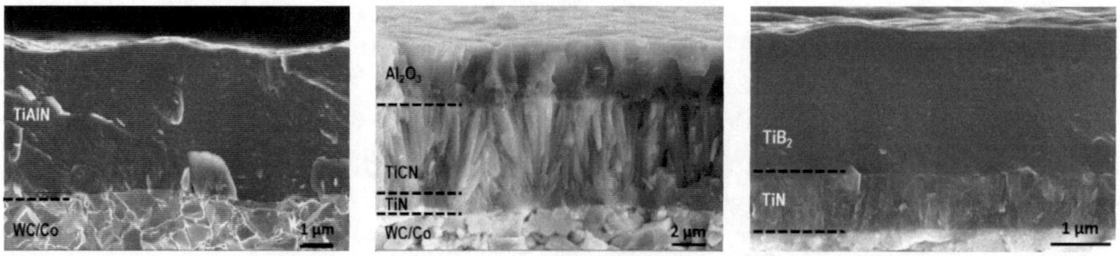


a) b) c)

**Figure 2.** Scanning electron microscope cross sections of the investigated a) PVD TiAlN ,b) CVD TiCN-$Al_2O_3$ and c) CVD TiN-$TiB_2$ coatings.

Due to the fact, that metal cutting operation takes place at elevated temperatures, the temperature dependent hardness of the outermost layer is important. Therefore, the coating hardness in the as deposited state as well as the hot hardness at 500 °C are given together with the corresponding coating thickness in

Table 2. The CVD TiCN-$Al_2O_3$ and PVD TiAlN coatings were post-treated in order to reduce the surface roughness and tailor the stresses in the CVD coating.

**Table 2.** Overview of coating properties

| Deposition method | Coating materials | Total thickness | Hardness RT | Hardness 500 °C |
|---|---|---|---|---|
| CVD | TiCN-$Al_2O_3$ | 10 µm | 26 GPa [15] | 16 GPa [15] |
| PVD | TiAlN | 5 µm | 28 GPa [15] | 9 GPa [15] |
| CVD | $TiB_2$ | 3.5 µm | 44 GPa [10] | - |

## 3. Results

The lifetime of the selected carbide and coating combinations is given as number of entries of the helirough process. The Figure 3 and Figure 4 showed the lifetime with increasing toughness of the cemented carbide substrate from ISO P20 toP45 for CVD TiCN-$Al_2O_3$ and PVD TiAlN, respectively.

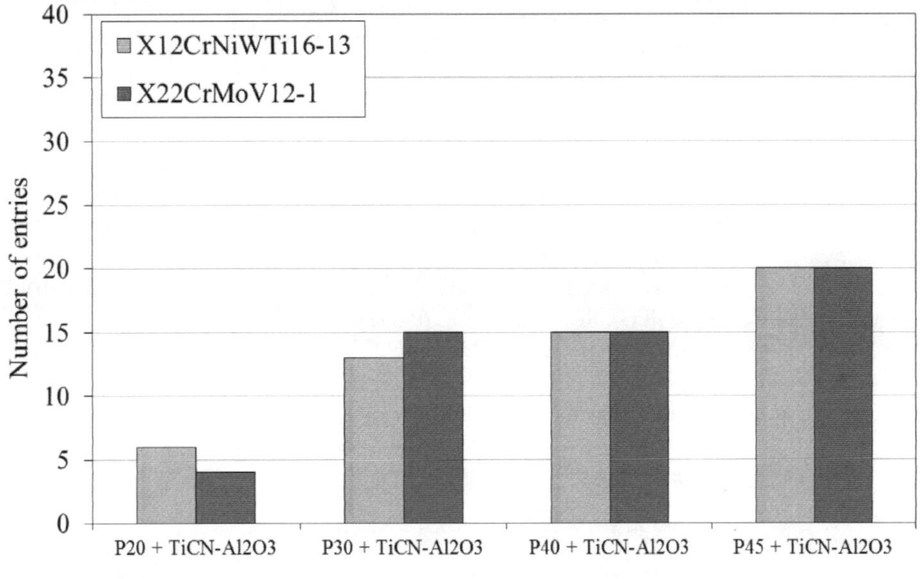


**Figure 3.** Lifetime of the cutting inserts with CVD TiCN-$Al_2O_3$ coatings, with increasing toughness of the cemented carbide from ISO P20 to P45 using F50 chipbreaker.

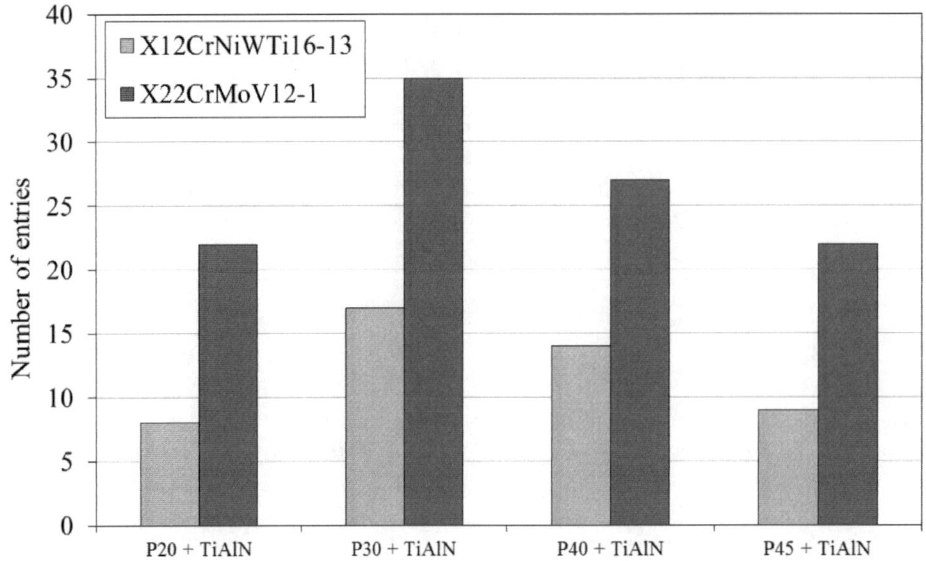


**Figure 4.** Lifetime of the cutting inserts with PVD TiAlN coatings, with increasing toughness of the cemented carbide from ISO P20 to P45 using F50 chipbreaker.

The CVD TiCN-$Al_2O_3$ coated substrates showed comparable lifetime for both work piece materials, which increases with increasing toughness of the carbide. The PVD TiAlN coated grades showed a maximum of performance for the P30 carbide and a similar trend for both work piece materials. The lifetime when machining martensitic X22CrMoV12-1 is approximately doubled compared to austenitic X12CrNiWTi16-13 work piece material.

The comparison of the lifetime of PVD TiAlN, CVD TiCN-$Al_2O_3$ and CVD $TiB_2$ coated P45 carbide is shown in Figure 5. The CVD $TiB_2$ coating could outperform the other coating materials on martensitic X22CrMoV12-1 work piece material, but not on austenitic X12CrNiWTi16-13.

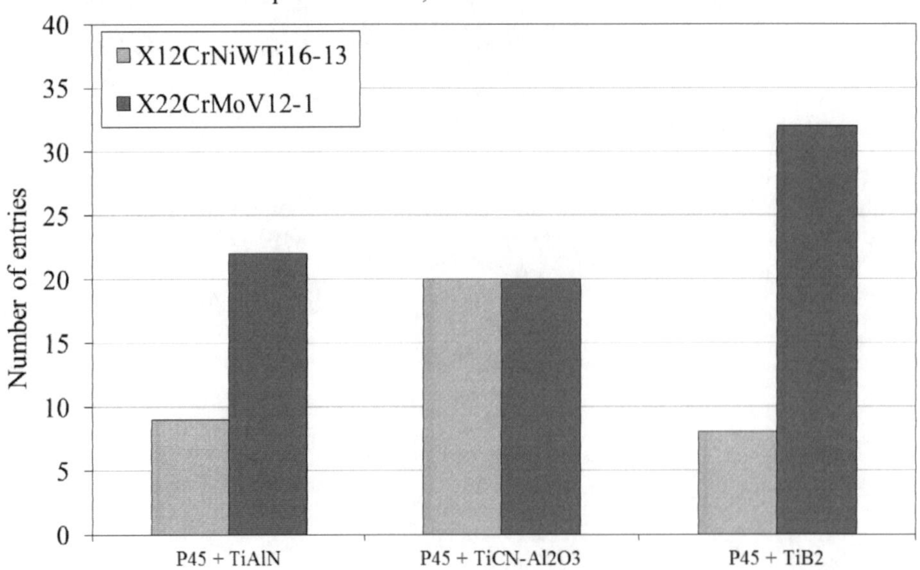


**Figure 5.** Lifetime of the cutting inserts with P45 carbide and varying coating material using F50 chipbreaker.

For the best performing carbide an coating combination on each work piece material, which means P45 carbide with CVD TiCN-$Al_2O_3$ coating on austenitic X12CrNiWTi16-13 and P30 carbide with PVD TiAlN coating on martensitic X22CrMoV12-1 the test was repeated using a more stable M50 chip breaker geometry, in comparison to the above used F50. The comparison of the number of entries is shown in Figure 6. A slight increase in lifetime could be reached for the CVD TiCN-$Al_2O_3$ coatings on both workpiece materials and for PVD TiAlN on austenitic X12CrNiWTi16-13. On martensitic X22CrMoV12-1 the F50 geometry with P30 carbide and PVD TiAlN coating showed the pest performance.

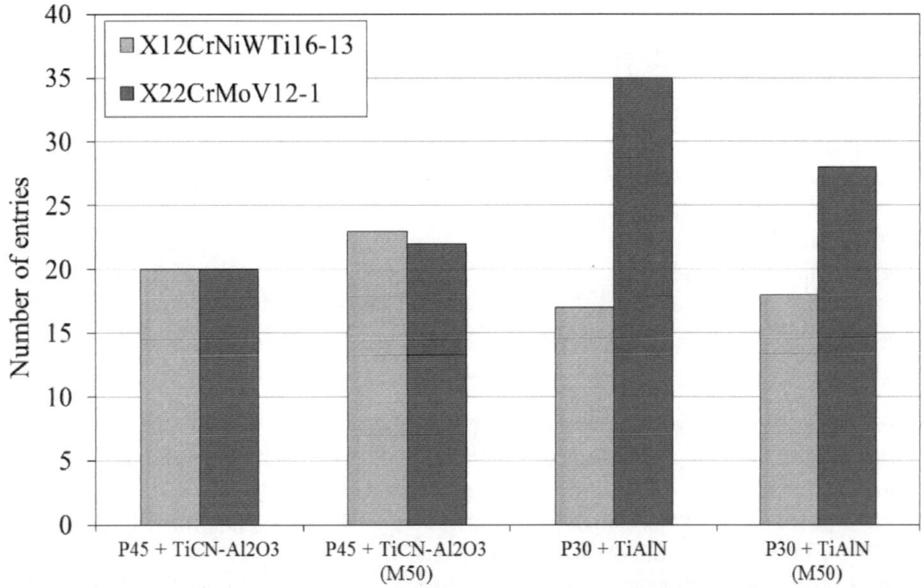


**Figure 6.** Lifetime of the cutting inserts with best performance P45 carbide and CVD TiCN-$Al_2O_3$ coating and P30 carbide with PVD TiAlN coating. Both where tested in F50 and M50 chip breaker geometry.

## 4. Discussion

The cutting performance of several carbide and coating combinations for machining turbine blade materials of high alloyed steels was compared. For a dry helirough process on martensitic X22CrMoV12-1 the combination P30 carbide with PVD TiAlN coating and F50 chip breaker showed the best performance of 35 entries. The second best combination was the P45 substrate with CVD $TiB_2$ coating with 32 entries. This result showed that the martensitic steels are more abrasive and the coatings with the highest hardness at room temperature showed the lowest wear rate.

When machining austenitic X12CrNiWTi16-13 the combination of P45 carbide and CVD TiCN-$Al_2O_3$ in M50 chip breaker geometry outperformed all other combinations with 23 entries. The second best version was P30 carbide with PVD TiAlN in M50 geometry, but showed with 18 entries only 78 % of the lifetime. The thermal load on the cutting tool seems to be much higher when machining austenitic work piece material, therefore a more stable geometry is necessary to avoid plastic deformation at the cutting edge, which could lead to coating delamination. For this application the insulation of the $Al_2O_3$ coating and its high temperature wear resistivity are crucial.

To gain a fundamental understanding of cutting operations and tool wear, the interaction of work piece material, cutting parameter, tool geometry and tool materials including carbide and coating has to be investigated more in detail. Especially the mechanical and thermal loads during application have to be determined, to set the right boundary conditions for the choice of cutting materials and their further development. An all-in-one solution suitable for every purpose is not possible when highest performance is desired.

## 5. Conclusion

For the best cutting performance on dedicated applications and work piece materials, the right choice of insert grade, which means the combination of carbide grade and coating, is crucial. As it turned out for different types of high alloyed stainless steels different insert grades have to be used. It could be demonstrated that ...

- P30 carbide with PVD TiAlN coating (CTPP235) in F50 geometry showed the best result on martensitic X22CrMoV12-1 work piece material and the best overall performance. This should be the first choice.
- P45 carbide with CVD TiCN-$Al_2O_3$ (experimental) in M50 geometry outperformed other combinations on austenitic X12CrNiWTi16-13 work piece material. The width of machining parameters is narrower, but in the optimum the difference in lifetime is significant.
- CVD $TiB_2$ coatings could be an alternative when machining abrasive materials with moderate thermal load.

A further optimization to improve the wear resistivity and especially the high temperature material properties is crucial to meet the requirements of future machining demands. Therefore detailed investigations and measurement methods have to be developed.

## References

[1] H.M. Ortner, P. Ettmayer, H. Kolaska, I. Smid, The history of the technological progress of hardmetals, International Journal of Refractory Metals and Hard Materials, 49 (2015) 3-8.
[2] C. Czettl, Design of CVD Coatings for Cutting Tools, Chair for Functional Materials and Material Systems, Montanuniversität, Leoben, 2013.
[3] S. Ruppi, Advances in chemically vapour deposited wear resistant coatings, Journal De Physique IV, 11 (2001) 847-859.
[4] U. Schleinkofer, C. Czettl, C. Michotte, 1.16 - Coating Applications for Cutting Tools, in: V.K. Sarin (Ed.) Comprehensive Hard Materials, Elsevier, Oxford, 2014, pp. 453-469.
[5] M. Tkadletz, J. Keckes, N. Schalk, I. Krajinovic, M. Burghammer, C. Czettl, C. Mitterer, Residual stress gradients in α-$Al_2O_3$ hard coatings determined by pencil-beam X-ray nanodiffraction: The influence of blasting media, Surface and Coatings Technology, 262 (2015) 134-140.
[6] N. Schalk, C. Mitterer, C. Czettl, B. Satory, M. Penoy, C. Michotte, Dry-Blasting of α- and κ-$Al_2O_3$ CVD Hard Coatings: Friction Behaviour and Thermal Stress Relaxation, Tribology Letters, 52 (2013) 147-154.
[7] C. Mitterer, 2.16 - PVD and CVD Hard Coatings, in: V.K. Sarin (Ed.) Comprehensive Hard Materials, Elsevier, Oxford, 2014, pp. 449-467.
[8] S. PalDey, S.C. Deevi, Single layer and multilayer wear resistant coatings of (Ti,Al)N: a review, Materials Science and Engineering A, 342 (2003) 58-79.
[9] EMUGE-Franken, Handbuch der Gewindetechnik und Frästechnik EMUGE-Franken, Erlangen, 2004.
[10] N. Schalk, J. Keckes, C. Czettl, M. Burghammer, M. Penoy, C. Michotte, C. Mitterer, Investigation of the origin of compressive residual stress in CVD $TiB_2$ hard coatings using synchrotron X-ray nanodiffraction, Surface and Coatings Technology, 258 (2014) 121-126.
[11] W. Wallgram, U. Schleinkofer, Synthesis, Structure, and Behaviour of a new CVD $TiB_2$ Coating with Extraordinary Properties for High Performance Applications, in: L.S. Sigl, P. Rödhammer, H. Wildner (Eds.) $17^{th}$ Int. Plansee Seminar, Plansee Group, Reutte, Austria, 2009, pp. HM32-31/HM32-37.

[12] M. Pfeiler, G.A. Fontalvo, J. Wagner, K. Kutschej, M. Penoy, C. Michotte, C. Mitterer, M. Kathrein, Arc evaporation of Ti-Al-Ta-N coatings: The effect of bias voltage and Ta on high-temperature tribological properties, Tribology Letters, 30 (2008) 91-97.
[13] M. Pfeiler, C. Scheu, H. Hutter, J. Schnoller, C. Michotte, C. Mitterer, M. Kathrein, On the effect of Ta on improved oxidation resistance of Ti-Al-Ta-N coatings, Journal of Vacuum Science & Technology A: Vacuum, Surfaces, and Films, 27 (2009) 554-560.
[14] C. Czettl, C. Mitterer, U. Mühle, D. Rafaja, S. Puchner, H. Hutter, M. Penoy, C. Michotte, M. Kathrein, CO addition in low-pressure chemical vapour deposition of medium-temperature $TiC_xN_{1-x}$ based hard coatings, Surface & Coatings Technology, 206 (2011) 1691-1697.
[15] M. Rebelo de Figueiredo, M.D. Abad, A.J. Harris, C. Czettl, C. Mitterer, P. Hosemann, Nanoindentation of chemical-vapor deposited $Al_2O_3$ hard coatings at elevated temperatures, Thin Solid Films, 578 (2015) 20-24.

IOP Conf. Series: Materials Science and Engineering **119** (2016) 012012 doi:10.1088/1757-899X/119/1/012012

# Synthesis and characterization of nano-structured molybdenum-iron intermetallics by gas-solid reaction technique

**A A El-Geassy[1] and S Seetheraman[2]**

[1]Central Metallurgical Research &Development Institute, P.O. Box 87-Helwan, Cairo, Egypt

[2]Royal Institute of Technology, KTH, SE-100 44 Stockholm, Sweden

E-mail: elgeassy@yahoo.com

**Abstract.** Ammonium molybdate and ferrous oxalate were used for the synthesis of nano-structured Mo-Fe intermetallics. Co-precipitation technique was applied to produce Mo/Fe precursors containing 58/42, 72/28 and 30/70 mass% respectively. The different phases formed were identified by XRD. The macro- and microstructures were microscopically examined by Reflected Light Microscope (RLM) and Scanning Electron Microscope (SEM) coupled with Electron Dispersion Spectroscopy (EDS). TG-DTA-DSC technique was applied to follow up the behavior of precursors up to 900°C (10K/min.). Endothermic peaks were detected at 97.8, 196.9 and 392.7°C due to the decomposition reactions to produce $MoO_3$ and $Fe_2O_3$. The exothermic peak resulted at 427.8°C is due to the solid state reaction between these oxides to form $Fe_2(MoO_4)_3$. Precursors were isothermally reduced at 600-850°C in a flow of purified $H_2$ and the $O_2$-weight loss from the reduction reactions was continuously recorded as a function of time. The isothermal reduction behavior of precursors was investigated. The structures of reduced products and the different phases formed were investigated and correlated with the reduction conditions. At >60% reduction extents, $Fe_2(MoO_4)_3$] phase was first reduced to $Fe_2MoO_3O_8$ before the formation of FeMo, while $FeMoO_4$ and $MoO_2$ were reduced to FeMo and Mo. In precursors containing higher content of $Fe_2O_3$, FeMo, $Fe_3Mo$ and Fe phases were produced. The visual observation of reduced samples showed that the volume was gradually shrinking with rise in temperature up to 800°C and beyond which a measurable swelling was observed reaching about 170% at 900°C.

## 1. Introduction

Gaseous reduction technique is considered a promising method for the synthesis of binary and tertiary metallic alloys powder because the applied reduction temperature is fairly low compared with the traditional thermal processes. Moreover, the phase composition and the morphology of the produced alloys can be controlled by adjusting the reaction parameters such as precursor composition, particle size of reactants, reduction temperature, reducing gas, etc. The production of tailor-made materials by gaseous reduction provides an excellent process which has a positive impact on the environment by applying green energy technology. The wide range of applicability of gas-solid reaction process has not been fully explored toward the synthesis of novel complex materials. Various physical properties of solid reactants play an important role in the manner in which the reduction reactions proceed. About 90% of the world's ferroalloys production is consumed by the steel industry.

These alloys are essential additives to steel for the inclusion of the alloying elements into the cast or to act as deoxidizers. A great attention is being devoted towards the use of intermetallic alloys as semi-conductors, photovoltaic cells, chemical manufacture and special types of ceramics [1,2]. Molybdenum-Iron alloys are used in different industries such as constructional steel, stainless steel, cast iron, tool and high speed steel and super alloys. Morales et al.[3] studied the production of $Fe_2Mo$ alloy from the reduction of $Fe_2MoO_4$ by $H_2$ in a fluidized bed reactor for the synthesis of Fe-Mo alloys having different Fe/Mo ratios[4]. Molybdenum is essential as an alloying/reinforcement agent and its properties maintain functional reliability under ever increasing temperatures for electronic and aerospace industries [5]. Fe-Mo alloy can be produced by mechanical alloying [6], by electro-deposition [7,8], by hydrogen reduction [3,9] or by ball-milling and subsequent hydrogen-reduction of $MoO_3$ powder[10]. Nanostructured Mo-Fe precursor can be synthesized by chemical co-precipitation, sol-gel, thermal, mechanical processes, etc. On the other hand, co-precipitation technique from the aqueous salts is applied for production of high purity precursors. It has been demonstrated that the co-precipitation method is most suitable compared with other routes due to its low cost in addition to its suitability for mass production. This technique offers considerable advantages, such as better mixing of the starting materials, better control over crystallite size and excellent chemical homogeneity.

The aim of the present work is to study the synthesis of Mo/Fe precursors having different Mo/Fe mole ratios from their aqueous salts using co-precipitation technique. Precursors were then isothermally reduced with $H_2$ at 650-850°C to produce nano-structured Fe-Mo intermetallics. Dry, roasted precursors and Mo/Fe intermetallics were physically and chemically characterized and correlated with the reaction mechanism.

## 2. Materials and experimental techniques

Wet chemical method was used for producing three Mo/Fe precursors from high purity (>99%) ammonium molybdate [$(NH_4)_6Mo_7O_{24}.4H_2O$] and ferrous oxalate ($Fe_2C_2O_4.2H_2O$) hot solutions. The mole ratios in the staff were calculated to synthesize precursors containing 58/42, 72/28 and 30/70% Mo/Fe as illustrated in Fig.1 [ 11] and named herein after as precursors *I, II* and *III* respectively. For a given precursor composition, pre-calculated weights of ammonium molybdate and ferrous oxalate were dissolved in hot distilled water and the aqueous solutions were mixed together, heated up to 80°C for 24hrs. with continuous stirring, then left to dryness. The solid residue was collected and mechanically ground in a ball mill for 5 hrs. The different phases in precursors were identified by X-ray diffraction (XRD). For a given precursor composition, pre-calculated weights of ammonium molybdate and ferrous oxalate were dissolved in hot distilled water and the aqueous solutions were mixed together, heated up to 80°C for 24hrs. with continuous stirring, then left to dryness. The solid residue was collected and mechanically ground in a ball mill for 5hrs. The different phases in precursors were identified by XRD (PW 1730 with Ni filtered Cu radiation at 40kV, 30mA). The morphology and grain structure were examined with Reflected Light Microscope (RLM) and Scanning Electron Microscope (SEM) coupled with Electron Dispersion Spectroscopy (EDS).

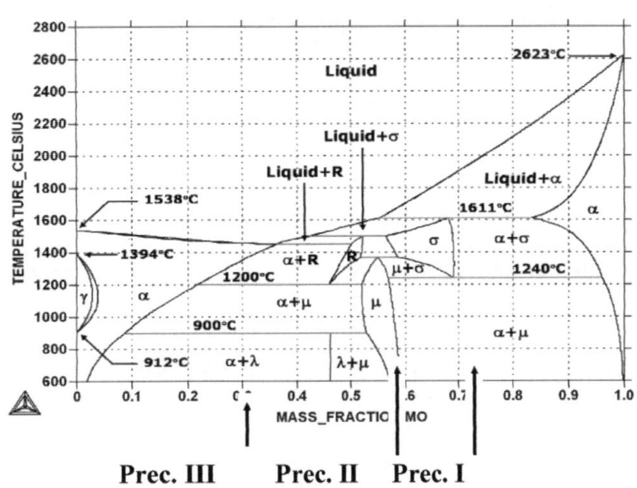


**Figure 1.** Molybdenum-iron phase diagram [11].

The thermal behaviour of dry precursors was examined by DTA-TG-DSC analyzer (Netzsch STA 409, Germany). The magnetic properties were tested at room temperature using a Vibrating Sample Magnetometer (VSM, 9600-1 LDJ,USA) and a maximum applied field of 15kOe. High pressure mercury pore sizer (Micromertics Pore-sizer, 9320) was used to measure total porosity, pore size distribution, bulk and apparent densities.

Precursors were isothermally reduced at 600-850°C in a flow of purified $H_2$ gas. The $O_2$-weight loss resulting from the reduction reactions was continuously recorded as a function of time using TG technique. The gas purification system, reduction apparatus and the reduction technique were previously described elsewhere [12]. The reduced products were physically and chemically characterized.

## 3. Results and Discussion

### *3.1. Characterization of precursors*

The different phases formed in the dry samples were examined by XRD and the patterns obtained showed the formation of less-crystalline materials. In these precursors, iron oxalate, ammonium molybdate, $MoO_3(H_2O)_{0.5}$ and FeO(OH) were produced. The extents of these phases (expressed as I/Io) are Mo/Fe mole-ratio dependant. The crystallite size was calculated from the X-ray data using TOPAS.2 software.

Dry precursors were thermally analyzed with TG-DTA-DSC analyzer to follow up their behavior during heating to 900°C (10 K/ min.) in air atmosphere. The obtained TG and DSC profiles are shown in Fig. 2 (a), (b) respectively. For a given precursor composition, the TG profile is characterized by the presence of mass loss at 97.8, 196.9 and 392.7°C and the total mass losses recorded were 39.3, 32.4 and 50.8% for precursor *I, II* and *III* respectively. At 97.8°C, the mass loss was due the removal of their moisture content, while that recorded at 196.9 and 392.7°C were account for the thermal dissociation reactions of iron oxalate and ammonium molybdate to produce $Fe_2O_3$ (reaction 1) and $MoO_3$ (reactions 2-4) respectively. The DSC profile given in Fig. 3b shows the presence of endothermic peaks at the above mentioned temperatures corresponding to the moisture and the dissociation reactions. This was confirmed by subjecting ferrous oxalate and ammonium molybdate individually to thermal analysis at the same conditions. The obtained TG and DSC profiles revealed the following:

a- For ferrous oxalate; the thermal decomposition reactions started at 222.3°C and ended at 315.1°C giving 55.96% total mass loss (reaction 1);

$$[2FeC_2O_4.2H_2O] + \tfrac{1}{2}\, O_2 \rightarrow Fe_2O_3 + 2\,CO + 2\,CO_2 + 4H_2O \quad (1)$$

b- For ammonium molybdate, the thermal decomposition reactions started at 130°C and ended at 348°C giving 17.77% total mass loss (reactions 2-4);

$$[(NH_4)_4Mo_7O_{24}.4H_2O] \rightarrow [(NH_4)_4Mo_5O_{17}] + NH_3 + 4\,H_2O \quad (2)$$

$$[(NH_4)_4Mo_5O_{17}] \rightarrow [(NH_4)_2Mo_5O_{13}] + NH_3 + H_2O \quad (3)$$

$$[(NH_4)_2Mo_5O_{13}] \rightarrow MoO_3 + NH_3 + H_2O \quad (4)$$

The exothermic peak obtained at 427.8°C (Fig.2b) is attributed to the solid state reaction between $Fe_2O_3$ (reaction 1) and $MoO_3$ ( reaction 4) to produce iron molybdate $Fe_2(MoO_4)_3$ as follow:

$$Fe_2O_3 + 3\,MoO_3 \rightarrow Fe_2(MoO_4)_3 \quad (5)$$

The XRD phase analysis obtained from the analysis of roasted precursors shows the presence of iron molybdate [$Fe_2(MoO_4)_3$] phase in precursor *II*, while $Fe_2(MoO_4)_3$ and $\alpha$-$Fe_2O_3$ were detected in precursors *I* and *III.* This was the result of the excess iron oxalate in precursors *I* and *III* rather than the stoichometric amount to produce [$Fe_2(MoO_4)_3$]. Table 1 summarizes the phases formed in roasted precursors, their molar ratios and average crystallite sizes.

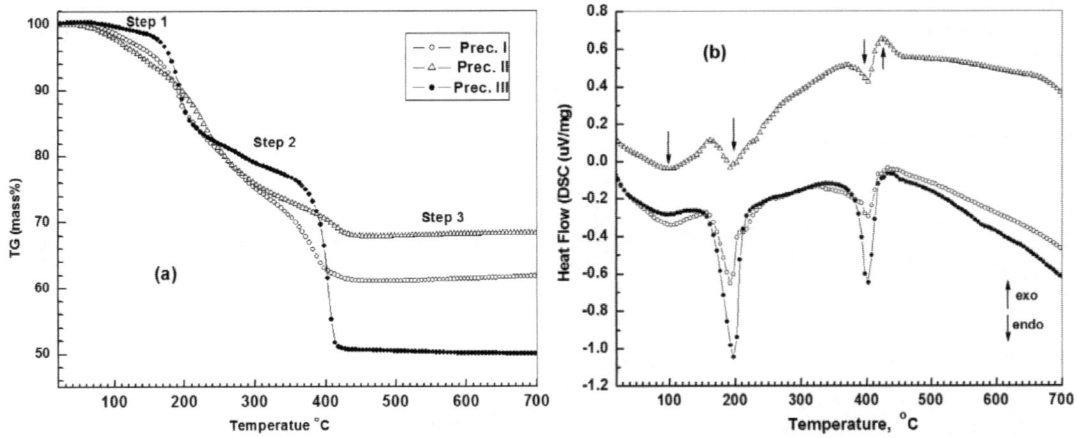


**Figure 2.** Thermal behavior of Mo/Fe dry precursors; (a) TG profile (b) DSC profile

Table1. XRD phase analysis for precursors roasted at 500°C for 3hrs.

| Prec. | Composition Mo% | Composition Fe% | Phases identified | Mole ratio (%) | Crystallite sizes (μm) |
|---|---|---|---|---|---|
| *I* | 58 | 42 | $Fe_2(MoO_4)_3$<br>$\alpha$-$Fe_2O_3$ | 72.2<br>27.8 | 3.5 |
| *II* | 72 | 28 | $Fe_2(MoO_4)_3$ | 100 | 5.8 |
| *III* | 30 | 70 | $Fe_2(MoO_4)_3$<br>$\alpha$-$Fe_2O_3$ | 46.68<br>53.32 | 3.0 |

The internal structure developed in the roasted precursors *I, II* and *III* is depicted in Fig. 3 (a), (b) and (c) respectively. Figure 3 (a) shows the presence of relatively small numbers of coarse iron molybdate grains (light gray) embeded in large number of fine hematite grains (gray) and pores (dark gray). The structure formed in precursor II (Fig. 3 b) is characterized by well defined elongated grains having different sizes. Each of these grains are composed of large number of iron molybdate particles coalesced to form aggregates in which large sizes of pores can be distinguished. For precursor *III* (Fig. 3 c), the structure shows the presence of very fine particles of hematite grains (gray) together with iron molybdate (light gray). It can be noticed that the higher the iron molybdate content in the precursor, the larger the grain structure formed. The grain structure of precursors *I, II* and *III* by SEM-EDS is given in Fig. 4 (a), (b) and (c) respectively. Figure 4 (a) shows the presence of well defined iron molybdate grains (345um) together with small grains of hematite. The structure of precursors *II* reveals the formation of large grains of iron molybdate (560 μm) connected together forming a relatively dense structure as shown in Fig.4 (b). The structure in precursor *III* (Fig. 4 c) shows the presence of large number of smaller hematite grains (194 μm) forming aggregates in which the original grains are still identified.

The total porosity and pore size distribution in roasted precursors were measured and the variation between cumulative intrusions (ml/g) , incremental intrusions (ml/g) and the corresponding pore diameter (μm) are given in Fig 5 (a) and (b) respectively. The cumulative intrusion values given in Fig. 5 (a) are slightly increased with the decrease in pore diameter up to about 8.5 μm then sharply increased to about 3.0 μm followed by a slight increase up to 0.42 μm. It also reveals that precursor *III* has finer pores than the other precursors. Fig. 5 (b) shows that the incremental pore volume slightly increased with the pore diameter in the range of macropores. In the mesopore range, the incremental value sharply increased with the decrease in pore diameter showing higher increase for precursor *II.*

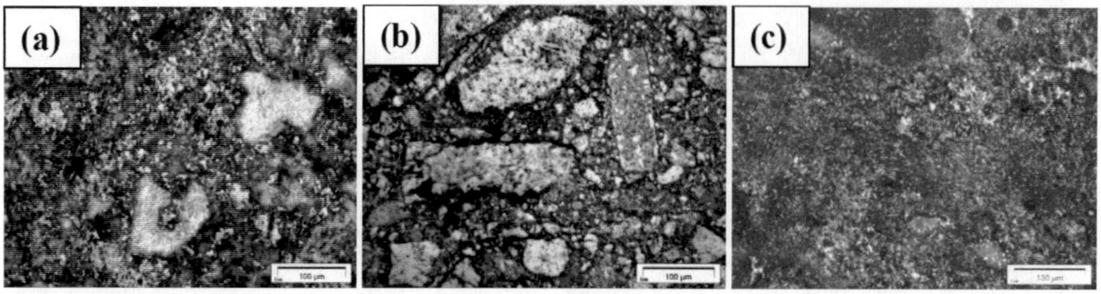


**Figure 3**. RLM structures for roasted precursors; (a) prec. *I*, (b) prec. *II*, (c) prec. *III*

**Figure 4**. SEM image for roasted precursors; (a) prec. *I* (b) prec. *II* (c) prec. *III*

Table 2 summarizes the measured values of pore area, average pore diameter, bulk and apparent densities measured in the precursors.

Table 2. Porosity measurements of precursors roasted at 500°C for 3hrs.

| Prec. No. | Composition Mo% | Fe% | Bulk density ($gml^{-1}$) | App. density ($gml^{-1}$) | Pore diameter (μm) | Pore area $m^2/g$ | Porosity % |
|---|---|---|---|---|---|---|---|
| *I* | 58 | 42 | 2.676 | 6.2947 | 0.0259 | 41.01 | 57.49 |
| *II* | 72 | 28 | 2.579 | 6.4278 | 0.0223 | 38.19 | 59.87 |
| *III* | 30 | 70 | 2.7622 | 5.3525 | 0.0315 | 43.45 | 48.39 |

The influence of Mo/Fe mole ratio on the magnetic properties of precursor was investigated. The M-H magnetic hysteresis loops of the roasted precursors are given in Fig.6. It can be seen that the higher saturation magnetization value ($M_s$ = 4.95 emu/g) was obtained in precursor *III* in which free hematite is present ($Fe_2O_3$ = 53.32%); while the lower value corresponding to precursor *II* ($M_s$ = 1.15 emu/g) was recorded.

*3.2. Synthesis of Mo/Fe intermetallics*

Gaseous reduction technique was applied for the synthesis of Mo/Fe intermetallics at >900°C. The reduction was carried out isothermally at 600-850°C in a flow of purified $H_2$ gas and the $O_2$-weight loss resulted from the reduction of $Fe_2(MoO_4)_3$ and $\alpha$-$Fe_2O_3$ was continuously recorded as a function of time using TG technique. The extents of reduction were calculated and plotted against the corresponding reaction time. Fig.7 (a) and (b) shows the isothermal reduction behaviour of precursors containing different Mo/Fe molar ratio at 700 and 800°C respectively. They indicated that the reduction was completed and all $O_2$-contents in the phases were removed. For each precursor, the rate of reduction increases with rise in temperature.

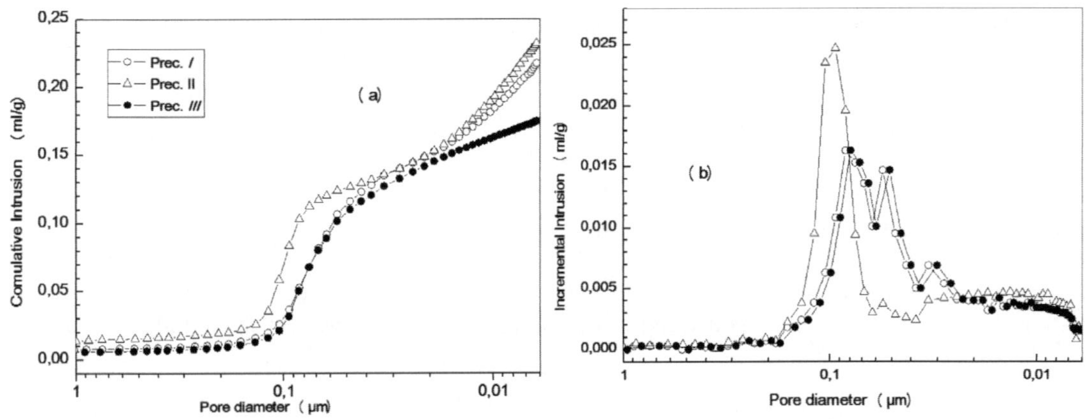


**Figure 5.** Pore Size analysis for the roasted precursors (500°C, 3h);
(a) Cumulative intrusion (ml/g) (b) incremental intrusion (ml/g)

**Figure 6.** B-H hysteresis loop of the roasted precursors.

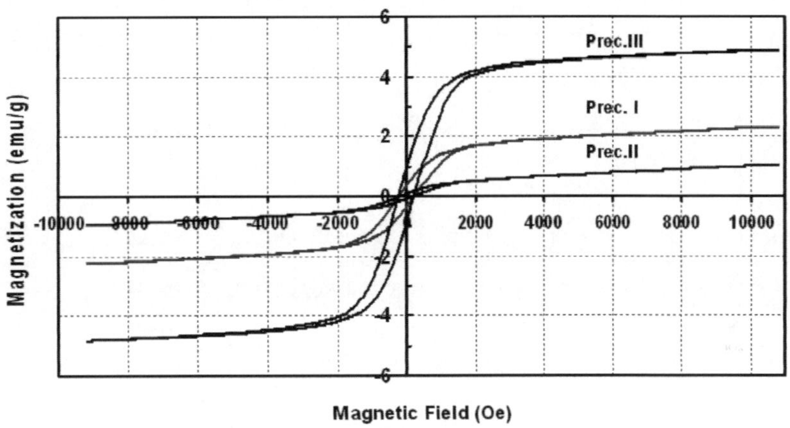


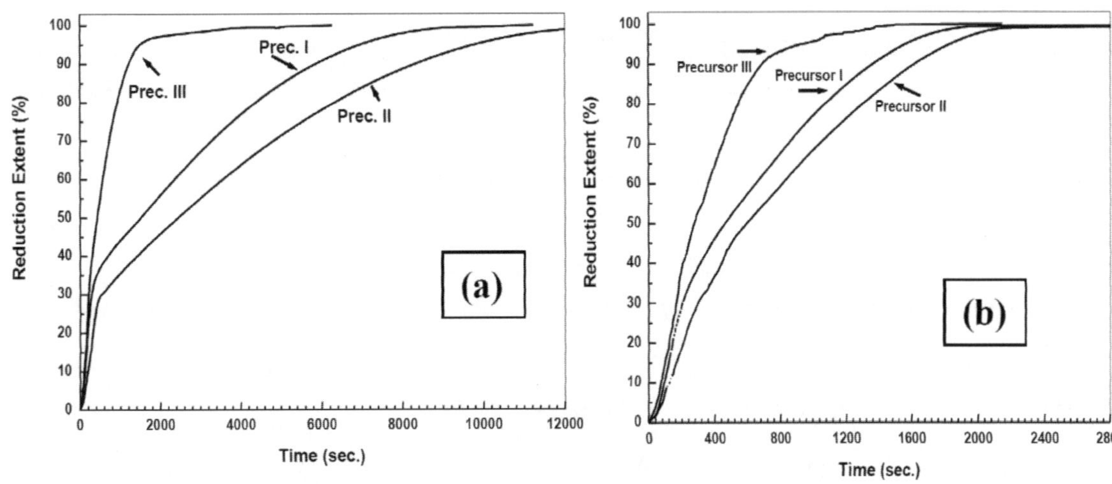


**Figure 7.** Isothermal reduction behavior of precursors; (a) at 700°C; (b) at 800°C

At a given temperature, the reduction rate of precursors increased in the order *III* >*I* >*II*. This indicated that the reduction rate of $Fe_2(MoO_4)_3$ is lower than $Fe_2O_3$ and the rate increases with the increase in α-

$Fe_2O_3$ content in precursors. The reduction kinetics and mechanism of precursors will be given and discussed in details elsewhere [13].

### *3.3 Characteristics of the Mo/Fe intermetallics*

Nano-structured Mo/Fe intermetallics were prepared from the reduction of precursors at 600-850°C. The reduction process was carried out isothermally in $H_2$ gas flow using TG technique. The reduced samples were examined by XRD, RLM and SEM-EDS. Figs. 8-10 show the XRD patterns for precursors I, II and III reduced at 700-850°C respectively. In precursor II, where $MoO_3$ and $\alpha$-$Fe_2O_3$ are in the stoichiometic ratio, only FeMo intermetallic was identified. In other precursors, where excess $\alpha$-$Fe_2O_3$ is present, Fe/Mo and $Fe_3Mo$ intermetallics were produced. With the increase in $\alpha$-$Fe_2O_3$ mole ratio, metallic iron was detected. The phases identified in the reduced precursors *I, II* and *III* in $H_2$ are presented in Table 4.

Table 4. Phases identified in roasted and reduced precursors at 700-850°C.

| Prec. | Composition Mo% | Fe% | Roasted samples | Reduced samples |
|---|---|---|---|---|
| *I* | 58 | 42 | $Fe_2(MoO_4)_3$, $\alpha$-$Fe_2O_3$ | FeMo, $Fe_3Mo$ |
| *II* | 72 | 28 | $Fe_2(MoO_4)_3$ | FeMo, Mo |
| *III* | 30 | 70 | $Fe_2(MoO_4)_3$, $\alpha$-$Fe_2O_3$ | FeMo, $Fe_3Mo$, Fe |

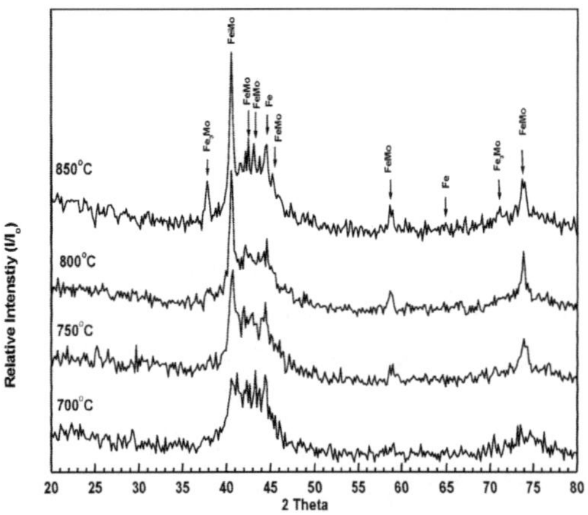


**Figure 8.** Phases identified in precursor *I* reduced at 700-850°C.

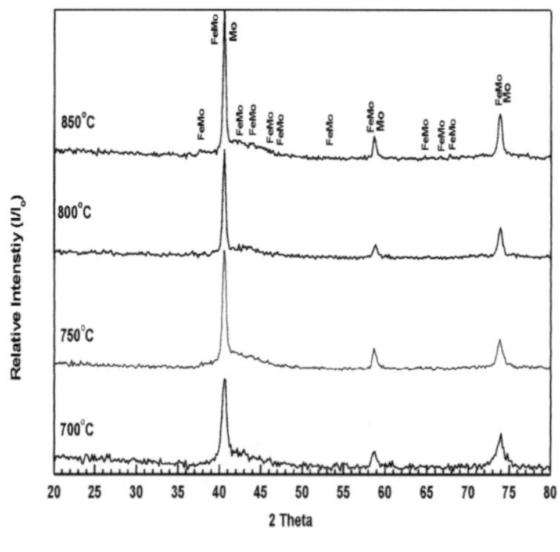


**Figure 9.** Phases identified in precursor *II* reduced at 700-850°C.

It can be seen that the Fe/Mo is the main phase identified and $Fe_3Mo$ is formed in precursors *I* and *III*. The presence of metallic iron in precursor *II* resulted from the excess mole ratio of $Fe_2O_3$. In order to predict the chemical reaction equations; precursors were reduced to different extents and reaction products were analyzed by XRD. Based on the phase analysis, the following chemical reaction equations were given and discussed elsewhere [13].

a- precursor *I*; $2\,Fe_2(MoO_4)_3 + 3\,Fe_2O_3 + 33\,H_2 \rightarrow 4\,FeMo + 2\,Fe_3Mo + 33\,H_2O$ (6)

b- precursor *II*; $Fe_2(MoO_4)_3 + 12\,H_2 \rightarrow 2\,FeMo + Mo + 12\,H_2O$ (7)

c- precursor *III*; $Fe_2(MoO_4)_3 + 2\,Fe_2O_3 + 18\,H_2 \rightarrow 2\,FeMo + Fe_3Mo + Fe + 18\,H_2O$ (8)

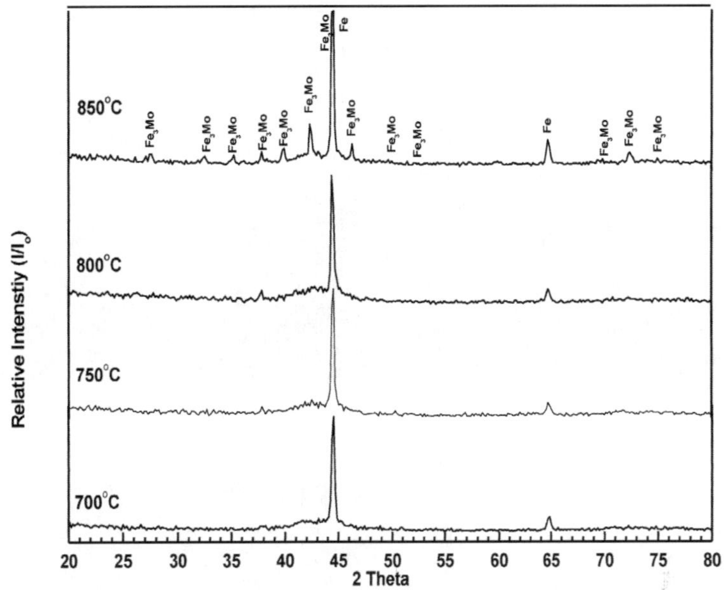


**Figure. 10** Phases identified in precursor *III* reduced at 700-850°C.

The photomicrographs showing the internal structures (RLM) formed in the reduced precursors *I, II* and *III* at 800°C in $H_2$ are shown in Fig.11(a), (b) and (c) respectively. The structure given in Fig.11(a) shows the presence of fine grains of hematite and larger elongated grains of Mo/Fe intermetallics.

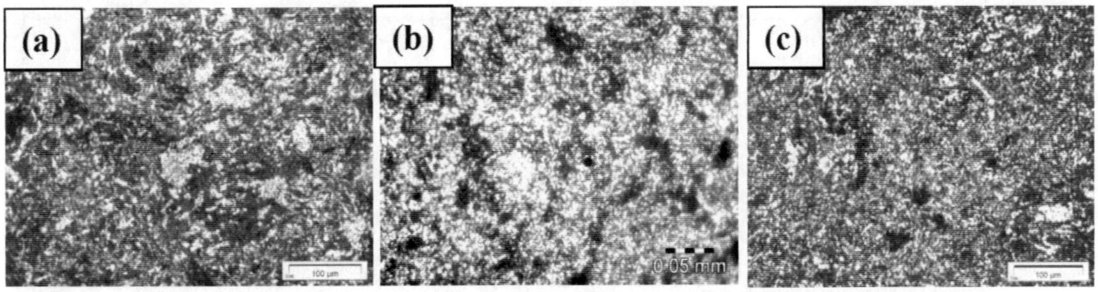


**Figure 11**. Internal structure of reduced samples; (a) Prec. *I* (b) Prec. *II* (c) Prec. *III*

The structure shown in Fig.11 (c) for precursor III is characterized by the presence of small amounts of metallic iron (light) and fine grains of intermetallics. The structures of grains in reduced samples I, II and III are shown in Fig.12 (a), (b) and (c) respectively. Unlike in precursors I and III, the grain structure in precursor *II* is characterized by the formation of large grains in which fibers or needles made from FeMo intermetallics as was defined by EDS analysis. In precursors *I* and *III*, small grains composed of $Fe_3Mo$ and larger FeMo grains can be distinguished. The grain structure formed in reduced precursor *II* at 850 and 900°C is shown in Fig.13 (a) and (b) respectively. The grains structure of the produced Fe/Mo intermetallics were changed from needle form formed at lower temperatures to grain and were composed of large numbers of smaller micrograins which are connected together forming big clusters. It can be seen that relatively smaller micrograins were formed at 850°C and enlarged by the effect of sintering at 900°C .
The visual external shape of compacts made from precursors *I, II* and *III* reduced at 650-900°C showed no remarkable changes in the visual shape of compacts from precursors *I* and *III*. For

precursor *II*, the samples showed deformation in shape at ≥ 850°C resulting in an increase in the volume by 175% at 875°C as shown in Fig.14.

**Figure 12.** SEM image for samples reduced at 800°C; (a) Prec. *I* (b) Prec. *II* (c) prec. *III*

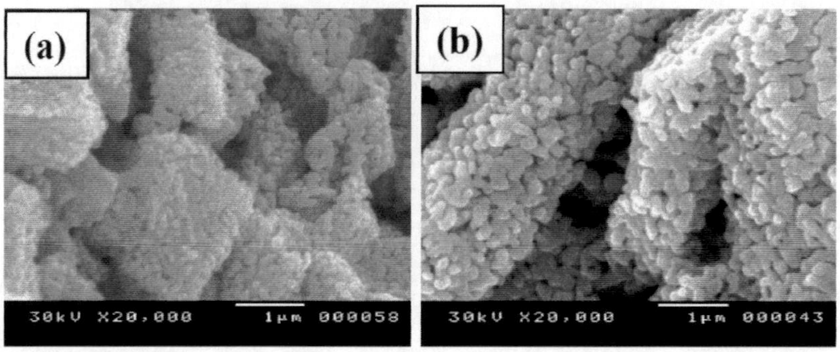


**Figure 13. SEM** image of precursor *II* reduced at;

(a) 850°C
(b) 900 °C

**Figure 14.** Visual observation of precursor *II* reduced at 650 - 900°C

## 1. Conclusion

Co-preciptation technique was applied for the preparation of precursors containing 58/42, 72/28 and 30/70 Mo/Fe mole% named as Prec. *I*, *II* and *III* respectively. The thermal behaviour of precursors was carried out up to 900°C (10K/min.) showing the presence of endothermic peaks at 97.8, 196.9 and 392.7°C in addition to an exothermic peak at 427.8°C. The endothermic peaks were due to the moisture content and the thermal dissociation reactions to produce $MoO_3$ and $Fe_2O_3$. The exothermic peak was resulted from the solid state reaction between $MoO_3$ and $Fe_2O_3$ to develop $Fe_2(MoO_4)_3$. The precursors were roasted at 500°C for 3hrs and were microscopically examined. In roasted samples, 72.2, 100 and 46.68 mole% of $Fe_2(MoO_4)_3$ were detected and the balance is being $\alpha$–$Fe_2O_3$. Nano-structured Mo/Fe intermetallics were obtained from the isothermal reduction tests at 600-850°C in $H_2$ gas flow. The reduced products showed the presence of FeMo intermetallic in precursors which were resulting from the reduction of $Fe_2(MoO_4)_3$. In samples containingg excess molar ratios of $Fe_2O_3$, $Fe_3Mo$ was detected and with the increase of $Fe_2O_3$, metallic iron was identified. At ≤800°C, the

Mo/Fe is characterized by the formation of clusters that contain well defined needles structure. At ≥ 850°C, the reduced samples were swelled and the measured ΔV values reached to 175% at 900°C.

**References**

[1] Tremel W, Kleinke H, Derstroff V and Reinser C 1995 *J. Alloys Compd.* **219** 73.
[2] Romankiw LT, Croll IM and Hatzakis M 1970 *IEEE Trans. Magn.* **6** 579.
[3] Morales R, Arvanitidis I, Sichen D, Seetheraman S 2002 *Metall Mater Trans* B **33B** 589.
[4] Morales R: 2003 *Scand. J. Metallurgy* **32** 263.
[5] Garg P, Park SJ and German RM 2007 *Int.J. Refract. Met. Hard Mater*, **25** 13-16.
[6] Moumeni H, Alleg S and Greneche JM 2006 *J Alloys Compd.* **419** 140-144.
[7] Zehn-Jiang N, Shi-Bing Yand Shao-Min Z 1998 *J. Electroanal. Chem.* **455** 205-207.
[8] Luciana S, Sanches, Claudia B, Mariano, Lucia H and Mascaro 2007 *J. Alloys Compd.* **439** 342-345.
[9] Morales R, Sichen D, Seetheraman S: 2003 *Metall. Mater. Trans.B*, **34B** 661-667.
[10] Gil-Su Kim, Young Jung Lee, Dae-Gun Kim and Young Do Kim 2008 *J. Alloys Compd.* **454** 327-330.
[11] T. Massalski T 1986 *Binary alloy phase diagrm, American Socity for Metals* 1080.
[12] El-Geassy AA and Rajakumar V 1985 *Trans Iron Steel Inst. Jap.* **25** 449.
[13] El-Geassy AA 2015 publication in progress.

# Thermodynamic optimization of individual steel database by means of systematic DSC measurements according the CALPHAD approach

**P Presoly[1], J Six[2] and C Bernhard[1]**
[1] Chair of Ferrour Metallurgy, Montanuniversitaet Leoben, Franz-Josef-Straße 18, A-8700 Leoben, Austria
[2] voestalpine Stahl GmbH, voestalpine-Straße 1, 4020 Linz, Austria

Email: peter.presoly@unileoben.ac.at

**Abstract.** Reliable thermodynamic data are essential information required for the design of new steel types and are a prerequisite to effective process optimization and simulation. Moreover, it is important to know the exact temperatures at which the high-temperature phase transformations ($T_{Liquid}$, $T_{Solid}$, $T_{Perit}$, $T_{\gamma\rightarrow\delta}$) occur in order to describe the solidification sequence and to describe further processing parameters. By utilizing DTA/DSC measurements, our earlier experimental studies of selected commercial DP, TRIP and high-Mn TWIP steels, have indicated that currently commercially available databases can often not be utilised to reliably describe the behaviour and microstructural development in such complex alloy systems. Because of these ostensible deficiencies, an experimental study was undertaken in an attempt to determine the pertaining thermodynamic data to analyse the behaviour of the important five-component Fe-C-Si-Mn-Al alloy system. High purity model alloys with systematic alloy variations were prepared and utilized in order to determine the influence of individual alloying elements in this complex, but industrially important alloy system. The present study provides new validated experimental thermodynamic data and analysis of the five-component Fe-C-Si-Mn-Al system, which will allow the construction of new phase diagrams, prediction of solidification sequences and the assessment of micro-segregation.

## 1. Introduction and Motivation

Thermodynamic databases are widely used and have become standard tools in research, development and production control in the steel industry. With user-friendly software programs, which access these databases, complex chemical reactions and phase equilibria that occur during steelmaking operations can be calculated over wide ranges of temperature, pressure and chemical compositions. A variety of Gibbs free energy minimizing commercially software packages such as ThermoCalc, FactSage, MTDat, PANDAT, JMatPro, IDS are freely available. The reliability of the predictions made by using these programs is limited by the accuracy of the thermodynamic data underpinning the binary, ternary and quaternary sub-systems. Some of the experimental data of important binary and ternary systems are quite old and have not been sufficiently verified by modern experimental techniques. Moreover, the purity of some of the older alloys are sometimes suspect. Computational thermodynamics are usually based on the CALPHAD approach [1,2]. The CALPHAD method has the potential to enable the calculation of complex, higher alloyed systems on the basic of well-established and reliable thermodynamic data, describing binary, ternary and quaternary sub-systems.

Of specific interest in these developments is steel, the main metallic construction material and hence, thermodynamic steel databases are continuously improved. Some of the most important developments in the design of steels in the last few decades have been the concepts of high strength and ductile steels for the automotive industry. Three of these alloying concepts are Dual-Phase (DP), TRansformation Induced Plasticity (TRIP) and Twinning Induced Plasticity (TWIP) steels. These steel types are considered as a new generation of automotive steels due to the advantages of weight reduction, energy saving and safety performance of automobiles. All these steel types are based on the iron-carbon-manganese system with additions of ferrite formers such as silicon and aluminium, in addition to micro-alloying elements. This means, that the alloy matrix consists of the Fe-C-Si-Al-Mn system (in different variations, often with or without Al, or with very high Mn contents up to 25 wt.-%, for example in TWIP steels). The main differences compared to previous steel grades are the significantly higher levels of the ferrite formers Si and Al.

Apart from research on the material and product properties at lower temperature, knowledge of the high-temperature phase transformation temperatures ($T_{Liquid}$, $T_{Solid}$, $T_{Perit}$, $T_{\gamma\rightarrow\delta}$) plays an important role in describing initial solidification phenomena and the control of subsequent production processes. As visualized in Figure 1a four different transformation sequences (I-IV) can be distinguished in the Fe-C system, characterized by the points $C_A$, $C_B$, and $C_C$ [3]. Figure 1b visualizes the strong influence of alloying elements, the effect of which can be distinguished as either austenite formers (such as Mn, Ni, Co, Cu, N...) or ferrite formers (such as Cr, Mo, W, Al, Si, Ti, V....).

Special features of this high-temperature system are the two different phases: δ- and γ-iron, showing significant differences in their density due to the different crystal structures (δ = face centred cubic / γ = body centred cubic) [4]. Steels which solidify according to the transformation sequence of range II are of special interest, because in this case the transformation from δ-Fe to γ-Fe (i.e. the peritectic phase transition $L+\delta \rightarrow \delta+\gamma \rightarrow \gamma$) starts in the solid-liquid two phase region and ends in the solid. The specific peritectic transformation range II is implied in an increased appearance of defects (such as hot tears, surface defects, depressions, and in the worst case breakouts) in the course of the continuous casting process [5,6].

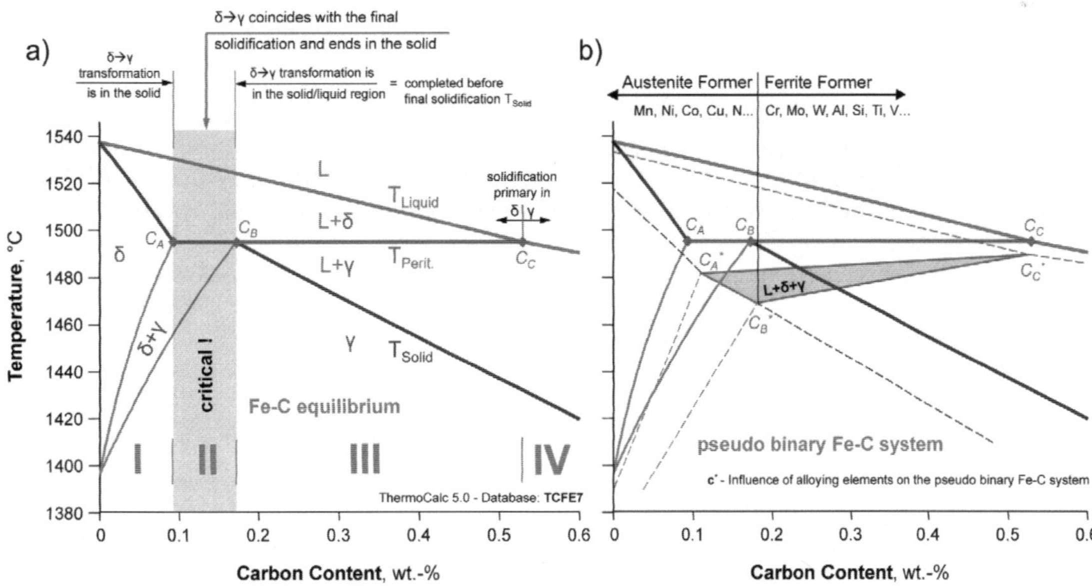


**Figure 1.** a) Fe-C equilibrium diagram with the critical peritectic range and b) influence of alloying elements on the Fe-C system in the high temperature range.

## 2. Demand for a database development

Previous experimental studies [3,6,7] have shown that if new alloying systems are considered, the range of validity of the thermodynamic database that is being used has to be carefully evaluated in each case. Investigations of 22 steel grades from the production program of voestalpine Stahl Linz in 2013 proved that by means of thermodynamic calculations with different databases about two-thirds of the steel grades are correctly identified with respect to the transformation range I to IV. Unfortunately, this also means that up to one-third of these steels were calculated inaccurately or incorrectly [6].

A recent investigation of selected DP-, TRIP- and TWIP-steels by means of Differential thermal analysis (DTA) and Differential scanning calorimetry (DSC) measurements explicitly indicated that current calculations of phase transformations in the Fe-C-Si-Mn system need to be reassessed [7]. Specifically, measurements on TWIP steels demonstrate that there exists a huge potential for further investigations of the higher alloyed Fe-C-Si-Al-Mn system. These arguments also apply to some other high alloyed systems. The need for the measurement of these systems – despite problems with the manufacturing of the samples and the measurement itself, is evident:

a) Verification of the results of available databases for high alloyed systems,
b) Creation of reliable databases for new alloy developments
c) Use of the data for the assessment of solidification kinetics (prediction of micro-segregation and formation of primary precipitates).

One important component of this work is the focus on segregated-areas which occur at the dendritic scale during the solidification. Inter-dendritic segregation, which can be calculated by micro-segregation models, are often responsible for the formation of defects during the casting process (e.g., hot tearing) and can also negatively affect product quality (inhomogeneous microstructure, and undesirable primary precipitates). In a recent paper the use of thermodynamic databases for the calculation of micro-segregation is described in detail for the case of a TRIP steel [8]. Figure 2 shows the measured and calculated equilibrium temperature of a medium-alloyed TRIP steel containing 0.77 wt.-% Al. The DSC measurement (heating rate corrected) on the left side of Figure 2 clearly shows a large peritectic peak between the solidus and liquidus temperature, which is typical for alloys between $C_B^*$ and $C_C^*$ (transformation sequences III). The position of the alloy in a pseudo-binary Fe-C phase is visualized on the right-hand side of Figure 2. The TRIP steel under investigation is well described by the TCFE7 (ThermoCalc) and the FSstel (FactSage 6.4) databases (only the peritectic temperature is calculated marginally too high). Since the original SGTE2014 database has no thermodynamic description of the ternary Fe-C-Al system, the peritectic range and the solidus temperature are calculated too low. The calculation with a newly optimized extract of the SGTE2014 database predicts all transformation temperatures very well. Although the initial concentration (Fe-0.21%C-0.54%Si-2.12%Mn-0.77%Al) of this commercial TRIP steel is well described by both the FSstel and SGTE databases, subsequent micro-segregation calculations show significant differences [8]. Only the optimized database extract could achieve realistic results, since the Fe-C-Si-Mn-Al system (and its subsystems) has been individually improved. The following section explains the method by which this specific database was systematically further developed by means of DSC measurements.

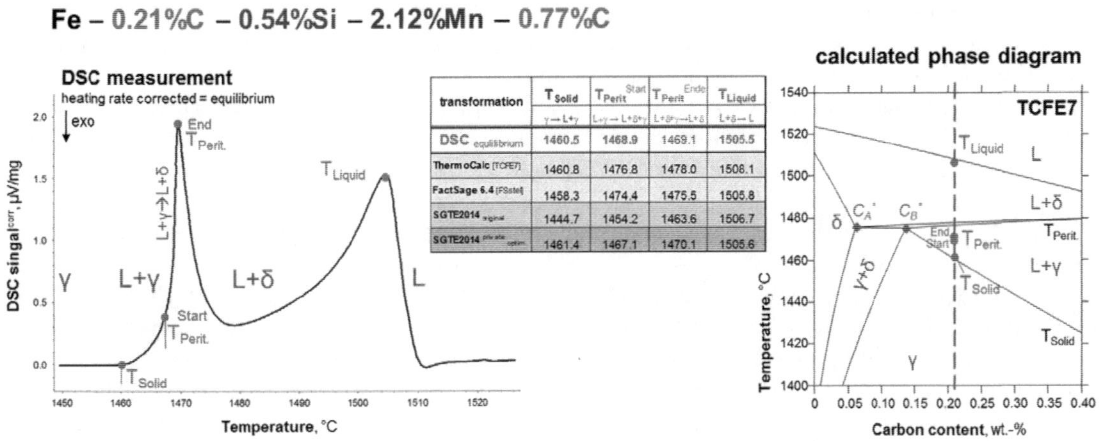


| transformation | $T_{Solid}$ | $T_{Perit}^{Start}$ | $T_{Perit}^{Ende}$ | $T_{Liquid}$ |
|---|---|---|---|---|
| | γ→L+γ | L+γ→L+δ+γ | L+δ+γ→L+δ | L+δ→L |
| DSC equilibrium | 1460.5 | 1468.9 | 1469.1 | 1505.5 |
| ThermoCalc [TCFE7] | 1460.8 | 1476.8 | 1478.0 | 1508.1 |
| FactSage 6.4 [FSstel] | 1458.3 | 1474.4 | 1475.5 | 1505.8 |
| SGTE2014 original | 1444.7 | 1454.2 | 1463.6 | 1506.7 |
| SGTE2014 private optim. | 1461.4 | 1467.1 | 1470.1 | 1505.6 |

**Figure 2.** Investigation of a TRIP steel.

## 3. Methodology to reassess thermodynamic databases

The most important step for database optimization is the identification of the affected subsystem which needs to be reassessed. Therefore, it is important to understand the construction of a CALPHAD database, which is illustrated in Figure 3. The Gibbs Energy $G^{\varphi}$ of each phase φ consists of three terms; ${}^{0}G^{\varphi}$ describes the standard Gibbs energy of a mechanical mixture of the constituents of the phase, ${}^{ideal}G^{\varphi}$ describes the entropy of mixing an ideal solution and ${}^{E}G^{\varphi}$ describes the so-called excess term. Based on experimental data, the excess term ${}^{E}G^{\varphi}$ is assessed to describe the interactions of the elements with each other and therefore the "real behaviour". In order to perform calculations in a five-component system, knowledge of 5 quaternary subsystems, 10 ternary subsystems and 10 binary subsystems are required. It is especially important to verify the sub-systems in the Fe-rich corner of the alloy system. Complex multi-component alloys and commercial steels cannot be used for database optimization, since the different interactions of the alloying elements overlap each other. Furthermore, undesirable trace elements in commercial steels negatively affect the determination of the equilibrium phase transformation temperatures by the formation of low melting phases. Consequently, high purity model alloys, with systematic alloy variations were used in the present investigation in order to determine the influence of specific alloying elements on the selected subsystems.

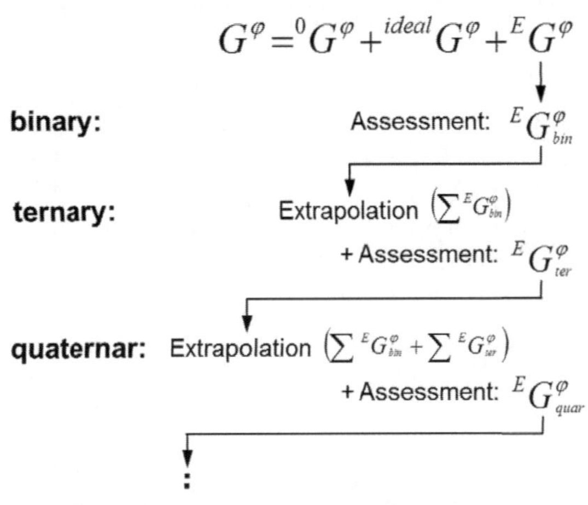


**Figure 3.** CALPHAD methodology [2].

In a previous study the need for a systematic reassessed of the quaternary Fe-C-Si-Mn system was clearly identified [7]. Since silicon and manganese are most important alloying elements, the description of their interaction is of fundamental importance and because interaction between alloying elements are stronger at higher concentrations, 2%Si and 2%Mn additions were chosen for further investigation. The test program with six alloys is visualized in Figure 4, whereby alloy III has been already investigated [7] and was clearly identified as peritectic (range II, between $C_A^*$ and $C_B^*$). This finding makes the whole investigation very exciting, since all calculations (TCFE7, FSstel, SGTE2014) provides nearly identical results, predicting the peritectic range II at significantly lower carbon contents.

On the right-hand side of Figure 4 the production of model alloys is illustrated. All Fe-C-2%Si-2%Mn model alloys (about 60g) were produced from high-purity raw-materials in an alumina crucible with a high-frequency re-melting and spin-casting machine under an argon atmosphere. The samples were very homogeneous due to the inductive melting, the spin casting and the rapid solidification in a copper mould. Chemical analyses were carried out by emission-spectroscopy and in addition, the carbon content was analysed by LECO combustion analysis since this method is more accurate.

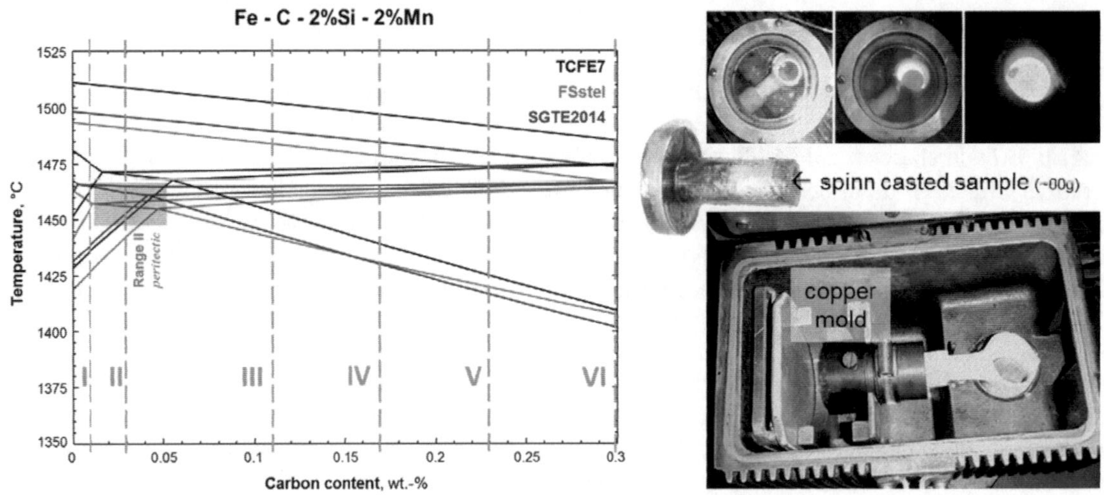


**Figure 4.** Fe-C-2%Si-2%Mn test program and the production of model alloys.

To assess the high temperature phase transformations ($T_{Liquid}$, $T_{Solid}$, $T_{Perit}$, $T_{\gamma \rightarrow \delta}$) DSC measurements were carried out under equilibrium conditions as described earlier [3,6]. The equipment used in that study was a NETZSCH DSC404F1 Pegasus with a Rhodium-furnace and a platinum DSC sensor with type-S thermocouples and $Al_2O_3$ sample cups. The experimental set-up was calibrated by measuring the well-known melting points of high purity In, Bi, Al, Ag, Au, Ni, and Co. In order to achieve the best equilibrium conditions, 50 mg samples and a heating rate of 10 K/minutes were used. In addition, the heating rate affected DSC measurements were corrected by the NETZSCH Tau-R software [9] regarding the thermal resistance and the time constants of the DSC sensor.

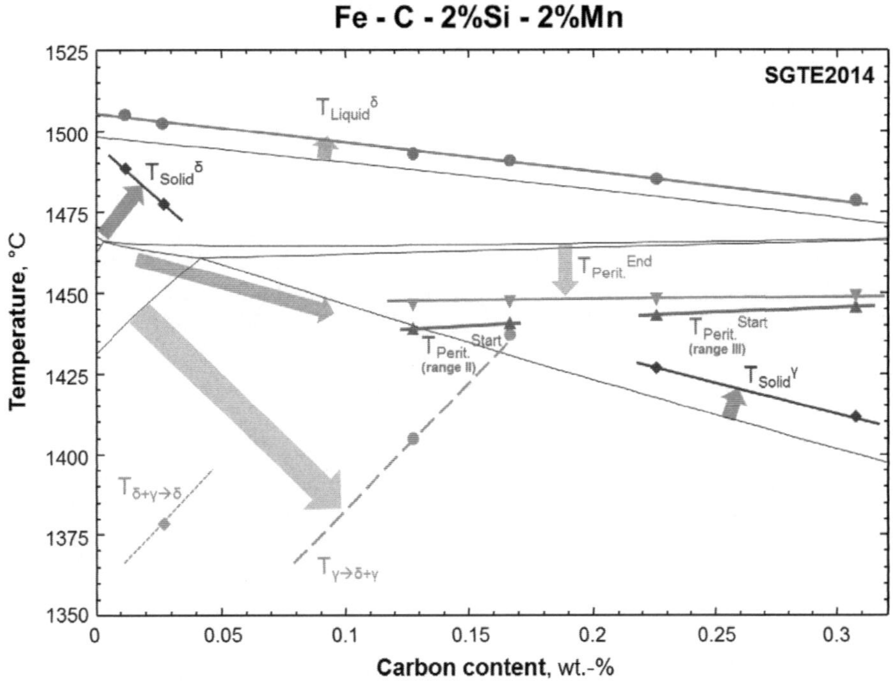


**Figure 5.** Fist measurement results of the Fe-C-2%Si-2%Mn.

The newly measured phase transformations are shown in Figure 5 as coloured dots, which are connected with simple lines so as to illustrate the new interfaces. The liquidus line can be determined very accurately since regression analysis shows a correlation of 98%. The newly determined phase transformations are in stark contrast to the calculated phase diagram using database SGTE2014. This simple comparison clearly shows that the delta-ferrite range is significantly larger than predicted by SGTE2014. A higher solidus temperature ($T_{Solid}^{\delta}$) is predicted, the gamma-to-delta phase transformation occurs at much higher carbon content and lower temperatures ($T_{\gamma \rightarrow \delta+\gamma}$) and the peritectic range (range II) occurs at higher carbon content. These results imply that silicon is a much stronger ferrite stabilizer for this system than calculated by SGTE2014. Furthermore, the liquidus temperature and the solidus temperature in the gamma range are higher than calculated by SGTE2014. The peritectic temperature, however, is clearly reduced in comparison to the by SGTE2014 calculation.

The current investigation demonstrates that by the use of model alloys (alloy variation) and DSC measurements, complete phase diagrams can be reassessed. In this case, however, six model alloys proved to be not enough to accurately predict the peritectic range and the points of interest $C_A^*$ and $C_B^*$. It was difficult to determine the $\gamma \rightarrow \delta$ phase transformation by using the current DSC-setup (50mg samples), due to the wide range of the $\gamma+\delta$ two-phase region and the resulting low enthalpy change. Therefore, further investigation of this solid-solid transition is of the essence.

The compilation in Figure 5 also shows that the biggest differences between the new measurements and the calculation with SGTE2014 occur at very low carbon concentrations. This finding clearly indicates that it is not only the Fe-C-Si-Mn system affected, but also the ternary Fe-Si-Mn system. Based on this result, carbon free Fe-Si-Mn alloys were prepared. In addition to the Fe-C-2%Mn-2%Si system, other quaternary combinations (high/low - Si/Mn) were studied, using the same method as described above, as part of a continuing comprehensive test program. Particularly important are Si-Mn combinations with higher Mn contents since steels of industrial importance are alloyed with significantly more Mn than Si and because Mn segregates much stronger than Si.

In addition to these experimental findings, an analysis of the thermodynamic databases (TCFE7, FSstel and SGTE2014) indicates that no thermodynamic description of the quaternary Fe-C-Si-Mn system is included. Therefore, the reason for the divergent calculations must be in an associated subsystem. The ternary Fe-C-Mn system (source [10]) appears well described, however, the ternary Fe-C-Si (source [11,12]) system shows a need for improvement according a previous study [7]. However, the deviations in the Fe-C-Si system are too small to explain the large discrepancies in the quaternary system. Therefore, the ternary Fe-Si-Mn system [Fe-Mn-Si:*Va*] (source [13]) has to be considered, since this system is particularly responsible for the low carbon region of the Fe-C-Si-Mn system, as visualized in Figure 5. It is perhaps also necessary to re-assess the binary Si-Mn subsystem, which describes the interaction of these important ferrite and austenite formers. This binary system was recently reassessed by Paek [14] and Zheng [15] and is visualized in Figure 6.

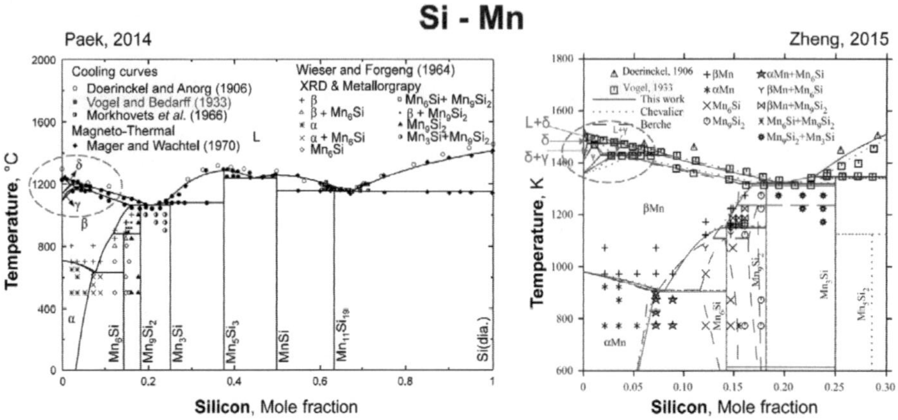


**Figure 6.** Reassessments of the binary Mn-Si system [14,15].

It is important to note that the data contained in these new publications have not yet been implemented in the databases. Both authors predicted similar phase diagrams, using the data of Voegel [16], which was more than 80 years ago. Hence, it would be instructive to repeat and verify these results, especially the $\delta \Leftrightarrow \gamma$ transition and the peritectic range, as shown by the red circle in Figure 6. Further justification for continuing these studies are found in the suspicion that silicon is a strong ferrite (bcc) stabilizer, as indicated in the binary Si-Mn phase diagram, the dearth of published experimental measurements, especially with respect to the $\gamma \rightarrow \delta$ phase transformation and the sometimes very old raw data.

In addition to the systematic studies of the Fe-C-Si-Mn system, identical studies (experimental and literature reviews) were carried out for the Fe-C-Si-Al and Fe-Mn-Al-C systems. Only by following the careful experimental procedure, outlined above, is it possible to assess the multicomponent Fe-C-Si-Al-Mn system. By utilizing the comprehensive newly developed experimental techniques; an extract from the SGTE2014 database is further developed and optimized by using the OptiSage tool, an integral part of the program FactSage. Moreover, by the use of the verified thermodynamic data developed in this study, new phase diagrams can be predicted, micro-segregation can be better analysed and calculations can be made with respect to the sequence of solidification [8]. In conclusion, it should be emphasized that it is only possible to calculate the solute enrichment of the TRIP steel, shown in Figure 3, by the use of the newly, improved databases outlined in this study. Even if it were possible to describe the initial elemental distributions in the Fe-0.2%C-0.5%Si-2%Mn-0.8%Al system by the use of existing databases, it is not possible to predict the consequences of micro-segregation where the individual concentrations might be close to ~1%C, ~1-2%Si, ~4-6%Mn, ~0,6%Al (Al segregates negative) respectively (which are significantly different from the initial concentrations).

## 4. Summary and Outlook

The current study illustrates the importance of experimental thermodynamic investigations for the re-assessment of selected Fe-base systems. The current study clearly shows that the actual thermodynamic calculations of the quaternary Fe-C-Si-Mn system provide sufficient justification for further improvement. A very important finding is that in the Fe-C-Si-Mn system, which is one of the most industrially important alloy systems for the design of high strength and high ductility steels, the position of the peritectic range should be at higher carbon content and the δ-region should be larger (bcc more stable) than predicted by existing thermodynamic databases.

The methodology to re-assess thermodynamic databases by means of self-constructed phase diagrams is highly efficient, especially if the alloy production, the sample manufacture, the DSC measurements, the construction of phase diagrams and the database optimization takes place within the same research group. By the systematic approach taken in this study and the use of high purity model alloys, it has been possible to identify relevant subsystems and their interactions. The CALPHAD method has the potential to calculate complex highly alloyed systems by using well investigated and secured binary, ternary and quaternary subsystems. Once if this thermodynamic data are available, new phase diagrams can be composed, micro-segregation can be quantified and calculations can made of the sequence of solidification events.

## 5. Acknowledgement

Financial support by the Austrian Federal Government (in particular from Bundesministerium für Verkehr, Innovation und Technologie and Bundesministerium für Wirtschaft, Familie und Jugend) represented by Österreichische Forschungsförderungsgesellschaft mbH and the Styrian and the Tyrolean Provincial Government, represented by Steirische Wirtschaftsförderungsgesellschaft mbH and Standortagentur Tirol, within the framework of the COMET Funding Programme is gratefully acknowledged. This work is part of the ongoing K2-MPPE project A4.18.

## 6. References

[1] Jung, I-H 2010 *Calphad* **34** 332,
[2] Lukas H L, Fries S G and Sundman B 2007 Computational Thermodynamics - The Calphad Method, Cambridge University Press, New York.
[3] Presoly P, Pierer R Bernhard C 2013 *Metallurgical and Materials Transactions* **44A** 5377.
[4] Jablonka A, Harste H and Schwerdtfeger K 1991 *Steel Research* **62** 24.
[5] Xia G, Bernhard C, Ilie S and Fürst W 2008 *6th ECCC* (Riccione, Italy).
[6] Presoly P, Xia G, Reisinger P and Bernhard C 2014 *BHM* **159** 430.
[7] Presoly P and Bernhard C 2015 LMPC2015 (Leoben, Austria).
[8] You D, Bernhard C, Michelic S, Wieser G and Presoly P 2015 MPPE2015 (Leoben, Austria) – *in press.*
[9] Software: NETZSCH Tau-R Kalibration, Version 6, Netzsch-Gerätebau, 2012.
[10] Huang W 1990 *Metallurgical and Materials Transactions* **A21** 2115.
[11] Lacaze J and Sundman B 1991 *Metallurgical and Materials Transactions* **22A** 2211.
[12] Miettinen J and Hallstedt B 1998 *Calphad* **22** 231.
[13] Forsberg A and Agren J 1993 *Journal of Phase Equilibria* **14** 354.
[14] Paek M K, Pak J J and Kang Y B 2014 *Calphad* **46** 92.
[15] Zheng W, Lu X G, He Y, Cui Y and Li L 2015 *Journal of Alloys and Compounds* **632** 661.
[16] Vogel R and Bedarff H 1933 *Arch. Eisenhüttenwesen* **7** 423.

# Laser materials processing of complex components. From reverse engineering via automated beam path generation to short process development cycles.

**R Görgl[1], E Brandstätter**
JOANNEUM RESEARCH Forschungsgesellschaft mbH
MATERIALS – Institute for Surface Technologies and Photonics
Leobner Straße 94, 8712 Niklasdorf, Austria

E-mail: richard.goergl@joanneum.at

**Abstract.** The article presents an overview of what is possible nowadays in the field of laser materials processing. The state of the art in the complete process chain is shown, starting with the generation of a specific components CAD data and continuing with the automated motion path generation for the laser head carried by a CNC or robot system. Application examples from laser welding, laser cladding and additive laser manufacturing are given.

## 1. Introduction

In modern production technology one trend goes into producing complex parts in small batch sizes ore even as single part. Therefore the corresponding R&D times need to be kept as short as possible. Therefore in laser processing one is frequently confronted with the demands of developing laser processes for complex components in rather short times to keep the whole project within a reasonable economic frame.

When dealing with complex components without available CAD data, reverse engineering via laser-based scanner systems is a suitable solution. Such systems nowadays have no more problems with reflecting metallic surfaces, and also the resolution has become sufficient even for delicate laser processing applications.

Out of the CAD data complex laser beam paths for welding and/or cladding can be generated automatically using appropriate software, including the consideration of tool orientation and collision issues for any customary robotic system or CNC gantry unit.

Combining laser cladding of different materials with intermediate or post-process machining (milling and drilling) within one system leads to enhanced possibilities for additive manufacturing solutions.

## 2. Technology

The paper is dealing with the technologies of laser welding, laser cladding and laser additive manufacturing of metallic parts. Further information about these technologies is found elsewhere [1].

[1] To whom any correspondence should be addressed.

## 3. Challenges

### *3.1. Small batch sizes*
As mentioned in section 1, industrial demand in special components produced in small lot sizes is ever-increasing. This in turn leads to the necessity of low R&D times in order to make such specialized components economically reasonable. Therefore the corresponding process development times have to be short, which is only feasible when automating as much as possible even during R&D.

### *3.2. Lack of sufficient CAD data*
Although normally the CAD data of components to be processed is available, every now and then this may be not the case:

- If parts are transferred during the process several times from one company to another, there might be no easy access to the original manufacturers CAD data.
- Parts to be processed after being in use. Such parts may be deformed in a way that the original CAD data is no more valid.
- Parts which are on the market for a very long time. Manufacturing has started long time ago, before CAD was common, and the corresponding CAD data has never been generated later on.

### *3.3. Generating complex motion paths*
Laser material processing means that there is a laser head – mounted on a CNC or robot system – that needs to be moved along certain tracks above the component. If these tracks are in any way mathematically describable, the corresponding programming of the motion paths may get any complexity, but is always feasible. An example of such a complex motion path for laser cladding of an extruder screw is shown in figure 1.

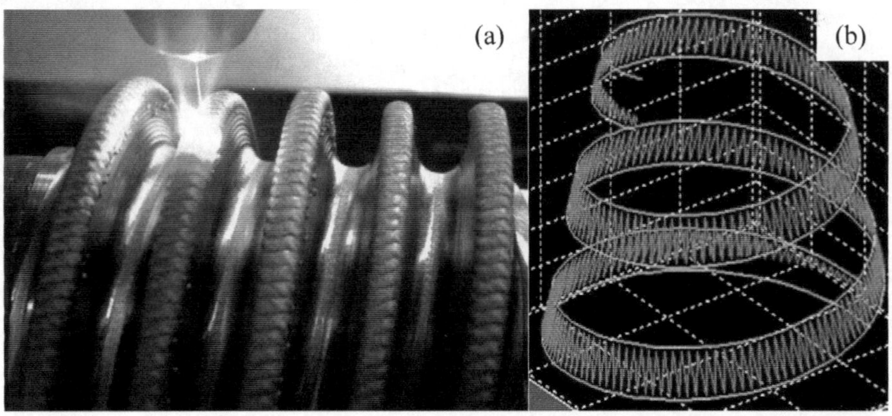


**Figure 1. (a)** Laser cladding of an extruder screw with decreasing diameter and truncated screws at the end. **(b)** Exaggerated display of corresponding motion path.

Exact programming of motion paths is not possible when dealing with mathematically not describable topologies, like free form surfaces. Such complex motion paths could only be teached, which means that the (robot) system is manually moved and the positions get stored. It is easily to understand that teaching soon exceeds any reasonable timeframes and is therefore not feasible in most cases.

## 4. Solutions

*4.1. Reverse engineering*

Reverse engineering refers to procedures that start with any real component and deliver its CAD data. One possibility is the use of laser-based scanner systems for 3D digitalization [2,3]. Such systems nowadays show sufficient spatial resolutions of less than 0.05 mm, and also the problem of highly reflecting surfaces that was common when dealing with metallic parts has been overcome. Furthermore, the corresponding software supports complex automated measurements. For example, the scanner system can be mounted onto a robot, and the measurement of large complex parts is done automatically. Figure 2 shows examples of the current state of the art.

**Figure 2. (a)** Robot-based automated measurement of a larger car body part [2], **(b)** 3D digitalization of a car tyre [2], **(c)** 3D digitalization of a coin [2], **(d)** Complex part generated via additive manufacturing and partly finished by subsequent milling, 3D digitalization of that part in **(e)** lower resolution of 0.1 mm and **(f)** higher resolution of 0.05 mm.

After such 3D digitalization the resulting CAD data can easily be imported in any standard CAD software for further adaptation, if desired.

*4.2. Automated motion path generation*

As described in section 3.3, general goal is the generation of corresponding motion paths for laser materials processing technologies, which is not mathematically solvable for components containing free form surfaces. Up to several years ago there was no universal and commercially available solution to this challenge, only several academic institutions were in possession of automation software tailored to specific applications and/or geometries. This situation is not satisfying when dealing with a wide range of different applications.

A few years ago a quite universal software solution was developed by the market leader of laser systems [4] and distributed along with their complete laser processing systems. In cooperation with another software company a special version of this software called *TopLas3D* [5], which can be adapted for all standard CNC and robot systems, is commercially available now.

We have implemented this software into an *ABB* robot system [6]. *TopLas3D* is capable of importing any standard CAD data format. In the next step the user selects the surfaces or edges that shall be processed. *TopLas3D* then gives a simulation of the motion path, where surface discontinuities (holes, edges, sharp folds,...) are considered. Since the complete workstation CAD data – including the mounted laser head – is also stored in the software, optimum tool orientation and automated collision detection is included. A lot of details like different motion strategies (rectangular, circular,...), basic CAD operations and many options for the user to set constraints or override the suggested simulation exist. A detailed description of all its capabilities would exceed this article by far, the interested reader is referred to the *TopLas3D* webpage [5]. When satisfied with the simulation results, NC data for the specific workstation is exported. Figures 3-5 show selected applications.

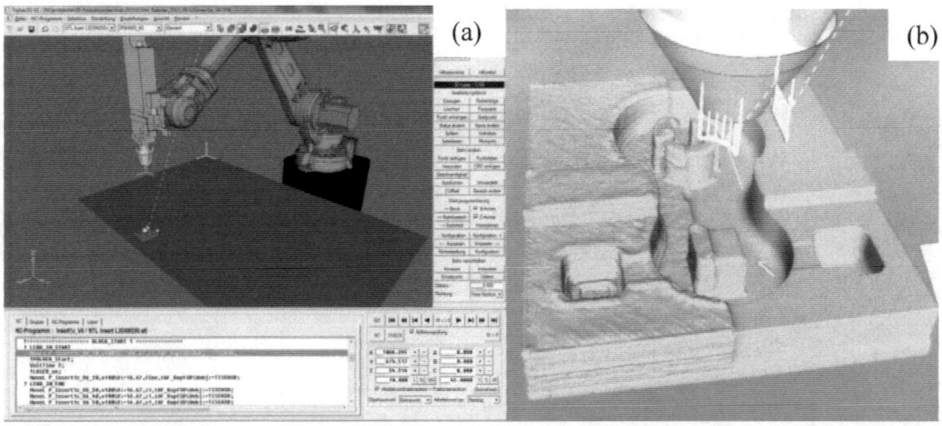


**Figure 3. (a)** View of the *TopLas3D* screen with the complete robot system and the CAD data of the component shown in figure 2d imported. **(b)** Screenshot of a motion path simulation on that component's surface.

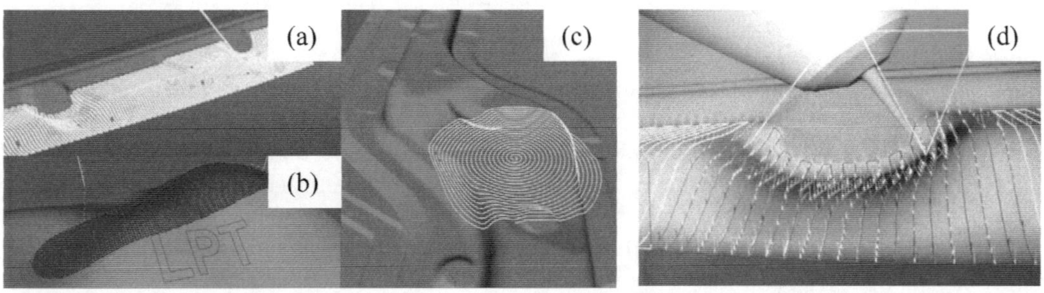


**Figure 4. (a-c)** Motion paths on various complex surfaces. **(d)** Screenshot of a motion path simulation of the path shown in figure 4a. The orange cone is the visualisation of the laser beam.

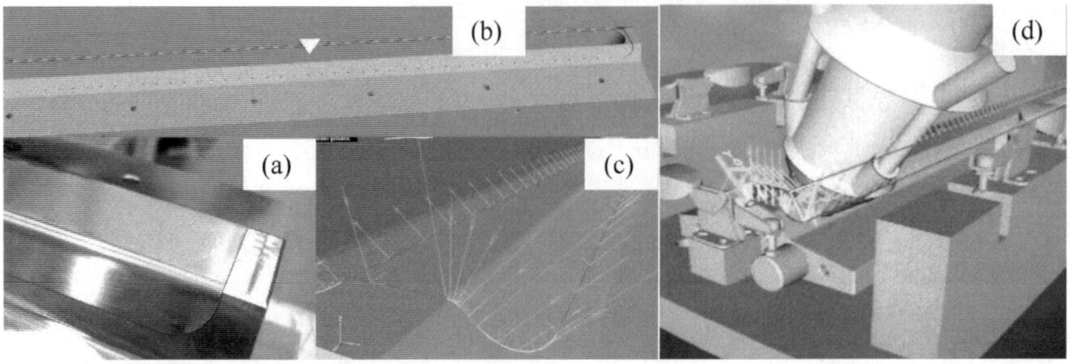


**Figure 5. (a)** Part to be laser welded. **(b)** CAD data of that part. **(c)** Motion path for the welding seam of that part. **(d)** Screenshot of the motion path simulation.

There are certain cases – especially when working with a robot station - when although the motion path is programmable in principle, its simulation is a good alternative. Consider for example a gyration movement, which origins in a complex interplay between the different robot axes to result in a smooth gyration movement of the tool. Figure 6 shows such a case.

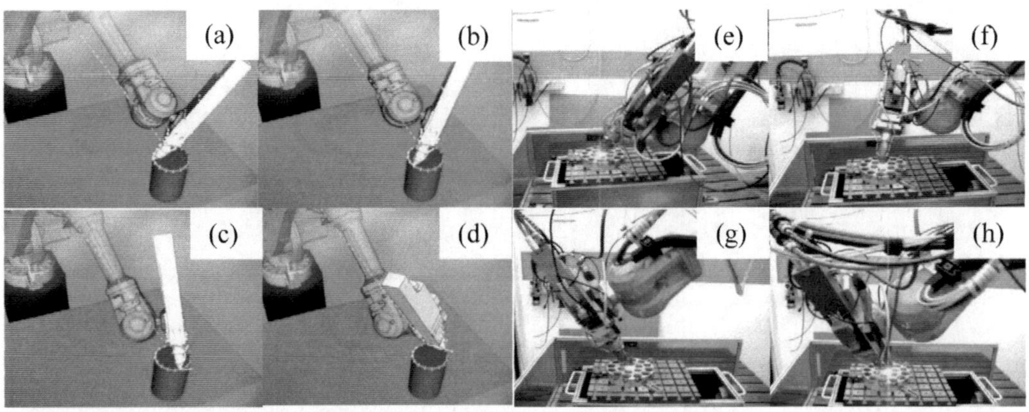


**Figure 6. (a-d)** Screenshots from the simulation of a gyration movement. **(e-h)** Images of a laser cladding process using a gyration movement in reality.

## 5. Application example

Nowadays oil field technology makes use of moving the drill strings in turns towards the oil field. Due to the sensors installed in immediate vicinity to the drilling head the base material has to be non-magnetic, which is fulfilled by high-alloyed Cr-Mn-steels which have high corrosion resistance and ductility. However, their wear resistance is quite low, which results in unacceptable high wear at field conditions. A laser cladding process was developed in cooperation with one of the global players in the drill string production market [7], whereby the following challenges had to be solved:

- Thermal load both to the cladding nozzle and to the cladded components because of cladding times up to several hours.
- Since subsequent machining should be avoided, a constant total layer thickness of $4.0^{\pm 0.2}$ mm was demanded. This was all the more a challenge, since the complete layer system consisted of four sublayers. Details of that layer system can be found elsewhere [8,9].

- Due to the complex surfaces of the drilling heads the motion paths for cladding had to be programmed by external software and fed into the machine. This was done at that time (2005) by tailoring available software to the specific needs. If a powerful software like *TopLas3D* had been available then, R&D times would have been drastically reduced.

Figure 7 shows an example of a programmed motion path and an image of the cladding process.

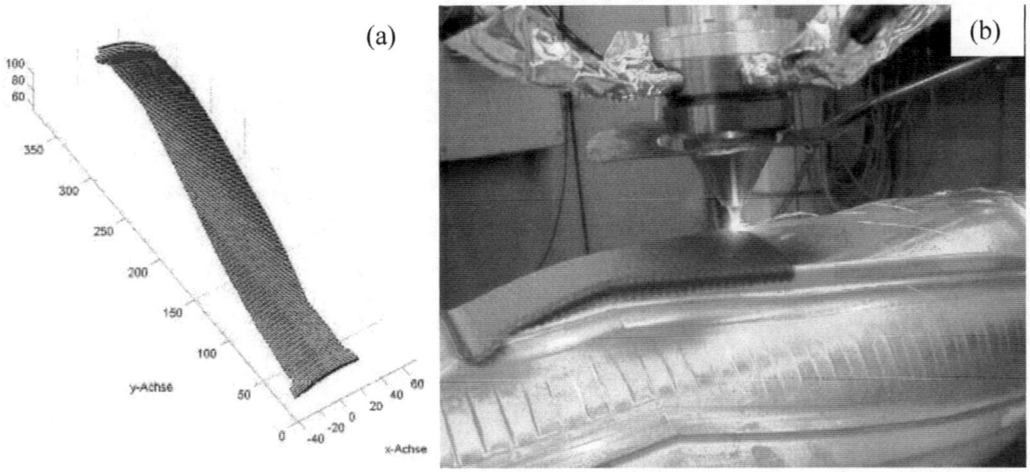


**Figure 7. (a)** Programmed motion path from 2005. **(b)** Image of the laser cladding process.

This process is still state of the art in oilfield technology and meanwhile running at the customers sites on four laser cladding stations performing multi-shift operations [7].

## 6. Additive manufacturing solutions

The technology of laser cladding is used mainly for the following applications:

- Deposition of protection layers (wear, corrosion,…) onto metallic components
- Repair of worn out metallic components
- Additive manufacturing of metallic components. In the context of additive manufacturing this technology is often referred to as LMD (laser metal deposition).

When using LMD for additive manufacturing there will always arise height differences in the generated part during the process. To level them out it would need very complex programming and/or fast and subtle online corrections of process parameters. A more expedient solution is frequent levelling out by intermediate milling, but which is only economically reasonable when it can be done without re-clamping of the component in formation.

Such a combination – but with standard wet milling – was brought to market by one of the big players in the production of tooling machines in 2014 [10], but was developed as special model – with high-speed dry milling – for *JOANNEUM RESEARCH* about 15 years ago. Figure 8 shows our combined laser and milling station as well as an additive manufactured part. In figure 2d already a part generated via LMD was shown, with one half remaining as it is immediately after LMD manufacturing and the other half finished by subsequent milling.

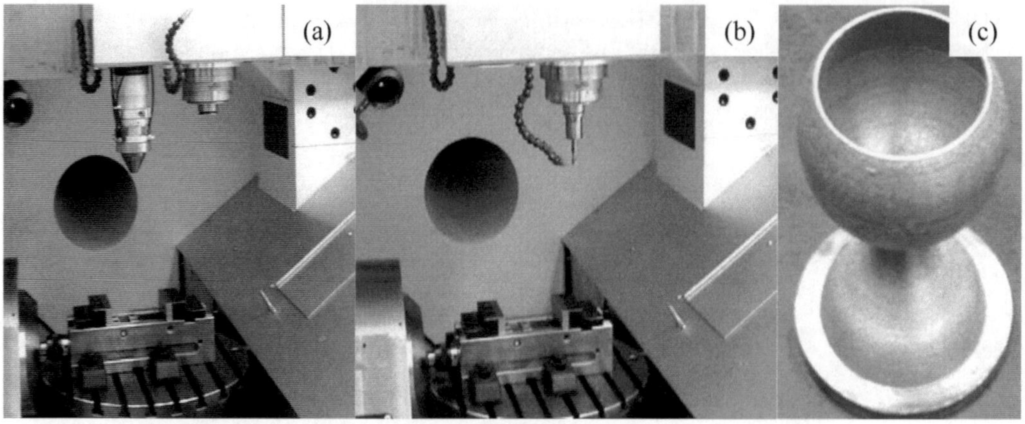


**Figure 8.** Combined 5-axes laser and milling station **(a)** with extended laser cladding nozzle and **(b)** with extended milling spindle. **(c)** Additive manufactured goblet. By making use of all five axes overhangs can be handled. In this case no further milling was applied.

During the last decade the technology of SLM (selective laser melting) has come to a considerable maturity level in the field of additive manufacturing [11,12]. Unjustly LMD and SLM are often regarded as competing technologies, which they clearly are not when looking at the details. Both technologies have different strengths and limitations, which in fact makes them complementary. Whereas LMD is not capable of producing such fine and complex structures that are feasible with SLM [13-16], SLM fails in the combination of different materials. Other drawbacks of SLM are the still extremely long fabrication times and limited maximum part sizes. Figure 9 shows examples of so-called "functional tools" consisting of different materials fabricated by LMD technology.

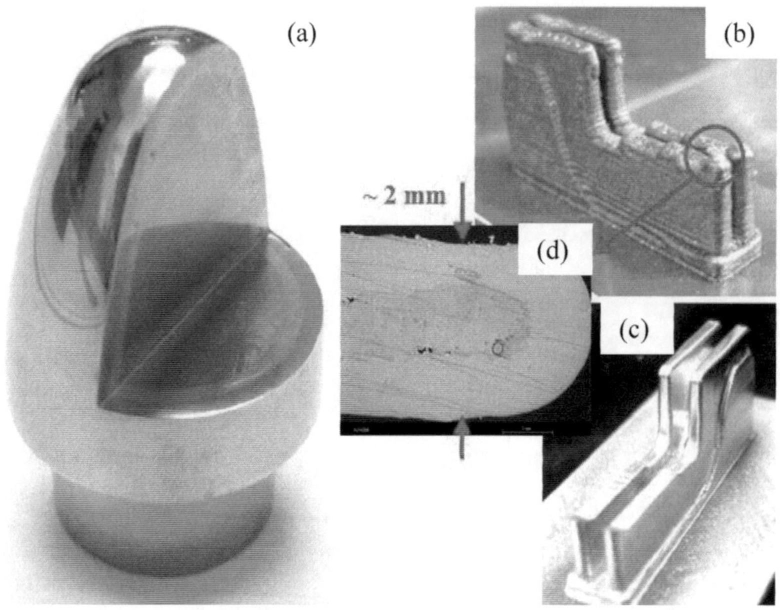


**Figure 9. (a)** Cooling insert made of inexpensive material for the most part, but with a copper layer beneath a special abrasion-resistant surface. **(b-d)** Functional complex cooling insert with copper beneath the surface **(b)** directly after LMD fabrication and **(c)** after final machining. **(d)** Microsection of the part.

## 7. Summary

In the field of laser materials processing of metallic materials the tools existing nowadays enable the automation of important process steps even during R&D phase, thus fulfilling the industrial needs for short development times.

These steps in particular consist of the generation of CAD data of any given metallic component and the automated generation of the corresponding motion paths for laser processing. Such automation tools can also be used for the production of additively manufactured components, even consisting of different materials.

## References

[1] Görgl R, Brandstätter E, 2011 Laser Cladding vs. Laser Alloying – A Comparative Study *Proc. 3rd IFHTSE "Heat Treatment and Surface Engineering of Tools and Dies" (Wels, Austria, March 2011)* **74-83**
[2] Carl Zeiss Optotechnik GmbH, COMET and T-SCAN series
[3] IPO.Plan GmbH, IPO.Eye
[4] TRUMPF GmbH + Co. KG, TruTopsCell software
[5] Mebitec - Meerbuscher Informationstechnik GmbH, TopLas3D software
[6] ABB AG, IRB 4600 series
[7] Schoeller-Bleckmann Oilfield Technology GmbH
[8] Brandstätter E, 2006 Laserauftragsschweißen am Beispiel von Werkzeugen für die Ölfeldtechnik *Lasermagazin* **1**
[9] Brandstätter E, 2006 Laserauftragsschweißen als innovatives Verfahren zur Panzerung von Werkzeugen der Ölfeldtechnik *Proc. 9. Werkstofftechnisches Kolloquium (Chemnitz, Germany, September 2006)* **370-376**
[10] DMG Mori AG, Lasertec 65 3D
[11] Gibson I, Rosen D W, Stucker B, 2010 *Additive manufacturing technologies: rapid prototyping to direct digital manufacturing* (New York, Heidelberg, Dordrecht, London: Springer ISBN ISBN: 978-1-4419-1119-3)
[12] Wohlers T, 2015 *Wohlers Report 2015, 3D Printing and Additive Manufacturing State of the Industry* (Fort Collins, CO: Wohlers Associates)
[13] Diatlov A, 2013 Manufacturing antenna components for satellites out of AlSi10Mg with SLM *Annual Report 2013* **74** (Fraunhofer Institute for Laser Technology)
[14] Merkt S J, Hinke C, 2013 Functional integrated rim support with internal lattice structure *Annual Report 2013* **75** (Fraunhofer Institute for Laser Technology)
[15] Santos Batista R J, Meiners W, 2013 Additive manufacturing of microstructures with SLM *Annual Report 2013* **76** (Fraunhofer Institute for Laser Technology)
[16] Rolink G, Weisheit A, 2013 Laser-based additive manufacture with iron-aluminide alloys *Annual Report 2013* **77** (Fraunhofer Institute for Laser Technology)

# Fatigue crack growth threshold as a design criterion – statistical scatter and load ratio in the Kitagawa-Takahashi diagram

**S.Kolitsch[1,2], H.-P. Gänser[1], J. Maierhofer[1], R. Pippan[2]**

[1] Materials Center Leoben Forschung GmbH, Roseggerstraße 17, A-8700 Leoben
[2] Erich Schmid Institute of Materials Science, Austrian Academy of Sciences, and Department of Materials Physics, University of Leoben, Jahnstraße 12,A-8700 Leoben

E-mail: stefan.kolitsch@mcl.at

**Abstract**. Cracks in components reduce the endurable stress so that the endurance limit obtained from common smooth fatigue specimens cannot be used anymore as a design criterion. In such cases, the Kitagawa-Takahashi diagram can be used to predict the admissible stress range for infinite life, at a given crack length and stress range. This diagram is constructed for a single load ratio *R*. However, in typical mechanical engineering applications, the load ratio *R* varies widely due to the applied load spectra and residual stresses. In the present work an extended Kitagawa-Takahashi diagram accounting for crack length, crack extension and load ratio is constructed. To describe the threshold behaviour of short cracks, a master resistance curve valid for a wide range of steels is developed using a statistical approach.

## 1. Introduction

Cracks reduce the endurable stress of components so that the endurance limit cannot be used anymore as a design criterion. In such cases, the Kitagawa-Takahashi (KT) diagram [4] is commonly used. In this diagram the admissible stress range depends on the crack length; this curve delimits the area of non-propagating cracks. By means of an intrinsic crack length $a_{0,\mathrm{H}}$ introduced by El Haddad [2], a transition from the threshold of long cracks to the endurance limit is given (Fig. 1). The intrinsic length $a_{0,\mathrm{H}}$ for the transition is calculated via

$$a_{0,\mathrm{H}} = \frac{1}{\pi}\left(\frac{\Delta K_{\mathrm{th,lc}}}{Y \cdot \Delta\sigma_{\mathrm{e}}}\right)^2, \quad (1)$$

and the admissible stress range depending on the intrinsic length can then be expressed as follows:

$$\Delta\sigma_{\mathrm{th,lc}}(a_0) = \frac{\Delta K_{\mathrm{th,lc}}}{Y \cdot \sqrt{\pi \cdot (a + a_{0,\mathrm{H}})}}. \quad (2)$$

Here, $\Delta K_{\mathrm{th,lc}}$ denotes the fatigue crack growth threshold stress intensity factor range for long cracks, $\Delta\sigma_{\mathrm{e}}$ the endurance limit stress range of polished specimens without flaws, and $Y$ is the geometry factor of the crack. It should be noted that the values $\Delta K_{\mathrm{th,lc}}$ and $\Delta\sigma_{\mathrm{e}}$ depend on the load ratio $R$ so that the curve is only valid for a single load ratio $R$.

This approach can be used for materials where the crack closure is built up completely. For other materials where the crack closure is not built up, El Haddad´s description is not conservative anymore. For this reason Chapetti used an exponential function to describe the crack closure effect for short cracks in the KT-diagram [1]. Nevertheless, in this function preexisting notches are not considered. Similar work was done by Tabernig [9].
The goal of the present work is to extend these concepts and take different load ratios into account.

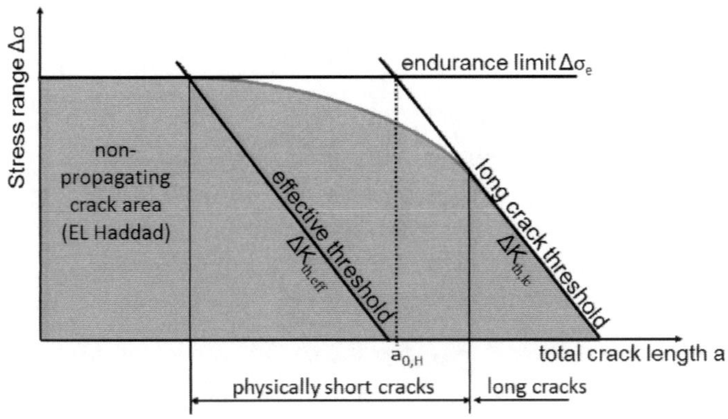


**Figure 1.** Kitagawa-Takahashi diagram, showing the area of non-propagating cracks according to El Haddad.

**Nomenclature**

| | | | |
|---|---|---|---|
| $\Delta a$ | crack extension length | $l_i$ | fictitious length scales |
| $a_0$ | notch depth | $R$ | load ratio |
| $a_{0,\mathrm{H}}$ | fictitious intrinsic length scale following El Haddad | $\Delta\sigma$ | stress range |
| $a$ | total crack length | $\Delta\sigma_\mathrm{e}$ | endurance limit stress range of polished specimens without flaws |
| $\Delta K$ | stress intensity factor range | $\Delta\sigma_\mathrm{th}$ | threshold stress range for crack propagation |
| $\Delta K_\mathrm{th}$ | threshold of stress intensity factor range for crack propagation | $\Delta\sigma_\mathrm{th,lc}$ | threshold stress range for crack propagation calculated using $\Delta K_\mathrm{th,lc}$ |
| $\Delta K_\mathrm{th,eff}$ | intrinsic (effective) threshold stress intensity factor range | $v_i$ | weighting factors |
| $\Delta K_\mathrm{th,lc}$ | long crack growth threshold stress intensity factor range | $l_{90}$ | fictitious length where 90 % of the crack closure is done |
| $\Delta K_{\mathrm{th,lc}(R=0)}$ | long crack growth threshold stress intensity factor range for load ratio equal to zero | $C$ | weighting factor |
| $\mathrm{f_{cl}}$ | function to describe the master resistance curve | $l_0$ | weighting factor |
| $\mathrm{f_{cext}}$ | function to describe the standardized crack extension | $\Delta\sigma_\mathrm{UTS}$ | ultimate tensile stress |
| $f_\mathrm{D}$ | tension and compression endurance strength factor | $M_\sigma$ | mean stress sensitivity |
| $K_\mathrm{D}$ | mean stress factor | $a_\mathrm{m}$ | material constant |
| $b_\mathrm{m}$ | material constant | | |

## 2. Build-up of crack closure

In Fig 2 the combination of the notch depth and the crack length is illustrated.

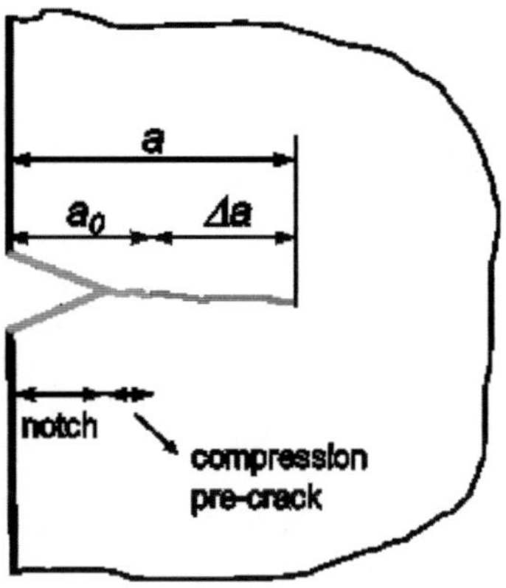


**Figure 2.** Schematic illustration of the proposed mechanical model: emanating from a deep sharp notch $a_0$, a crack of extension $\Delta a$ grows. Only on this crack extension $\Delta a$ the build-up of crack closure is possible [6].

The stress intensity factor range is defined by

$$\Delta K = Y \cdot \Delta\sigma \cdot \sqrt{\pi \cdot a}\ , \tag{3}$$

where the total crack length

$$a = a_0 + \Delta a \tag{4}$$

is the sum of notch depth $a_0$ and crack extension $\Delta a$.

For the description of the crack closure effect only the crack extension $\Delta a$ is important and not the total crack length. To consider also the notch depth, Maierhofer et al. [5] describe the crack resistance curve by a double exponential function (Eq (5)). The different closure effects and their contributions to the total effect are shown schematically in Fig. 3, the plasticity-induced and roughness-induced closure contributions being represented each separately by an exponential function with a characteristic length scale $l_i$.

$$\Delta K_{\text{th}} = \Delta K_{\text{th,eff}} + (\Delta K_{\text{th,lc}} - \Delta K_{\text{th,eff}}) \cdot \left[1 - \sum_{i=1}^{2} \nu_i \cdot \exp\left(-\frac{\Delta a}{l_i}\right)\right] \tag{5}$$

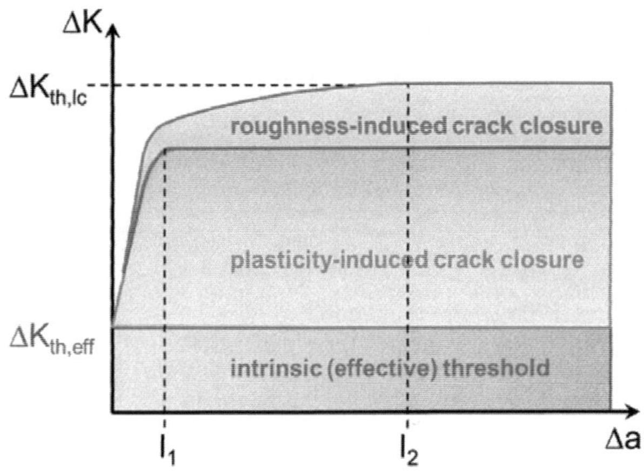


**Figure 3.** Illustration of the crack resistance curve caused by two different closure mechanisms; each closure mechanism is built up completely over a specific crack extension (described by the fictitious length scales $l_i$) [5]

## 3. Cyclic fracture mechanics experiments

Constant amplitude load increasing tests [5] have been conducted on single edge notched bending (SENB) specimens for five different materials to obtain the cyclic crack resistance curves shown in Fig. 4. Most of the materials have been tested for two different load ratios $R = -1$ and $R = 0.1$. By using the experimental data points and the experimentally determined threshold for long cracks $\Delta K_{\mathrm{th,lc}}$, the parameters $\nu_i$ and $l_i$ in Eq (5) can be fitted and an estimate for the crack resistance curve can be calculated (the intrinsic threshold is for all iron alloys $\Delta K_{\mathrm{th,eff}} \approx 2.5$ MPa√m; this is also clearly visible from Fig. 4). Those estimates are shown by the solid and dashed lines in Fig. 4.

It can be seen that Eq (5) describes the experimentally observed behavior well. However, the contributions from plasticity-induced and roughness-induced crack closure are different between the materials and are not comparable to each other. The threshold for long cracks changes due to the load ratio, but also due to differences in the closure mechanisms. This means that the resistance curve can be described by an exponential function whose parameters have to be determined separately for each material and load ratio.

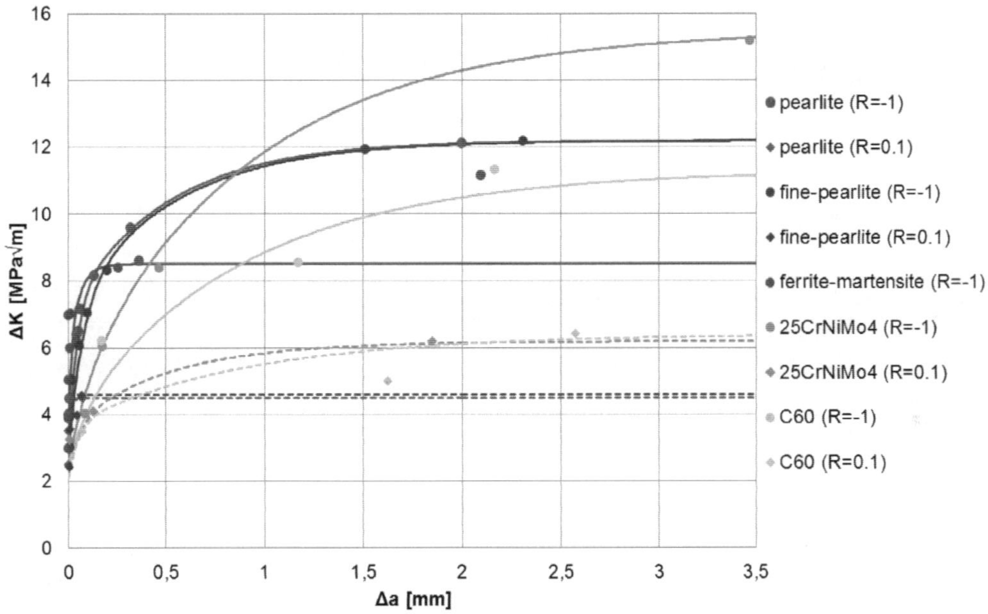


**Figure 4**. Crack resistance curve for $\Delta K_{\mathrm{th}}$: experimental data points and analytically estimation curve from Eq. (5).

## 4. Proposal for resistance curves for $\Delta K_{\mathrm{th}}$

To design a more general diagram to compare the resistance curves of each material, the curves have to be converted to a standardized form. To this purpose, we relate the threshold increase due to closure effects at a given crack extension, $\Delta K_{\mathrm{th}} - \Delta K_{\mathrm{th,eff}}$, to the total threshold increase due to closure for long cracks, $\Delta K_{\mathrm{th,lc}} - \Delta K_{\mathrm{th,eff}}$. This crack closure factor is thus calculated by

$$f_{\mathrm{cl}} = \frac{\Delta K_{\mathrm{th}} - \Delta K_{\mathrm{th,eff}}}{\Delta K_{\mathrm{th,lc}} - \Delta K_{\mathrm{th,eff}}} \quad (6)$$

To bring the length scales over which the build-up of closure occurs into a standardized form, the crack extension $\Delta a$ is rendered dimensionless via division by the crack extension $\Delta a$ where 90% of crack closure is reached, denoted by $l_{90}$:

$$f_{\text{cext}} = \frac{\Delta a}{l_{90}} \,. \tag{7}$$

In Fig. 5 the standardized form for the resistance curves of all experimental data is plotted with the standardized range of the stress intensity factors and the normalized crack length. Based on this proposal the material and physical behavior of the different material is standardized and they are comparable. A simple analytical approximation of this curve is given, similarly as in Eq (5), by

$$\frac{\Delta K_{\text{th}} - \Delta K_{\text{th,eff}}}{\Delta K_{\text{th,Ic}} - \Delta K_{\text{th,eff}}} = 1 - e^{\frac{\Delta a}{l_{90}}\frac{1}{l_0}+C} \,, \tag{8}$$

$$f_{\text{cl}}\left(f_{\text{ext}}\right) = 1 - e^{-\left(\frac{1}{l_0} f_{\text{ext}} + C\right)} \,, \tag{9}$$

where $l_0$ and $C$ denote constants to fit the experimental data points.

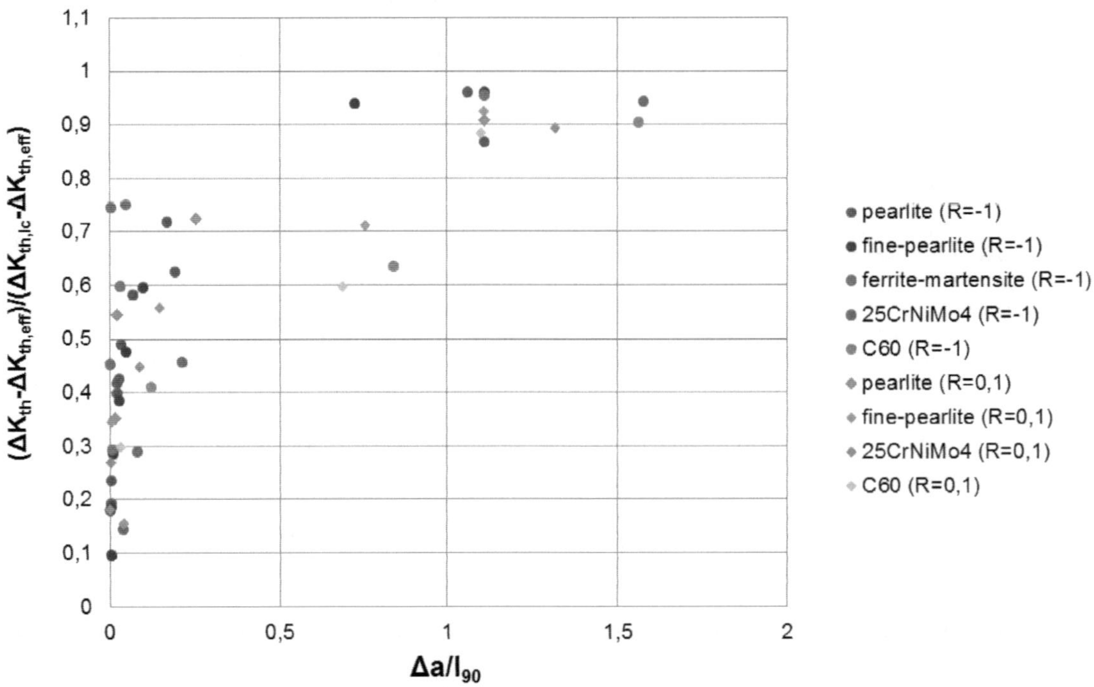


**Figure 5.** Standardized resistance curve of $\Delta K_{\text{th}}$ obtained from Fig. 4 by normalizing the individual data points via Eqns (6) and (7).

To provide a general resistance curve for $\Delta K_{\text{th}}$ of all tested materials a statistical analysis of the standardized experimental data is necessary.

## 5. Statistical approach

By determining the constants $l_0$ and $C$ by means of statistical regression, a general resistance curve can be obtained. A linear model for the data is obtained by transforming Eq (8) into

$$-\ln\left(1 - \frac{\Delta K_{\text{th}} - \Delta K_{\text{th,eff}}}{\Delta K_{\text{th,Ic}} - \Delta K_{\text{th,eff}}}\right) = \frac{\Delta a}{l_{90}} \cdot \frac{1}{l_0} + C \,, \tag{10}$$

$$y(f_{ext}) = -\ln(1-f_{cl}) = \frac{1}{l_0}\cdot f_{ext} + C\,. \tag{11}$$

The linear regression of $y = -\ln(1-f_{cl})$ vs. $f_{ext}$ to the mean as well as the lines corresponding to ±1 coefficient of variation (COV) of the regression parameters $1/l_0$ and $C$ are plotted in Fig. 6.

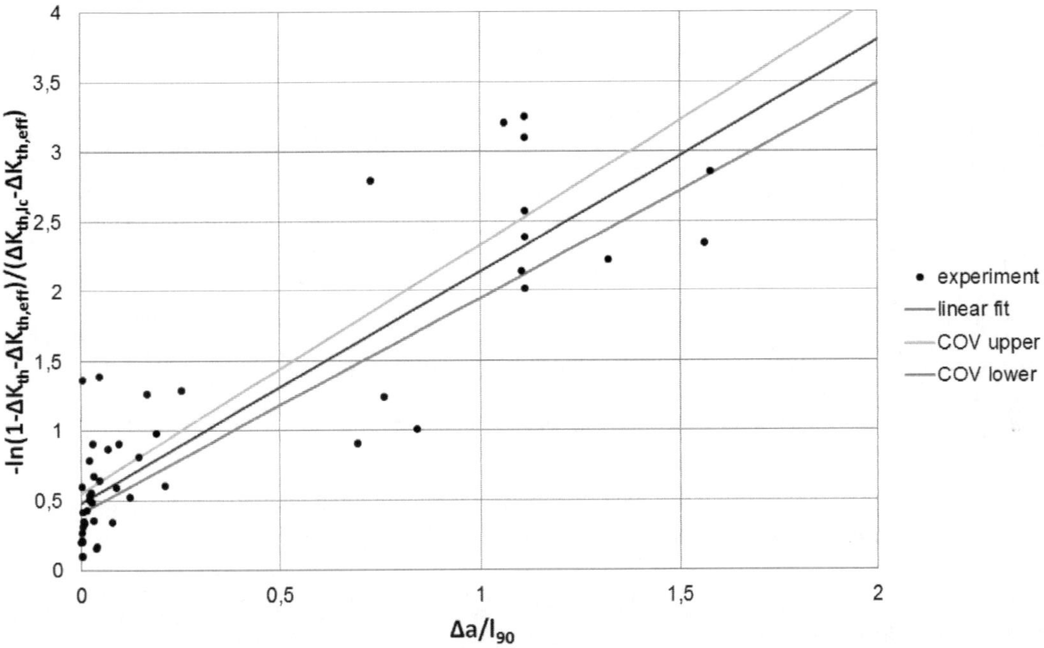


**Figure 6**. Linear regression analysis of experimental data.

The general resistance curve obtained by back-transformation is plotted in Fig. 7. The restriction of this fitting curve is that it does not start anymore at zero. This means that the closure estimate is not conservative for very short cracks in particular. However, such very short crack extensions are of minor importance for damage tolerance issues, given the current detection limits in non-destructive testing.

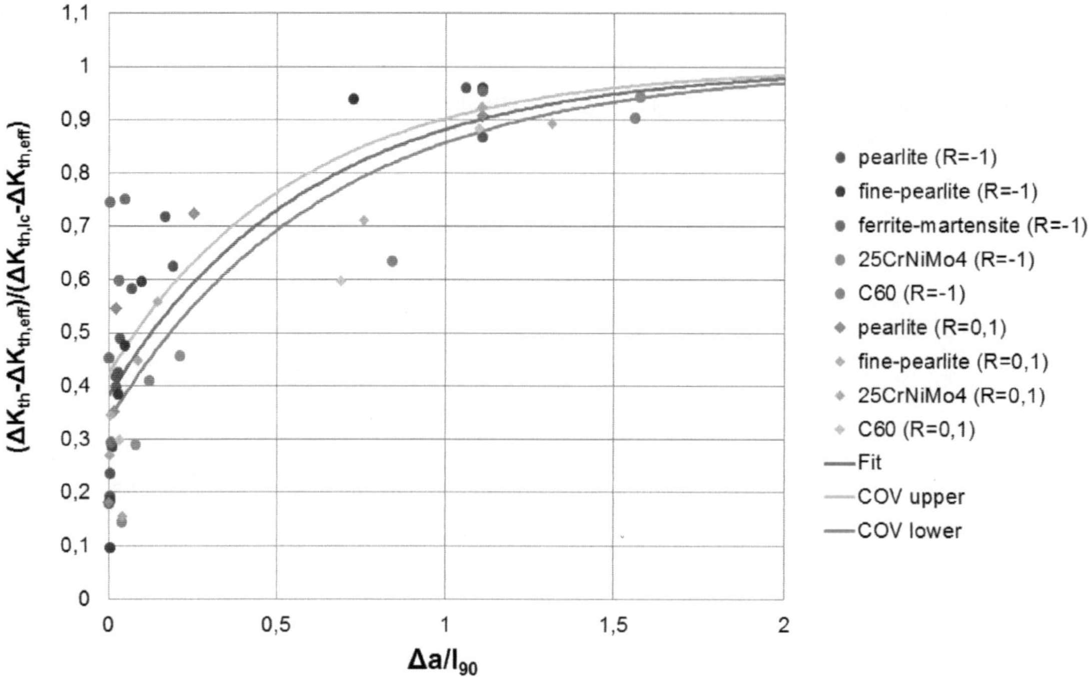


**Figure 7.** Master resistance curve with statistical regression analysis.

## 6. Consideration of load ratio *R*

Using the proposed resistance curve for $\Delta K_{\text{th}}$, the Kitagawa-Takahashi diagram can be calculated for different materials. In addition, the influence of the load ratio $R$ is estimated by the simple bi-linear approximation shown in Fig 8:

$$\Delta K_{\text{th,lc}}(R) = \begin{cases} \Delta K_{\text{th,lc}(R=0)} \cdot (1-R) & \text{for} \quad \Delta K_{\text{th,lc}(R=0)} \cdot (1-R) > \Delta K_{\text{th,eff}} \\ \Delta K_{\text{th,eff}} & \text{for} \quad \Delta K_{\text{th,lc}(R=0)} \cdot (1-R) \leq \Delta K_{\text{th,eff}} \end{cases} \tag{12}$$

Here $\Delta K_{\text{th,lc}(R=0)}$ denotes the range of the stress intensity factor for long cracks at a load ratio $R = 0$.

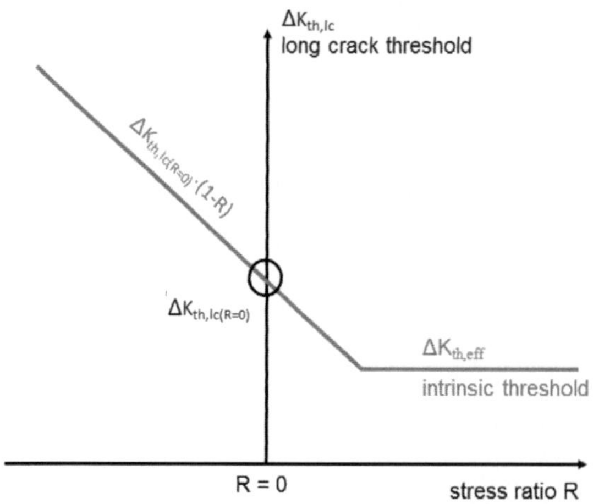


**Figure 8.** Dependency of long crack threshold and the load ratio.

Due to the linear connection between the threshold for long cracks and the load ratio, the range for any stress intensity factor can be calculated. Combining Eq (8) with the master resistance curve, Eq (9), one obtains the threshold stress intensity factor range as a function of crack extension $\Delta a$ and load ratio $R$,

$$\Delta K_{\mathrm{th}}(\Delta a, R)=\begin{cases}\Delta K_{\mathrm{th,eff}}+\left[\Delta K_{\mathrm{th,lc}(R=0)}\cdot(1-R)-\Delta K_{\mathrm{th,eff}}\right]\cdot \mathrm{f}_{\mathrm{cl}} & \text{for } \Delta K_{\mathrm{th,lc}}(R)>\Delta K_{\mathrm{th,eff}}\\ \Delta K_{\mathrm{th,eff}} & \text{for } \Delta K_{\mathrm{th,lc}}(R)\leq\Delta K_{\mathrm{th,eff}}\end{cases} \quad (13)$$

The admissible applied stress range $\Delta\sigma$ is then obtained by equating the crack growth resistance to the applied load,

$$\Delta K_{\mathrm{th}}(\Delta a, a_0)=\Delta\sigma\cdot Y\cdot\sqrt{\pi\cdot(a_0+\Delta a)}\,. \quad (14)$$

The admissible stress range $\Delta\sigma$ thus depends on three variables, $\Delta a$, $a_0$ and $R$, and is calculated as follows:

$$\Delta\sigma(\Delta a, R, a_0)=\min\left(\frac{\Delta K_{\mathrm{th}}(\Delta a, R)}{Y\cdot\sqrt{\pi\cdot(a_0+\Delta a)}}, \Delta\sigma_{\mathrm{e}}(R)\right). \quad (15)$$

Not only $\Delta\sigma$, also the stress endurance limit $\Delta\sigma_{\mathrm{e}}$ depends on the stress ratio $R$ [3]:

$$\Delta\sigma_{\mathrm{e}}(R)=2\cdot\sigma_{\mathrm{UTS}}\cdot f_{\mathrm{D}}\cdot K_{\mathrm{D}}(R). \quad (16)$$

Here, $\sigma_{\mathrm{UTS}}$ denotes the ultimate tensile stress of the material, $f_{\mathrm{D}}$ the endurance limit factor depending on the material class, and $K_{\mathrm{D}}(R)$ the mean stress factor accounting for the mean stress sensitivity of the material. The mean stress sensitivity is considered by using the Haigh-diagram (Fig. 9).

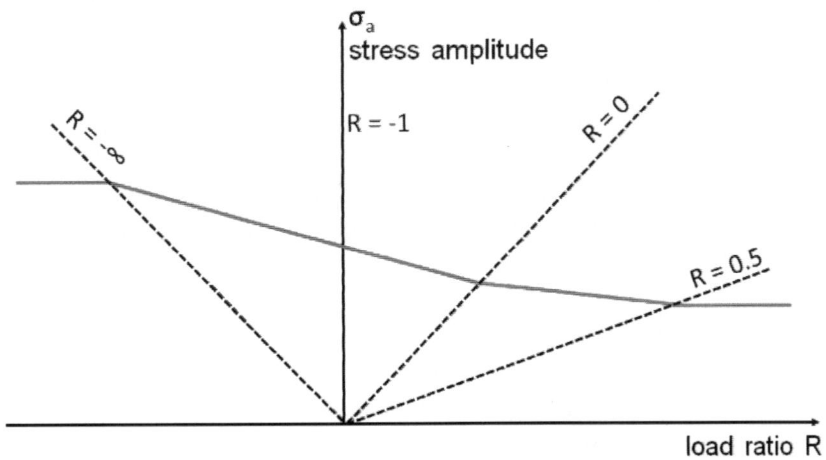


**Figure 9.** Haigh diagram

The mean stress factor depending on the stress ratio can be calculated as follows [3]:

$$K_D(R)=\begin{cases}\dfrac{1}{1+M_\sigma\cdot(1+R)/(1-R)} & \textit{for} \quad R<0\\[2ex] \dfrac{1}{(1+M_\sigma)\cdot[3+M_\sigma\cdot(1+R)/(1-R)]} & \textit{for} \quad 0<R<0.5\\[2ex] \dfrac{3+M_\sigma}{3\cdot(1+M_\sigma)^2} & \textit{for} \quad R>0.5\end{cases} \tag{17}$$

$M_\sigma$ describes the mean stress sensitivity depending on the ultimate tensile stress $\sigma_{\mathrm{UTS}}$ with constants $a_{\mathrm{m}}$, $b_{\mathrm{m}}$ depending on the material class [3]:

$$M_\sigma = a_{\mathrm{m}}\cdot 10^{-3}\cdot\frac{\sigma_{\mathrm{UTS}}}{\mathrm{MPa}}+b_{\mathrm{m}} \tag{18}$$

So the stress range can be calculated for different stress ratios. In Fig. 10 the stress range $\Delta\sigma$ ($\Delta a, a_0, R$) is plotted for a constant initial notch depth $a_0 = 0.1$ mm depending on the crack extension $\Delta a$. The stress decreases with increasing load ratio $R$.

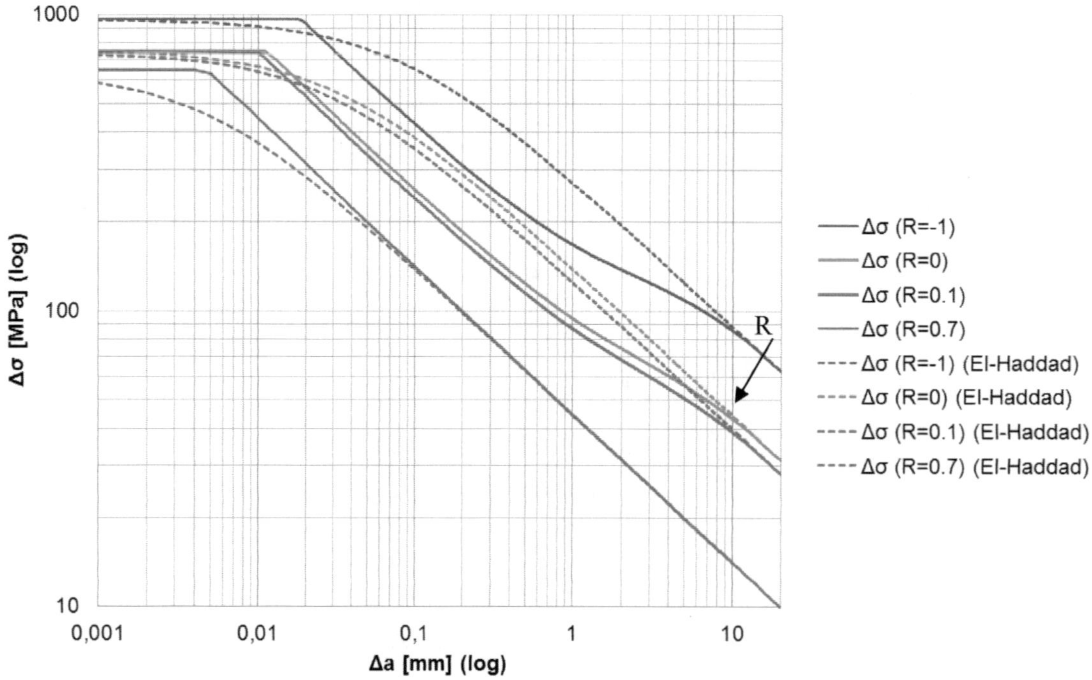


**Figure 10**. Admissible stress range $\Delta\sigma$ vs. crack extension $\Delta a$ for different load ratios $R$ and a constant initial notch depth $a_0 = 0.1$ mm.

Overall, Eq (15) describes for a given constant load ratio $R$ a three-dimensional surface [6], which is shown in Fig 11. This surface represents the stress range depending on the crack extension $\Delta a$, the initial crack (notch) depth $a_0$ and is plotted for different load ratios $R$. A different load ratio $R$ will create an offset and scaling of the failure surface which is shown for four different load ratios in Fig. 11.

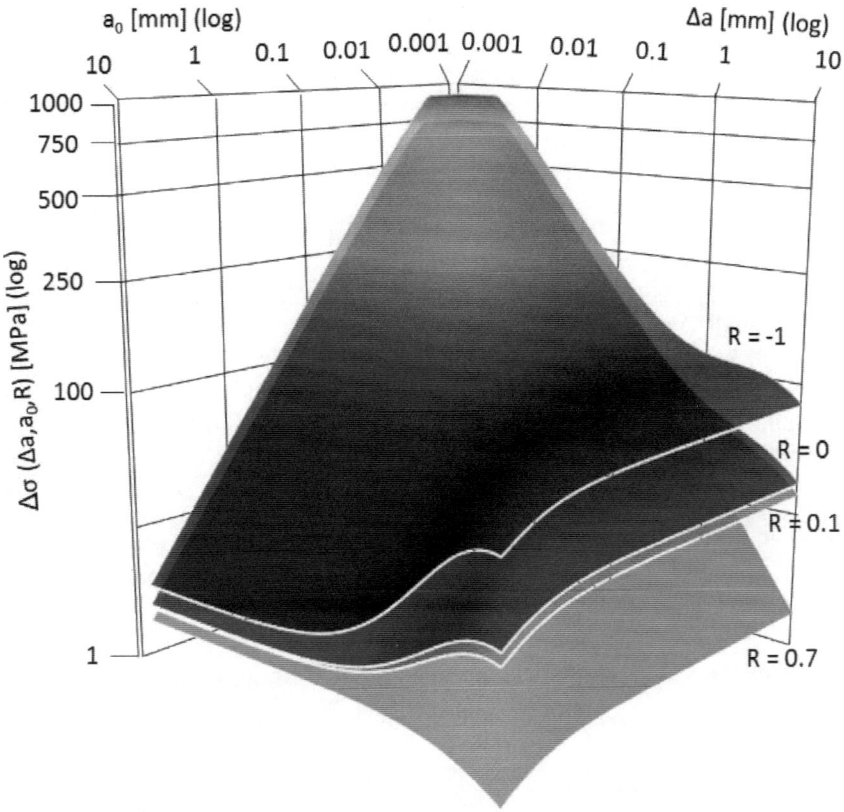


**Figure 11.** Three-dimensional plot of admissible stress range $\Delta\sigma$ depending on $a_0$, $\Delta a$, $R$

## 7. Conclusion

A master resistance curve for fatigue crack closure has been proposed; its parameters have been determined for medium-strength steels by statistical regression. With this standardized resistance curve, a Kitagawa-Takahashi diagram depending on the initial notch/crack depth $a_0$ and the crack extension $\Delta a$ can be computed for any load ratio $R$. This extended Kitagawa-Takahashi diagram is readily applicable to the damage tolerant design of components.

## 8. Acknowledgements

Financial support by the Austrian Federal Government (in particular from Bundesministerium für Verkehr, Innovation und Technologie and Bundesministerium für Wirtschaft, Familie und Jugend) represented by Österreichische Forschungsförderungsgesellschaft mbH and the Styrian and the Tyrolean Provincial Government, represented by Steirische Wirtschaftsförderungsgesellschaft mbH and Standortagentur Tirol, within the framework of the COMET Funding Programme is gratefully acknowledged.

## 9. References

[1] Chapetti MD. Fatigue propagation of short cracks under constant amplitude loading. Int J Fatigue 2003;25(12): p. 1319-26

[2] El Haddad MH, Topper TH, Smith KN. Prediction of non-propagating cracks. Eng Fract Mech 1979;11: p. 573-584.

[3] Forschungskuratorium Maschinenbau (FKM): Rechnerischer Festigkeitsnachweis für Maschinenbauteile. $6^{th}$ Ed., VDMA 2012

[4] Kitagawa H, Takahashi S. Applicability of fracture mechanics to very small cracks or cracks in the early stage. In: Proceeding of the second international conference on mechanical behavior of materials, ASM; 1976. p. 627-631.

[5] Maierhofer J, Pippan R, Gänser HP. Modified NASGRO equation for physically short cracks. Int J Fat 2014;59: p. 200-207.

[6] Maierhofer J, Gänser HP, Pippan R. Modified Kitagawa-Takahashi diagram accounting for finite notch depths. Int J Fat 2015;70: p. 503-509.

[7] R. Pippan, M. Berger, H.P. Stüwe. The influence of crack length on fatigue crack growth in deep sharp notches. Metall Trans 18A (1987), 429-435.

[8] Tanaka K, Akiniwa Y. Resistance-curve method for predicting propagation threshold of short fatigue crack at notches. Eng Fract Mech 1988;30(6): p. 863-76.

[9] Tabernig B, Powell P, Pippan R. Resistance curves for the threshold of fatigue crack propagation in particle reinforced aluminium alloys. Fatigue crack growth thresholds, endurance limits, and designs, ASTM STP 1372, J.C. Newman, JR. and R.S. Piascik, Eds., American Society for Testing and Materials, West Conshohocken, PA.

IOP Conf. Series: Materials Science and Engineering **119** (2016) 012016 doi:10.1088/1757-899X/119/1/012016

# Modeling of variant-interaction during bainitic phase transformation

**U Ehlenbröker[1], R Mahnken[1,3], M Petersmann[2] and T Antretter[2]**
[1]Chair of Engineering Mechanics, University of Paderborn,
Warburger Straße 100, 33100 Paderborn, Germany
[2]Institute of Mechanics, Montanuniversität Leoben,
Franz-Josef Straße 18, A-8700 Leoben, Austria

**Abstract.** In our research, we develop a thermodynamically consistent multi-scale model for phase transformations from austenite into $n$ possible bainite variants. Each material point of the macroscopic configuration represents a polycrystal which describes the mesoscopic configuration. The microscopic configuration consists of an agglomeration of variants which is attached to each single crystal of the mesoscopic configuration. In addition, the model allows simulation of the macroscopic effects of volume change due to phase transformation as well as transformation-induced plasticity (TRIP). In this paper we present the results of recent work in the context of this model, which is concerned with the extension of the model for an effect of variant-interaction between the different crystallographic variants of bainite. For this reason we make use of the theory of transformation hardening. Thereby we are able to include an effect of preferential variant formation. This leads to the simultaneous formation of a selection of variants while the evolution of crystallographic unfavorable variants is handicapped or even completely suppressed. This extension of the model aims for the fact that in general not every crystallographic variant of bainite forms within a single austenite grain.

## 1. Introduction

The bainite phase transformation is of specific industrial relevance, especially in manufacturing processes which include hot-forming and quenching of large components. During this treatment most parts of the workpiece are subjected to cooling rates which are appropriate to yield a substantial amount of bainite in the final product. This leads to a notable influence on the product properties.

For a good overview on the research concerning the issues relevant to this work, we explicitly refer to the introduction of [1] and all references therein, since the present work is an extension to this publication. As to the nature of bainitic transformation it says "The crystallographic structure of the bainitic product phase features many different allotropic variants similar to what is known for martensite. In fact, the same crystallographic framework as for martensite can also be used in the case of bainite. A viable material model describing the mechanics of bainitic transformation must capture both the diffusive nature of the process as well as the multi-variant product phase formation. In addition, as one of the macroscopic characteristic features of low alloyed steels, phase transformations under moderate deviatoric stresses far from the yield stress lead to a permanent macroscopic inelastic deformation commonly termed transformation induced plasticity (TRIP)."

[3] Email: mahnken@ltm.upb.de, phone: (+49) 5251 60-2283

In this paper we are concerned with the effect of variant-interaction during bainite transformation on the micro-scale of the material. During this transformation, preferential variant formation and variant interaction can be observed. To include these effects in our material model, we resort to the theory of transformation hardening. Concerning these topics of variant-interaction, interaction matrices and transformation hardening, we exemplarily refer to [2-6] among others. As a consequence the volume-fractions of the already evolved bainite variants have an influence on the subsequent bainite transformation. This ansatz may lead to the partial suppression of several variants and further to the remaining of retained austenite at the end of the transformation. This is reasonable, because in general, the bainite transformation is not complete and the different variants do not always evolve in equal volume-fractions, due to several influencing factors, e.g. chemical composition of the considered steel or thermo-mechanical loading.

The general and thermodynamic framework presented in this paper has already been published in [1]. Main aspects of the modeling section have been transferred to this article in order to provide a sufficient basis and the main equations of the material model. We skip the derivation of a special form of the Clausius-Duhem inequality, the heat-conduction equation and prove of thermodynamic consistency. For closer details related to the aforementioned topics as well as to the modeling part in general, we refer to [1].

The structure of this paper is as follows: Section 2 presents the general constitutive framework, the main equations and the extension of the material model to the simulation of variant-interaction during the bainitic phase-transformation. In section 3, we give a brief numerical example in order to show the effect of the variant-interaction during the bainite-transformation in a crystal-grain. We conclude the paper with a brief summary and outlook in section 4.

## 2. Modeling and Multi-scale framework

### *2.1. Thermodynamic framework on the macro-scale*

The constitutive equations in this work are formulated within a geometrically linear theory. To this end the displacement gradient $\boldsymbol{u} \otimes \nabla$ is introduced at each material point $P$ with position vector $x$ of the macroscopic configuration $\mathcal{B}$ in figure 1. Here, within a cartesian frame with base vectors $\boldsymbol{e}_i$ , $i = 1, 2, 3$ and coordinates $x_i$ , $i = 1, 2, 3$ , the nabla operator is defined as $\nabla := \partial/\partial x_i \boldsymbol{e}_i$. This defines the strain tensor of the geometrically linear theory

$$\boldsymbol{\varepsilon} := \mathrm{sym}\,\{\boldsymbol{u} \otimes \nabla\}\,. \tag{1}$$

In our multi-scale model, illustrated in figure 1, the material is regarded as a coexisting mixture of its polycrystals as well as of its phases, here bainite and austenite. The macroscopic temperature $\theta$ in a body point is the average temperature inside the polycrystal. In this way we assume the same temperature in each single grain of a polycrystal on the mesoscale.

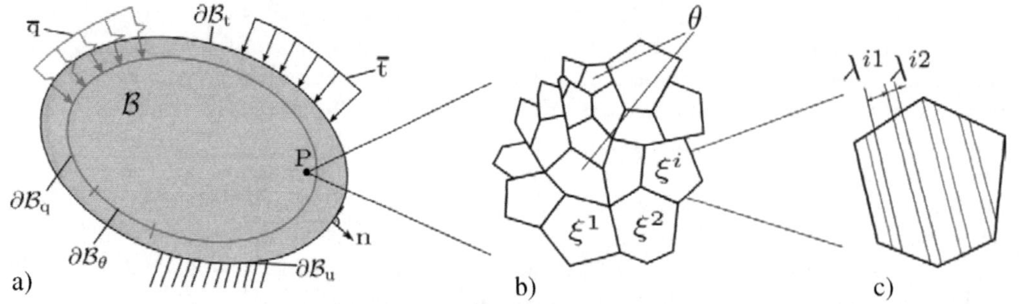


**Figure 1.** Multi-scale model: a) Macroscopic configuration $\mathcal{B}$, b) Mesoscopic configuration consisting of a polycrystal with volume fractions $\xi^i$ and c) Microscopic configuration of a single crystal grain with variants $\lambda^{ij}$; the temperature $\theta$ is regarded as a homogenized quantity.

The carbon diffusion at the mesoscopic and microscopic levels, i.e. inside of a single crystal and between grains of the polycrystal during the phase transformation (see e.g. [7]), is not taken into account by the model under consideration. The balance relations are (e.g. [8])

$$\begin{aligned}
&\mathbf{1.}\ \varrho_0 \ddot{\boldsymbol{u}} - Div[\boldsymbol{\sigma}] = \varrho_0 \boldsymbol{f}, && \text{(linear momentum)}\\
&\mathbf{2.}\ \varrho_0 \dot{e} + Div[\boldsymbol{q}_0] = \boldsymbol{\sigma} : \dot{\boldsymbol{\varepsilon}} + \varrho_0 r_\theta\,, && \text{(energy)}\\
&\mathbf{3.}\ -\varrho_0 \dot{e} + \varrho_0 \theta \dot{\eta} + \boldsymbol{\sigma} : \dot{\boldsymbol{\varepsilon}} - \frac{1}{\theta}\boldsymbol{q}_0\, Grad\,\theta \geq 0. && \text{(entropy)}
\end{aligned} \tag{2}$$

In this notation we use: $\varrho_0$ - density, $\boldsymbol{u}$ - displacement vector, $\boldsymbol{\varepsilon}$ – strain tensor of equation (1), $\theta$ - absolute temperature, $\boldsymbol{\sigma}$ - (symmetric) Cauchy stress tensor, $\boldsymbol{f}$ - mass density of external forces, $e$ - mass density of the internal energy, $\eta$ - entropy, $\boldsymbol{q}_0$ - heat-flux density vector, $r_\theta$ - mass density of heat supply. The dot above an argument denotes its time derivative. We also recall that the inequality (2.3) is known as the Clausius-Duhem inequality. These equations are completed by suitable boundary and initial conditions for the displacement vector $\boldsymbol{u}$ and for the temperature $\theta$ (for details see [1]).

Using the Legendre-transform $\Psi = e - \theta\eta$ between the internal energy $e$ and the Helmholtz free energy $\Psi$, the Clausius-Duhem inequality (2.3) becomes

$$\theta\gamma = \frac{1}{\varrho_0}\mathcal{P} - \dot{\Psi} - \dot{\theta}\eta - \frac{1}{\varrho_0\theta}\boldsymbol{q}_0\, Grad\,\theta \geq 0. \tag{4}$$

$\theta\gamma$ is the (mass density of) internal dissipation, $\gamma$ is the (mass density of) entropy-production rate. The stress-power $\mathcal{P}$ is given as a dual-pairing of the Cauchy stress $\boldsymbol{\sigma}$ and the strain-rate $\dot{\boldsymbol{\varepsilon}}$

$$\mathcal{P} = \boldsymbol{\sigma} : \dot{\boldsymbol{\varepsilon}}\,. \tag{5}$$

### *2.2. Meso-micro relations*

According to figure 1.b), a mesoscopic configuration, consisting of a polycrystal, is attached to each material point of the macroscopic configuration. Following [9] the polycrystal consists of $N$ crystals. The volume fraction of the crystal with orientation $i$ is denoted by $\xi^i$, and all polycrystals on the mesoscale satisfy the normalization condition $\sum_{i=1}^{N} \xi^i = 1$.

In the following we assume constant volume fractions for all crystals, which means $\dot{\xi}^i = 0$. Each crystal on the microscale has $n$ variants, and the index $j$ ranges from 0 for the austenite to the number $n$ of bainitic variants. The volume fractions of austenite $\lambda^{i0}$ and bainitic variants $\lambda^{ij}, j = 1, \dots, n, i = 1, \dots, N$ satisfy mass conservation, and each variant ranges between zero and one. In summary we have the constraints for all $i \in \{1, \dots, N\}$ and $j \in \{1, \dots, n\}$

$$\mathbf{1.}\ \sum_{j=0}^{n} \lambda^{ij} = 1\,, \qquad \mathbf{2.}\ \lambda^{ij} \geq 0\,. \tag{6}$$

The austenitization of a low alloyed steel is possible only above the equilibrium temperature $A_{C1}$. Therefore, for a temperature below $A_{C1}$ and above $M_S$ (martensite start temperature) austenite can transform into ferrite, pearlite and bainite. A reverse transformation in that temperature regime is not to be expected. To focus on the prototypical situation, we only consider the transformation from austenite to bainite. The evolution of its variants is subjected to the following constraints:

$$\mathbf{1.}\ \sum_{j=0}^{n} \dot{\lambda}^{ij} = 0 \qquad \mathbf{2.}\,\text{Austenite: } \dot{\lambda}^{i0} \leq 0 \qquad \mathbf{2.}\,\text{Bainite: } \dot{\lambda}^{ij} \geq 0\,. \tag{7}$$

In addition, we assume the following additive decomposition of the total mesoscopic strain tensor in the $j$th variant of the $i$th crystal:

$$\boldsymbol{\varepsilon}^{ij} = \boldsymbol{\varepsilon}^{ij}_{el} + \boldsymbol{\varepsilon}^{ij}_{th} + \boldsymbol{\varepsilon}^{ij}_{pl} + \boldsymbol{\eta}^{ij}_{tr}, \tag{8}$$

so that $\boldsymbol{\varepsilon}^{ij}$ consists of an elastic part $\boldsymbol{\varepsilon}^{ij}_{el}$, a thermal part $\boldsymbol{\varepsilon}^{ij}_{th}$, a plastic part $\boldsymbol{\varepsilon}^{ij}_{pl}$ and a rotated transformation part $\boldsymbol{\eta}^{ij}_{tr}$. $\boldsymbol{\varepsilon}^{ij}_{th}$ represents the thermal expansion and is given by $\boldsymbol{\varepsilon}^{ij}_{th} = (\theta - \theta_0)\,\boldsymbol{A}^{ij}$, where $\theta_0$ is a reference temperature and $\boldsymbol{A}^{ij}$ are the thermal expansion tensors associated with the $i$th crystal and with the $j$th variant (bainite: $j > 0$, austenite: $j = 0$). $\boldsymbol{\varepsilon}^{ij}_{pl}$ in equation (8) refers to the inelastic strains arising from the viscoplastic material behavior. The associated evolution equations are given in section 2.6. The rotated transformation strain tensor $\boldsymbol{\eta}_{tr}^{ij}$ in (8) is obtained from

$$\textbf{1.}\ \boldsymbol{\eta}_{tr}^{ij} = (\boldsymbol{R}^i)^T \boldsymbol{\eta}_{tr}^{j} \boldsymbol{R}^i, \quad \text{where} \quad \textbf{2.}\ \boldsymbol{\eta}_{tr}^{j} = \boldsymbol{U}_j - \boldsymbol{I}\,. \tag{9}$$

Here, $\boldsymbol{R}^i \in \boldsymbol{SO}_3$, $i = 1, \dots, N$, is a rotation tensor, $\boldsymbol{\eta}_{tr}^{j}$ is the linearized transformation strain tensor from austenite to the $j$th bainite variant, $\boldsymbol{U}_j$ is the symmetric transformation tensor and $\boldsymbol{I}$ is the identity tensor. The austenite is taken as a reference configuration with transformation strain $\boldsymbol{\eta}_{tr}^{i0} = \boldsymbol{0}$.

### 2.3. Macro-meso relations

For later exposition we assume the following relations between macroscopic and mesoscopic strains:

$$\begin{gathered}\textbf{1.}\ \boldsymbol{\varepsilon} := \sum_{i=1}^{N} \xi^i \sum_{j=0}^{n} \lambda^{ij} \boldsymbol{\varepsilon}^{ij}, \quad \textbf{2.}\ \boldsymbol{\varepsilon}_{el} := \sum_{i=1}^{N} \xi^i \sum_{j=0}^{n} \lambda^{ij} \boldsymbol{\varepsilon}_{el}^{ij}, \quad \textbf{3.}\ \boldsymbol{\varepsilon}_{th} := \sum_{i=1}^{N} \xi^i \sum_{j=0}^{n} \lambda^{ij} \boldsymbol{\varepsilon}_{th}^{ij} \\ \textbf{4.}\ \boldsymbol{\varepsilon}_{pl} := \sum_{i=1}^{N} \xi^i \sum_{j=0}^{n} \lambda^{ij} \boldsymbol{\varepsilon}_{pl}^{ij}, \quad \textbf{5.}\ \boldsymbol{\eta}_{eff} := \sum_{i=1}^{N} \xi^i \sum_{j=0}^{n} \lambda^{ij} \boldsymbol{\eta}_{tr}^{ij}\,.\end{gathered} \tag{10}$$

Here, $\boldsymbol{\varepsilon}$ is the macroscopic strain tensor from equation (1). $\boldsymbol{\varepsilon}_{el}$, $\boldsymbol{\varepsilon}_{th}$, $\boldsymbol{\varepsilon}_{pl}$ and $\boldsymbol{\eta}_{eff}$ are the effective elastic, thermal, the effective plastic strain and the effective eigenstrain tensor, respectively. We make the assumption of a constant stress $\boldsymbol{\sigma}$, then by inserting the effective strain (10.1) into (5), the stress power in (5) is obtained as

$$\begin{aligned}\mathcal{P} = \boldsymbol{\sigma} : \dot{\boldsymbol{\varepsilon}} &= \boldsymbol{\sigma} : \frac{d}{dt}\left(\sum_{i=1}^{N} \xi^i \sum_{j=0}^{n} \lambda^{ij} \boldsymbol{\varepsilon}^{ij}\right) \\ &= \sum_{i=1}^{N} \xi^i \sum_{j=0}^{n} \lambda^{ij} \boldsymbol{\sigma} : \dot{\boldsymbol{\varepsilon}}^{ij} + \sum_{i=1}^{N} \xi^i \sum_{j=0}^{n} \dot{\lambda}^{ij} \boldsymbol{\sigma} : \boldsymbol{\varepsilon}^{ij}\,.\end{aligned} \tag{11}$$

### 2.4. Helmholtz free energy

We assume a quadratic functional for the Helmholtz free energy, where, analogously to equation (9), the elasticity constants $\mathbb{C}_{tuvw}^{j}$ of the $j$th variant of crystal $i$ are transformed as $\mathbb{C}_{pqrs}^{ij} = \boldsymbol{R}_{tp}^{i} \boldsymbol{R}_{uq}^{i} \boldsymbol{R}_{vr}^{i} \boldsymbol{R}_{ws}^{i} \mathbb{C}_{tuvw}^{j}$ such that the energy density of variant $j$ of crystal $i$ is defined as

$$\begin{gathered}\textbf{1.}\ \Psi^{ij}\left[\boldsymbol{\varepsilon}_{el}^{ij}, \boldsymbol{q}_k^{ij}, \theta\right] = \psi_{el}^{ij} + \psi_{pl}^{ij} + \psi_{ch}^{ij}, \quad \text{where } \textbf{2.}\ \psi_{el}^{ij}\left[\boldsymbol{\varepsilon}_{el}^{ij}\right] = \frac{1}{2\varrho_0} \boldsymbol{\varepsilon}_{el}^{ij} : \mathbb{C}^{ij} : \boldsymbol{\varepsilon}_{el}^{ij}, \\ \textbf{3.}\ \psi_{pl}^{ij}\left[\boldsymbol{q}_k^{ij}\right] = \frac{1}{2\varrho_0} Q_0^j b^j (q_1^{ij})^2 + \frac{1}{2\varrho_0} H^j (q_2^{ij})^2 \text{ and } \textbf{4.}\ \psi_{ch}^{ij}[\theta] = \phi_{ch}^{ij} + \Delta\phi_{ch}^{ij}\theta\end{gathered} \tag{12}$$

and the total Helmholtz free energy is defined as a weighted sum

$$\begin{aligned}&\textbf{1.}\ \Psi\left[\boldsymbol{\varepsilon}_{el}^{ij}, \boldsymbol{q}_k^{ij}, \theta, \lambda^{ij}, \xi^i\right] = \Psi_{el}^{eff} + \Psi_{pl}^{eff} + \Psi_{ch}^{eff} + \Psi^{\theta}, \text{ where} \\ &\textbf{2.}\ \Psi_{el}^{eff}\left[\boldsymbol{\varepsilon}_{el}^{ij}, \lambda^{ij}, \xi^i\right] = \sum_{i=1}^{N} \xi^i \sum_{j=0}^{n} \lambda^{ij} \psi_{el}^{ij}\left[\boldsymbol{\varepsilon}_{el}^{ij}\right], \\ &\textbf{3.}\ \Psi_{pl}^{eff}\left[\boldsymbol{q}_k^{ij}, \lambda^{ij}, \xi^i\right] = \sum_{i=1}^{N} \xi^i \sum_{j=0}^{n} \lambda^{ij} \psi_{pl}^{ij}\left[\boldsymbol{q}_1^{ij}, \boldsymbol{q}_2^{ij}\right], \\ &\textbf{4.}\ \Psi_{ch}^{eff}\left[\theta, \lambda^{ij}, \xi^i\right] = \sum_{i=1}^{N} \xi^i \sum_{j=0}^{n} (\lambda^{ij} - \lambda_0^{ij}) \psi_{ch}^{ij}[\theta] \quad \text{and} \\ &\textbf{5.}\ \Psi^{\theta}[\theta] = \int_{\theta_0}^{\theta} c_d[\bar{\theta}] d\bar{\theta} - \theta \int_{\theta_0}^{\theta} \frac{c_d[\bar{\theta}]}{\bar{\theta}} d\bar{\theta}\,.\end{aligned} \tag{13}$$

$H^j$, $Q_0^j$ and $b^j$ are material parameters associated to the hardening variables, $q_1^{ij}$ and $q_2^{ij}$, used for the modelling of combined linear and nonlinear hardening. Further, $\phi_{ch}^{ij}$ describes a constant chemical energy, $\Delta\phi_{ch}^{ij}$ a temperature dependent part and $c_d$ the specific heat capacity.

### 2.5. Thermodynamic forces

The thermodynamic forces are obtained by standard procedure from the Clausius-Duhem inequality in (4) under use of (8) and (11). With the Helmholtz free energy from (13), the macroscopic Cauchy stress tensor for variant $j$ of crystal $i$ results as:

$$\mathbf{1.}\ \boldsymbol{\sigma}^{ij} = \xi^i\lambda^{ij}\boldsymbol{\sigma} = \varrho_0\frac{\partial\Psi}{\partial\boldsymbol{\varepsilon}_{el}^{ij}} = \xi^i\lambda^{ij}\mathbb{C}^{ij}:\boldsymbol{\varepsilon}_{el}^{ij} \quad\Rightarrow\quad \mathbf{2.}\ \xi^i\lambda^{ij}(\mathbb{C}^{ij})^{-1}\boldsymbol{\sigma} = \xi^i\lambda^{ij}\boldsymbol{\varepsilon}_{el}^{ij}\,. \tag{14}$$

We point out, that the assumption of a constant stress $\boldsymbol{\sigma}$ over the polycrystal is equivalent to the use of a Reuss-type lower bound. Taking the weighted sum of (14.2) over all crystals and variants results into

$$\sum_{i=1}^{N}\xi^i\sum_{j=0}^{n}\lambda^{ij}(\mathbb{C}^{ij})^{-1}\boldsymbol{\sigma} = \sum_{i=1}^{N}\xi^i\sum_{j=0}^{n}\lambda^{ij}\boldsymbol{\varepsilon}_{el}^{ij}\,. \tag{15}$$

Then, by use of the definitions (10) for the effective strains and the effective compliance tensor $\left(\mathbb{C}_{eff}\right)^{-1} := \sum_{i=1}^{N}\xi^i\sum_{j=0}^{n}\lambda^{ij}(\mathbb{C}^{ij})^{-1}$, equation (15) is written as

$$\boldsymbol{\sigma} = \mathbb{C}_{eff}:\boldsymbol{\varepsilon}_{el}\,. \tag{16}$$

We further identify the hardening stresses and chemical forces for phase-transformation as:

$$\begin{aligned}
&\mathbf{1.}\ \boldsymbol{Q}_1^{ij} = \varrho_0\frac{\partial\Psi}{\partial q_1^{ij}} = Q_0^j b^j q_1^{ij},\quad \boldsymbol{Q}_2^{ij} = \varrho_0\frac{\partial\Psi}{\partial q_2^{ij}} = H^j q_2^{ij},\ \text{ and}\\
&\mathbf{2.}\ \boldsymbol{\Lambda}^{ij} = \varrho_0\frac{\partial\Psi}{\partial\lambda^{ij}} - \xi^i\boldsymbol{\sigma\varepsilon}^{ij}\\
&\quad = -\xi^i\left(\tfrac{1}{2}\boldsymbol{\sigma}:(\mathbb{C}^{ij})^{-1}:\boldsymbol{\sigma} + \boldsymbol{\sigma}:(\boldsymbol{\varepsilon}^{ij} + \boldsymbol{\varepsilon}_{th}^{ij} + \boldsymbol{\varepsilon}_{pl}^{ij} + \boldsymbol{\eta}_{tr}^{ij}) - \varrho_0\psi_{pl}^{ij} - \varrho_0(\phi_{ch}^{ij} + \Delta\phi_{ch}^{ij}\theta)\right).
\end{aligned} \tag{17}$$

*2.6. Evolution equations: Phase transformation combined with variant interaction*
Contrary to martensitic displacive transformations, bainitic phase transformations are accompanied by diffusion processes making the bainite formation time dependent. Concerning the formulation of evolution equations for phase transformation, we resort to the methods of viscoplasticity. Therefore, we formulate evolution equations for the bainitic variants of Perzyna-type as follows:

$$\begin{aligned}
&\mathbf{1.}\ \mathit{Evolution\ of\ bainitic\ variants} && \dot{\lambda}^{ij} && = \frac{1}{v_{tr}}\langle\Phi_{tr}^{ij}\rangle^{n_{tr}}\left(H_s[\lambda^{i0},\lambda^{ij}]\right)^{n_{tr}},\ \ i,j>0\\
&\mathbf{2.}\ \mathit{Evolution\ of\ austenite} && \dot{\lambda}^{i0} && = -\sum\nolimits_{j=1}^{n}\lambda^{ij}\\
&\mathbf{3.}\ \mathit{Overstress\ function} && \Phi_{tr}^{ij} && = (\boldsymbol{\Lambda}^{i0} - \boldsymbol{\Lambda}^{ij}) - (f_c + \alpha_{vi}V^{ij})\\
&\mathbf{4.}\ \mathit{Saturation\ function} && H_s[\lambda^{i0},\lambda^{ij}] && = \left(\lambda^{i0}\right)^{s_1}\left(\lambda^{ij}\right)^{s_2}.
\end{aligned} \tag{18}$$

The function $\langle x\rangle$ is defined as $\langle x\rangle = x$ for $x > 0$, $\langle x\rangle = 0$ for $x \le 0$ ("Macaulay brackets"). It guarantees that the relation $\dot{\lambda}^{ij} \ge 0$, is always fulfilled. In addition, the overstress function $\Phi_{tr}^{ij}$ in (18.1) ensures that the bainitic variants only start to form out of the austenite if the difference between the free enthalpies of austenite and bainite exceeds an energetic barrier $f_c$. The evolution of austenite is calculated from the constraint (7.1) as the negative sum over the rates of bainite fractions. The speed of growth for bainite is dependent on the saturation $H_s$ in equation (18.4). On the one hand it ensures that for a decreasing amount of austenite the formation of bainite is slowed down. On the other hand small amounts of bainite lead to a slow formation of bainitic variants in the beginning of the simulation. This approach adopts the main ideas of the classical theory by Johnson-Mehl-Avrami-Kolmogorov for diffusion-controlled isothermal phase transformations (see e.g. [10]). The typical S-shape of the curves can be seen in the results of the numerical example in section 3.

For modeling of variant interaction, an influence factor $V^{ij}$ is introduced, which can raise the energetic barrier in $\Phi_{tr}^{ij}$ needed for the evolution of the different variants. Further, the parameter $\alpha_{vi}$ is used to control the overall strength of the interaction effect. For $\alpha_{vi} = 0$ the variant interaction is deactivated. The influence factor is defined as

$$V^{ij} := \sum_{k=1}^{n}\lambda^{ik}e_{vi}^{kj}\,, \tag{19}$$

where $e_{vi}^{kj}$ is a factor which describes the influence of variant $k$ on the evolution of variant $j$. Therefore, if variant $k$ hinders the evolution of variant $j$, the value of $e_{vi}^{kj}$ will be larger compared to the value of another variant, which is more likely to form besides variant $j$. The influence of variant $k$ is then automatically controlled with help of its volume-fraction $\lambda^{ik}$ in equation (19). At this point it has to be discussed, which values are suitable for the factors $e_{vi}^{kj}$ in order to lead to a realistic result.

All material parameters for the description of the evolution of bainite variants are summarized as $\kappa_{tr}^{j} = [\nu_{tr}, n_{tr}, f_c, s_1, s_2, \phi_{ch}^{ij}, \Delta\phi_{ch}^{ij}, \alpha_{vi}]$. At this point we use the same set of material-parameters for all variants, resulting in only one set of parameters for the evolution of variants.

*Remark.* For simulation of inelastic behavior of the constituents, time dependent viscoplasticity combined with non-linear isotropic hardening is used. Since the focus of this paper is on the variant-interaction, we do not further elaborate on the specific issue (details in: [1] and [11]).

## 3. Numerical Example

### *3.1. Numerical implementation*

For the numerical implementation of the constitutive equations in section 2, we make use of a two-step staggered algorithm (see figure 2). It consists of the following two steps:

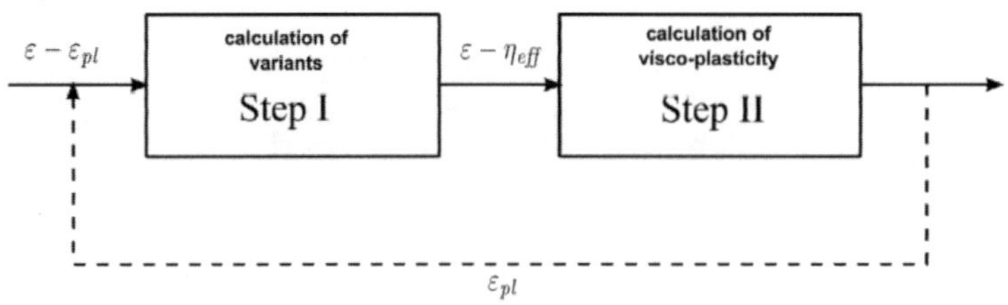


**Figure 2.** Numerical implementation: two-step-staggered algorithm.

- **Step I** Calculation of variants $\lambda^{ij}$ with a projected Newton algorithm, combined with an active-set-strategy, where the viscoplastic strains $\boldsymbol{\varepsilon}_{pl}^{ij}$ are constants.
- **Step II** Calculation of viscoplasticity with a Newton iteration and radial return method, where the variants $\lambda^{ij}$ are constants.

A description of the algorithm in Step I can be found in [12]. Basic principles of the algorithm in Step II can be found e.g. in [13].

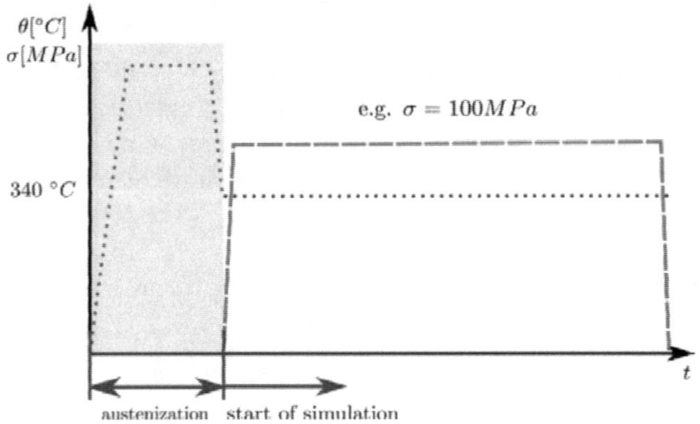


**Figure 3.** Variant selection: schematic of thermo-mechanical loading (dotted line: temperature, dashed line: stress).

### *3.2. Variant selection: Variant-interaction during bainitic phase transformation in a single crystal*

In this example the calculations are performed for a single grain, subjected to an external stress field and a constant temperature, as shown exemplarily in figure 3. We note that the grain counter $i$ is now equal to one while the variant counter $j$ may be any number from 1 to 24. The zone marked with cyan color in figure 3 represents the process of austenitization and the following cooling-phase needed for initiation of the bainitic transformation.

We assume that the material has been cooled down to a temperature of 340 °C with a moderate cooling-rate, making the formation of bainite possible. This is the point where our simulation starts. The temperature is supposed to be at the constant value of 340 °C and external stresses are applied to the crystal. It should be noted that the single element level considerations mainly serve the purpose of demonstrating the algorithm's capability of capturing the effect of variant-interaction. Therefore viscoplasticity has not been taken into account at this stage which would be difficult to deal with in a stress-controlled analysis at a single material point for a weakly hardening material. However, viscoplasticity would be accounted for at the next level, i.e. at the level of a polycrystal, e.g. by means of a representative volume element. A respective example has been given in [1].

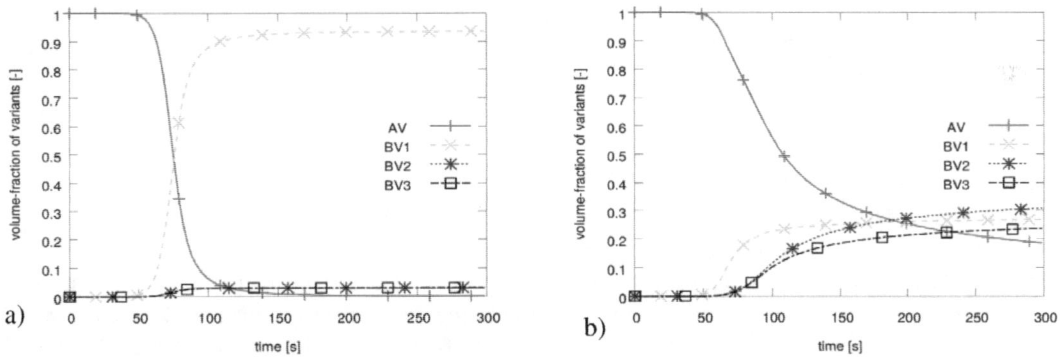


**Figure 4.** Variant selection: for a mechanical loading of $\sigma_x = 250\ MPa$; a) without variant-interaction ($a_{vint} = 0$); b) with variant-interaction ($a_{vint} = 125$)

For reasons of clarity, we present results of simulations considering the evolution of only three bainitic variants. The material parameters are chosen as in [1], except for $n_{tr}$, which is set to 2.5. The variant-interaction factors are chosen as follows, using matrix formulation:

$$\underline{\underline{E}}_{vi} := \begin{bmatrix} e_{vi}^{11} & e_{vi}^{12} & e_{vi}^{13} \\ e_{vi}^{21} & e_{vi}^{22} & e_{vi}^{23} \\ e_{vi}^{31} & e_{vi}^{32} & e_{vi}^{33} \end{bmatrix} = \begin{bmatrix} 0.01 & 0.5 & 0.75 \\ 0.5 & 0.01 & 0.25 \\ 0.75 & 0.25 & 0.01 \end{bmatrix}.$$

The transformation strain tensors for the variants are set to:

| | $U_{xx}$ | $U_{yy}$ | $U_{zz}$ | $U_{xy}$ | $U_{xz}$ | $U_{yz}$ |
|---|---|---|---|---|---|---|
| $U_0$ | 1.0 | 1.0 | 1.0 | 0.0 | 0.0 | 0.0 |
| $U_1$ | 1.0296 | 1.0 | 1.0 | 0.186 | 0.0 | 0.0 |
| $U_2$ | 1.0 | 1.0296 | 1.0 | 0.0 | 0.0 | −0.186 |
| $U_3$ | 1.0 | 1.0 | 1.0296 | 0.0 | 0.186 | 0.0 |

Figure 4 a) shows the results for variant evolution without variant-interaction, i. e. for $a_{vint} = 0$, under an external mechanical loading of $\sigma_x = 250\ MPa$. Here, the loading in x-direction leads to a preferred transformation of the austenite into the first bainitic variant (BV1). The other two variants (BV2 and BV3) only evolve to low, equal amounts. In comparison to this, figure 4 b) presents results for a simulation considering variant-interaction ($a_{vint} = 125$). The results show a more balanced evolution of the different variants. Although the variant BV1 again is the first one to evolve, the variant-interaction factor $e_{vi}^{11}$, in combination with the relatively high overall strength of the interaction $a_{vint} = 125$, hinders the evolution at the later stages of the transformation, allowing the evolution of the other two variants. Further, we point out, that the model is now capable of simulating transformations with a remaining volume-fraction of austenite. This is a reasonable extension, since general experiments often show a specific amount of untransformed retained austenite.

In figure 5, we compare several simulations for different values of $a_{vint}$. These results show the influence of the overall strength of the variant-interaction on the evolution of the bainitic variants.

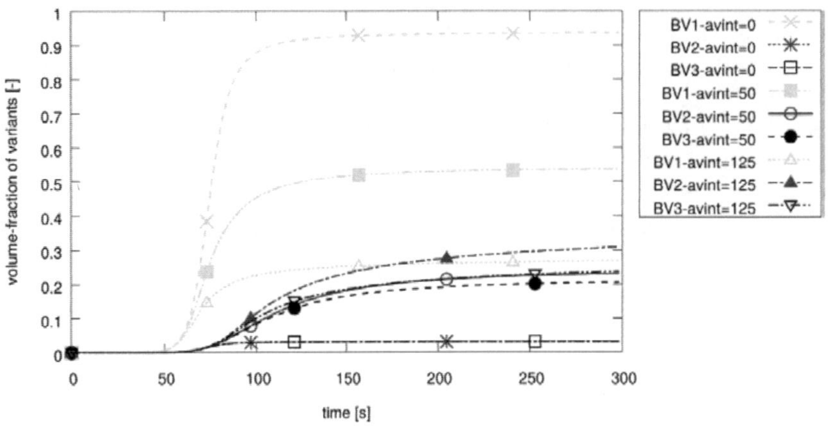


**Figure 5.** Variant selection: influence of the overall strength of the variant-interaction on the evolution of the bainitic variants

## 4. Summary and Outlook

In this paper, we extended a model for bainitic variant evolution by a mechanism for variant-interaction. This is achieved with an ansatz adopted from the theory of transformation hardening. With this extension, we obtain an effective tool for simulating the influence of the different variants belonging to each other, potentially hindering or preventing the evolution of crystallographically unfavorable variants.

The variant-interaction factors in the presented numerical example have been merely chosen to clarify the capabilities of the model. Future work should include the determination of proper values, reflecting the real variant-interaction. Preceding this task, we will have to identify the true transformation strain tensors of the bainitic variants. Further steps could be the modeling of mesoscopic carbon diffusion and the dissolution of austenite into bainitic ferrite and carbides.

## 5. References

[1] Mahnken R, Schneidt A, Antretter T, Ehlenbröker U and Wolff M 2015 *Int. J. Solids Struct.* **54** 156-71
[2] Franciosi P 1985 *Acta Metall.* **33** 1601-12
[3] Siredey N, Patoor E, Berveiller M and Eberhardt A 1999 *Int. J. Solids Struct.* **36** 4289-315
[4] Niclaeys C, Ben Zineb T, Arbab-Chirani S and Patoor E 2002 *Int. J. Plast.* **18** 1619-47
[5] Kundu S and Bhadeshia H K D H 2007 *Scripta Mater.* **57** 869-72
[6] Gérard C, Cailletaud G and Bacroix B 2013 *Int. J. Plast.* **42** 194-212
[7] Bhadeshia H K D H 2003 *Bainite in Steels, Transformations, Microstructure and Properties* (Cambridge: University Press)
[8] Haupt P 2002 *Continuum Mechanics and Theory of Materials* (Berlin: Springer)
[9] Hackl K and Heinen R 2008 *Continuum Mech. Thermodyn.* **19** 499-510
[10] Avrami M 1940 *J. Chem. Phys.* **8** 212-24
[11] Lemaitre J and Chaboche J L 1990 *Mechanics of Solid Materials* (Cambridge: University Press)
[12] Mahnken R and Wilmanns S 2011 *Comput. Mater. Sci.* **50** 2535-48
[13] Simo J C and Hughes T J R 1998 Computational inelasticity In: *Interdisciplinary Applied Mathematics. Mechanics and Materials* (Berlin: Springer)

**Acknowledgements**
Financial support by the Austrian Federal Government (in particular from Bundesministerium für Verkehr, Innovation und Technologie and Bundesministerium für Wissenschaft, Forschung und Wirtschaft) represented by Österreichische Forschungsförderungsgesellschaft mbH and the Styrian and the Tyrolean Provincial Government, represented by Steirische Wirtschaftsförderungsgesellschaft mbH and Standortagentur Tirol, within the framework of the COMET Funding Programme is gratefully acknowledged.

# Experimental characterisation and modelling of deformation-induced microstructure in an A6061 aluminium alloy

**J F Kreyca[1], A Falahati[1] and E Kozeschnik[1]**

[1]Institute of Materials Science and Technology, TU Wien, Getreidemarkt 9, 1060 Wien, Austria

E-mail: johannes.kreyca@tuwien.ac.at

**Abstract.** For industry, the mechanical properties of a material in form of flow curves are essential input data for finite element simulations. Current practice is to obtain flow curves experimentally and to apply fitting procedures to obtain constitutive equations that describe the material response to external loading as a function of temperature and strain rate. Unfortunately, the experimental procedure for characterizing flow curves is complex and expensive, which is why the prediction of flow-curves by computer modelling becomes increasingly important. In the present work, we introduce a state parameter based model that is capable of predicting the flow curves of an A6061 aluminium alloy in different heat-treatment conditions. The model is implemented in the thermo-kinetic software package MatCalc and takes into account precipitation kinetics, subgrain formation, dynamic recovery by spontaneous annihilation and dislocation climb. To validate the simulation results, a series of compression tests is performed on the thermo-mechanical simulator Gleeble 1500.

## 1. Introduction

Plastic forming behavior can be divided into different stages. A stage is defined by a change in the dominant microstructural mechanism leading to a characteristic stress-strain relation. For polycrystalline materials, mainly stages III and IV (eventually stage V) are relevant [1] [2]. The stress response of a material caused by external loading is closely related to the mean free path (MFP) for dislocations [3]. This MFP is influenced by obstacles. Dominant obstacles define the dominant hardening mechanism. As dislocations work as dominant obstacles for low strains, the overall dislocation density and its evolution over strain $\rho(\varepsilon(t))$ (i.e. over time for constant strain rate) is a convenient model state parameter. This is the case in stage III. In order to reduce their total energy, dislocations tend to agglomerate to a honeycomb like structure. The mechanism of subgrain formation [4] becomes dominant in stage IV. A model that describes stages III and IV, therefore, needs two state parameters, one for the evolution of the dislocation density and one for subgrain formation. In this paper, the dislocation density inside subgrains $\rho_i$ and the dislocation density stored in subgrains walls $\rho_w$ are used as state-parameters. The following evolution equations are used in this paper, see also refs. [5] [6].

$$\sigma = \sigma_0 + \sigma_i + \sigma_w = \sigma_0 + \alpha M b \mu \sqrt{\rho_i} + \alpha M b \mu \sqrt{\rho_w} \tag{1}$$

$$\frac{d\rho_i}{d\varepsilon} = \frac{M}{bA}\sqrt{\rho_i} - \frac{Bd_{crit}M}{b}\rho_i \tag{2}$$

$$\frac{d\rho_w}{d\varepsilon} = \frac{M}{bD}\sqrt{\rho_w} \tag{3}$$

where $\alpha$ is a constant, $M$ is the Taylor factor, $b$ is Burgers vector, $\mu$ is the shear modulus, $d_{\text{crit}}$ is a critical distance for spontaneous dislocation annihilation [7], $\nu$ is Poisson's ratio and $A$, $B$ and $D$ are hardening parameters. $\sigma_0$ is the yield stress. Eq. (1) describes the link between dislocation density and stress response and was first suggested by Taylor [8]. Eq. (2) and Eq. (3) are equivalent to the evolution equations defined in the Kocks-Mecking-model (KM-model) [9]. For a derivation of this model, see ref. [6].

## 2. Parameters A, B and D

Eqs. (1-3) present a convenient physical model that is capable of describing the main influences on flow curve behaviour. Parameter $A$ is related to dislocation storage, parameter $B$ to dislocation annihilation and parameter $D$ is related to relaxation processes due to subgrain formation. These mechanisms combined with yield stress define the deformation behaviour of a material. Parameters A, $B$ and $D$ fully characterize the hardening characteristics of a material under static loading conditions, where $A$, $B$ and $D$ must be expressed in relation to temperature, strain rate and the microstructural state. For further reading, see e.g. ref. [10]. The aim of this paper is to describe $A$, $B$ and $D$ based on comprehensive experimental testing. That means to derive these values for experimental flow curves in different heat treatment conditions, temperatures and strain rates as a basis for further physical modelling. Especially, the relation between yield stress and the storage and annihilation terms is closely investigated.

The relations between hardening parameters and the shape of a stress strain curve are the following. Parameter $A$ is indirectly correlated with the initial hardening rate ($d\sigma/d\varepsilon$) that is the angle between the stress-strain curve and the strain-axis. A low value of $A$ corresponds to a high hardening rate after yielding. Together with parameter $B$, the curvature in stage III and its saturation stress are defined. The saturation stress remains constant, if the product of $A$ and $B$ are constant [6]. A lower saturation stress correlates to a lower saturation dislocation density. Parameter $D$ is a measure for the hardening rate in stage IV. A high $D$ parameter correlates to a low hardening rate in stage IV.

## 3. Experiments and modelling

Experiments are carried out on a Gleeble 1500 Thermo-Mechanical Simulator. Cylindrical specimens, made from an A6061 aluminium alloy, are used in the dimensions of 15x10mm. The chemical composition is summarized in Table 1.

Table 1. Chemical Composition of A6061 alloy

| Element | Al | Si | Fe | Cu | Mn | Mg | Cr | Zn | Ti |
|---|---|---|---|---|---|---|---|---|---|
| Concentration [wt %] | 97.35 | 0.69 | <=0.7 | 0.23 | 0.11 | 0.85 | 0.18 | <=0.25 | <=0.15 |

The specimens are solution heat treated for 1h at 540°C, water quenched and aged at 170°C (artificial ageing). This procedure follows the suggestions of the norm EN 2700 [11]. The ageing time is varied between 1h, 5h, 10h and 96h, with all tests performed twice. Mo foils are used for friction control and as separation between the specimen and the W stamps. The stress is calculated under the assumption of constant specimen volume and no barrelling. Table 2 summarizes the parameters used in the calculations.

Table 2. Constants used in modelling

| Abbr. | Name | Value | Unit |
|---|---|---|---|
| $b$ | Burger's vector | $2.84\ 10^{-10}$ | [m] |
| $M$ | Taylor factor | 3.06 | [-] |
| $G$ | Shear modulus | $2.6\ 10^{10}$ | [Pa] |
| $\upsilon$ | Poisson's ratio | 0.3 | [-] |
| $Q^{fv}$ | Vacancy formation energy | 0.67 | eV |
| $\alpha$ | Strengthening coefficient | 0.5 | [-] |

The coefficients $A$, $B$ for inner dislocation density and $D$ for dislocation density in subgrain walls are calculated based on Eqs. (1), (2) and (3).

## 4. Results

*4.1 Temperature and strain rate behavior.*

Figure 1 shows the true stress over true strain behavior for 5h of ageing time. Figure 2 shows the corresponding hardening rate (d$\sigma$/d$\varepsilon$) for stage III over true stress $\sigma$ in a Kocks-plot. The following characteristics can be observed:

- higher yield strength for lower temperature and higher strain-rates (i.e. no negative strain rate sensitivity in that temperature and strain rate range). Interestingly, the slopes of the curves for both temperatures in stage IV are lower for higher strain-rates.
- The hardening rate at the yield point (y-intercept in Figure 2) is higher for lower temperatures.
- The lines in the Kocks-plot are steeper for higher temperatures and lower strain-rates (less pronounced for 25°C). That means that the stress-strain curves decrease faster into stage IV.

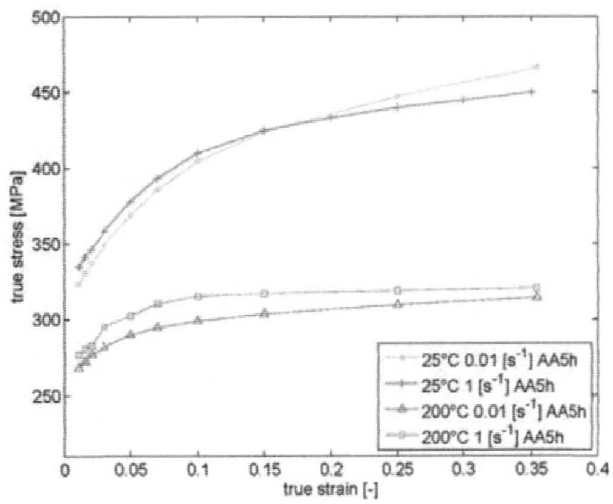


**Figure 1.** shows the true stress over true strain behaviour for different temperatures and strain rates for 5h of artificial ageing at 170°C.

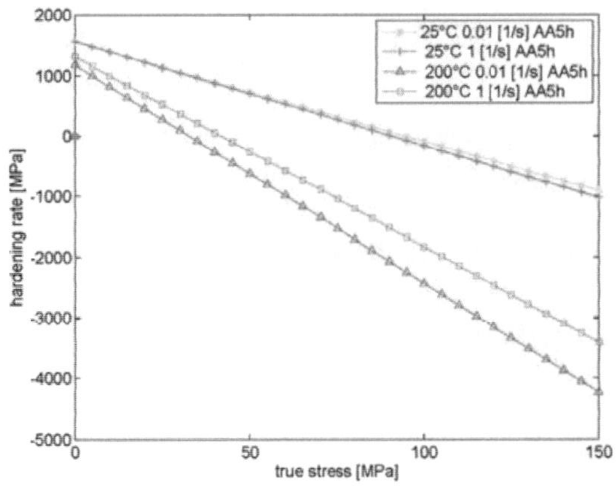


**Figure 2.** shows the hardening behaviour for the curves shown in Figure 1.

*4.2 Hardening parameters*

Figure 3 and Figure 4 show the parameters A, B and D for different ageing times and strain-rates over the yield stress. The values of the parameters vary between 35 and 65 for *A*, between 3 and 7 for *B* parameters and between 3 and 6 for *D*. Obviously there is a tendency for hardening parameters to increase with increasing yield strength. For the shape of the flow-curve, this means that a high yield strength leads to a lower initial hardening rate and lower hardening in stage IV, i.e. the curves are flatter.

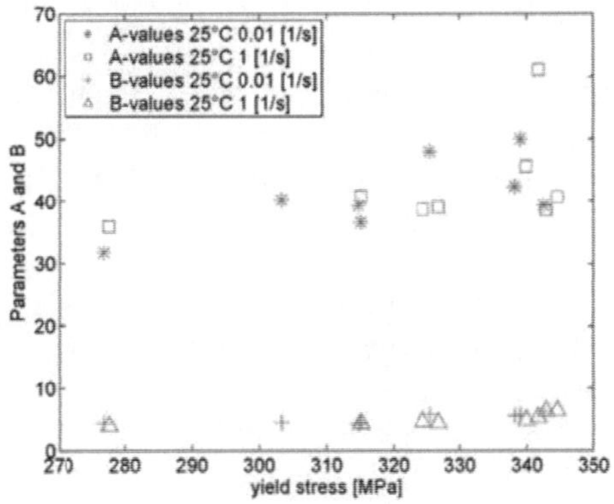


**Figure 3.** shows the A and B values for 25°C, two different strain rates and different ageing times.

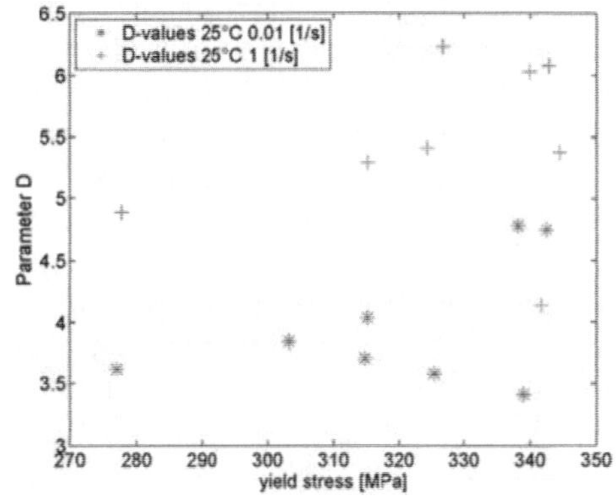


**Figure 4.** shows the D values for 25°C, two different strain rates and different ageing times.

Figure 5 shows the 0.2% proof stress for all tested samples for different strain-rate, temperatures and artificial ageing times.

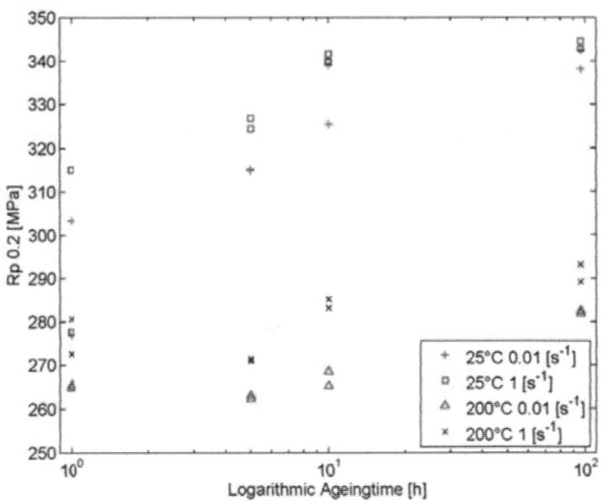


**Figure 5.** 0.2% proof stress for all samples.

## Discussion and Conclusion

The deformation behaviour of an A6061 aluminum alloy is investigated experimentally and modelled by a state-parameter-based Kocks-Mecking type approach. The following conclusions are obtained:

- *A*, *B* and *D* as defined above are useful parameters to define and compare hardening behaviour of a flow curve.
- Strain-rate, temperature and initial microstructure have an influence on yield stress and hardening behaviour. Within the range of artificial ageing at one temperature the testing temperature has the highest influence on yield stress and hardening.
- For a good agreement between model and experiment, all three parameters need to be adjusted. The original paper from Kocks [9] on this topic assumes a constant initial hardening rate (i.e. a constant *A* parameter) for pure metals. Parameter *A* is related to the obstacle spacing solely defined by dislocation arrangement in pure metals but also influenced by precipitation and solid solution in industrial alloys.
- A clear correlation between yield stress and hardening behaviour (Figures 3 and 4.) is observed. It appears that the influence of precipitation, solid solution and grain boundaries on the yield stress and the hardening behaviour are related through dislocation dynamics. A combined observation for further research is therefore recommended.

The presented model is capable of describing microstructure state dependent flow curve data in a comprehensible physical form. Furthermore, in combination with additional physical modelling for the hardening parameters, a predictive modelling of flow curves is achievable.

## References

[1] Kocks U and Mecking H 2003 *Prog. Mater. Sci.* **48** 171–273
[2] Nes E 1998 *Prog. Mater. Sci.* **41** 129–193
[3] Bergstro. Y 1970 *Mater. Sci. Eng.* **5** 193
[4] Holt D 1970 *J. Appl. Phys.* **41** 3197–01
[5] Kreyca J, Falahati A and Kozeschnik E 2015 *Materials Today: Proceedings* **2** 107–112
[6] Kreyca J, Falahati A and Kozeschnik E 2015 *Key Engineering Materials* **651–653** 587–591
[7] Brinckmann S 2011 *Int. J. Fat.* **33** 1369–75
[8] Taylor G I 1943 *Proc. R. Soc. Lond.* A **145** 362–87
[9] Kocks U 1976 *J. Eng. Mater. Tech.* **98** 76–85
[10] Galindo-Nava E I, Sietsma J and Rivera-Diaz-del-Castillo P 2012 *Acta Mater.* **60** 2615–2624
[11] DIN EN 2700 1989
[12] Montheillet F, Piot D, Matougui N, and Fares M 2014 *Metall. Mater. Trans.* A **45** 4324–32

# Characterization of the effect of cryogenic treatment on the tempering behavior of a secondary hardening high Co-Ni steel

**Marina Gruber [1], Gerald Ressel [1], Sarah Ploberger [2], Stefan Marsoner [1] and Reinhold Ebner [1]**

[1]Materials Center Leoben Forschung GmbH, Roseggerstraße 12, A-8700 Leoben, Austria
[2]Böhler Edelstahl GmbH & Co KG, Mariazeller Straße 25, A-8605 Kapfenberg, Austria

E-mail: marina.gruber@mcl.at

**Abstract.** For high Co-Ni steels sub-zero treatments are conducted to reduce the retained austenite phase fraction for obtaining excellent fracture toughness properties, but in general, cryogenic treatment has a great impact on the microstructural evolution of steels during tempering. Hence, the aim of this work was to analyze the influence of cryogenic treatments on the microstructural evolution of high Co-Ni steels, including carbide precipitation kinetics and austenite phase fraction evolution, during heating to elevated temperatures. In order to study the formation properties of carbides, the heating processes of cryogenically and non-cryogenically treated specimens were analyzed by dilatometer measurements. Furthermore, for determining the evolution of austenite phase fraction and hardness due to tempering, dilatometer investigations were combined with X-ray diffraction analyses and hardness measurements. It is revealed that sub-zero treated samples exhibit much stronger carbide precipitation signals. This was ascribed to the lower phase fraction of retained austenite, as more carbon is available for carbide precipitation.

## 1. Introduction

For tool steels, the application of a cryogenic treatment during the heat treatment, i.e. before tempering, has a massive impact on mechanical and microstructural properties such as enhancement of wear resistance and hardness, reduction of retained austenite phase fraction and refinement of carbide structure [1, 2, 3, 4, 5]. Nevertheless, a cryogenic treatment is also applied before tempering of the martensitic high Co-Ni steel Aermet® 100 for obtaining excellent toughness properties. Generally, high Co-Ni steels contain retained austenite, which is present after quenching from austenitization, and reverted austenite, which forms during tempering [6, 7]. As retained austenite in Aermet® 100 would lower toughness values because of its low stability, a sub-zero treatment is conducted to reduce the retained austenite content [6, 7, 8]. However, the effect of cryogenic treatment on carbide precipitation kinetics in high Co-Ni steels has not been precisely characterized so far.

Hence, this work presents experimental results on the influence of sub-zero treatments on the carbide precipitation kinetics and the austenite evolution during tempering of a high Co-Ni steel. The microstructural evolution during heating was characterized by dilatometer investigations on specimens, which were quenched to room temperature or cryogenically treated. Furthermore,

X-ray diffraction (XRD) analyses and hardness measurements in combination with dilatometer investigations were conducted to determine the austenite phase fraction and the hardness after various tempering treatments. It is shown that cryogenically treated samples present a more pronounced carbide precipitation reaction during heating compared to the non-cryogenically treated samples. It could be revealed that the austenite phase fraction affects the precipitation kinetics, causing further microstructural differences between cryogenically and non-cryogenically treated samples.

## 2. Experimental

The used material in this work was a high Co-Ni steel with 11 wt.% Ni, 13.5 wt.% Co, 2.9 wt.% Cr, 1.2 wt.% Mo and 0.22 wt.% C. The material was forged to round bars with a diameter of 200 mm and austenitized in a vacuum furnace at 885°C for 1 h and cooled to room temperature with 10°C/ min. Half of the samples were cryogenically treated (CT) in a freezing unit at −73°C for 1 h.

Dilatometer investigations were carried out by using a DIL 805A dilatometer from TA Instruments (formerly BAEHR). The samples for dilatometer investigations had a diameter of 4 mm and a length of 10 mm. The differential length change signals from the dilatometer measurements were obtained by deriving the relative length change with respect of time. CT and non-CT samples were heated (1st run) in the dilatometer to a maximum temperature of 620°C with 20°C/ min and immediately cooled to room temperature (20°C/ min) and additionally reheated (2nd run) with the same parameters. Heating parameters were chosen in a way that no significant reverted austenite formation occurs.

Austenite phase fraction and hardness measurements were carried out on CT and non-CT samples with a diameter of 10 mm and a length of 15 mm, which were heated in the dilatometer with 20°C/ min to various temperatures up to 620°C and immediately cooled to room temperature with 100°C/ min. XRD measurements for austenite phase fraction determination were performed by using a D8 Discover diffractometer from Bruker AXS. Cr-$K_\alpha$ radiation with a wavelength of 2.29 Å was used for measurements and for a quantitative determination of phase fractions the Rietveld method was applied [9]. The absolute inaccuracy of phase fraction determination was about 1-2 %. Transmission electron microscopy (TEM) investigations were carried out on a Philips CM12 microscope. Samples were conventionally prepared by grinding to a thickness of 80 $\mu$m and a subsequent electro polishing procedure.

## 3. Results

### *3.1. Characterization of the evolution of austenite*

To study the evolution of austenite content due to tempering, austenite phase fraction measurements were carried out on CT and on non-CT samples before heating and after heating to 200°C, 400°C, 500°C and 580°C and immediately cooling to room temperature. The results are shown in Fig. 1. The austenite content of the non-CT sample without tempering is approximately 6.5 %. By performing a cryogenic treatment the austenite content decreases to 3.5 %. After heating the cryogenically treated samples to different temperatures, the austenite phase fraction does not change significantly, whereas heating of the non-cryogenically treated samples beyond 400°C leads to a marked decrease of austenite content to about 4.5 %. For analyzing this decrease of phase content more precisely, the relative length change curves, obtained from dilatometer experiments, were studied. Fig. 2 depicts the dilatometer cooling curves of the non-cryogenically treated samples after tempering to 400°C, 500°C or 580°C. A distinct phase transformation can be observed for the samples, which were annealed to 500°C and 580°C, as indicated by the volume increase during cooling compared to the sample, which was annealed to 400°C. Since the phase transformation is accompanied with a volume increase, it is attributed to a transformation of austenite into a bcc phase, e.g. martensite. Furthermore, these dilatometer investigations also reveal that the reduction of austenite content of the non-CT samples occurs during cooling from the tempering treatment.

For characterizing the retained austenite more precisely, TEM investigations were carried out. Fig. 3 shows a TEM bright-field image of the microstructure of the investigated steel. The material exhibits a martensitic matrix after austenitizing and quenching to room temperature. Retained austenite films are present along martensite lath boundaries, as indicated by the arrows. It is estimated that austenite films have a thickness between 3 nm (detection limit) and 40 nm.

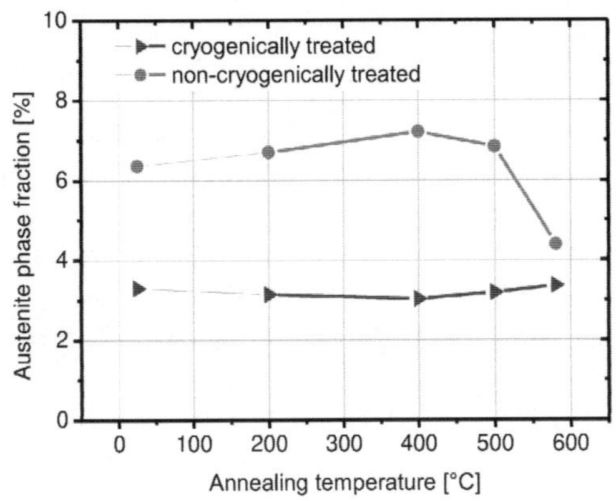


**Figure 1.** Austenite phase fraction of cryogenically and non-cryogenically treated samples after heating in the dilatometer to various temperatures up to 580°C (heating rate: 20°C/ min) and cooling down to room temperature (cooling rate: 100°C/ min). The non-cryogenically treated samples exhibit a decrease of austenite phase fraction, when tempering is conducted beyond 400°C.

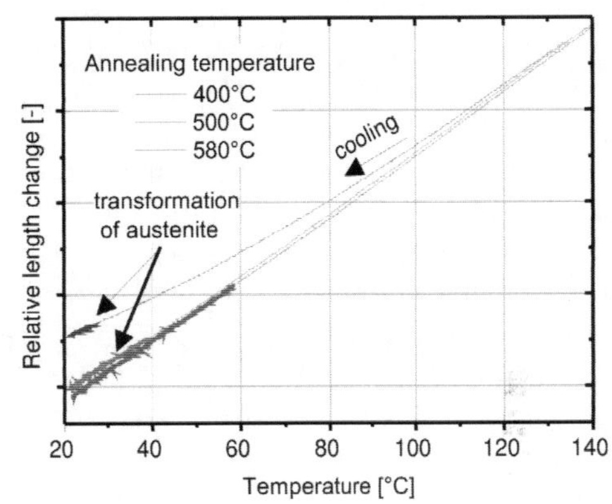


**Figure 2.** Relative length change of dilatometer cooling curves after heating of non-CT samples to 400°C, 500°C or 580°C and subsequent cooling (100°C/ min). A transformation of austenite occurs during cooling for the non-CT samples, which were annealed to 500°C or 580°C. For a better comparability, the curves were shifted along the y-axis. Therefore, no scaling of the y-axis is shown.

**Figure 3.** TEM bright-field image of the martensitic microstructure after austenitizing at 885°C for 1 h and quenching to room temperature. Retained austenite is present at martensite lath boundaries, as marked by the arrows.

### *3.2. Detailed analyses of the tempering behavior*

For a detailed characterization of the processes, taking place during tempering of CT and non-CT samples, dilatometer analyses were conducted. In Fig. 4 the relative length change during heating of CT and non-CT samples is plotted as a function of temperature. Generally, the heating curves of both conditions exhibit differences. The most significant difference is observed in the range between 250°C-350°C, as CT samples exhibit a more pronounced deviation from the linear expansion. For a more precise characterization of the phenomena, which occur during heating, the derivatives of the relative length change curves were calculated as depicted in Fig. 5. These curves show three main reactions, which are indicated by the numbers 1 - 3, respectively. The first effect during heating in the range between 100°C and 200°C may be attributed to transition carbide precipitation, as this process generally occurs in the temperature range between 80°C and 200°C and causes a reduction of volume [10, 11, 12]. The second effect, which also causes a reduction of volume, occurs in the temperature regime between 250°C and 400°C and can be assigned to cementite precipitation [10, 11, 12].

There are distinct differences between CT and non-CT samples regarding cementite precipitation properties, as precipitation is more enhanced for CT samples. This is also observed for the processes, which start at temperatures above 450°C - 500°C. These processes are ascribed to secondary hardening carbide precipitation, because the formation of these carbides generally proceeds in this temperature range and would cause an increase of length [6, 10, 13, 14, 15].

The dilatometer curves of both conditions do not show any increase of length in the range between 200°C and 350°C, which is typical for austenite transition into ferrite and carbides or bainite [10, 11, 12, 16, 17]. Hence, it is suggested that austenite stays stable and no austenite transformation occurs during heating. This is also supported by austenite phase fraction measurements, where no reduction of austenite phase fraction occurs up to temperatures of 400°C.

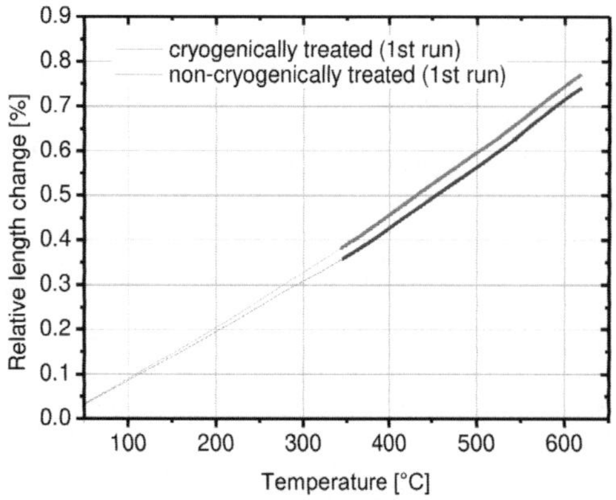


**Figure 4.** Relative length change curves of dilatometer heating experiments of cryogenically and non-cryogenically treated samples. The heating rate was 20°C/ min. Generally, the curves of CT and non-CT conditions exhibit a different behavior, which is indicated by the lower expansion for CT compared to non-CT samples during heating.

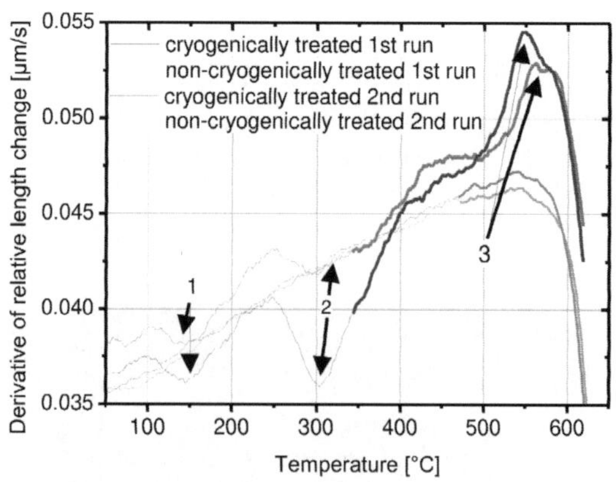


**Figure 5.** The derivatives of the relative length change curves obtained from dilatometer heating experiments of the 1st and the 2nd run of CT and non-CT samples. The heating rate was 20°C/ min. The effect 1 can be attributed to transition carbide precipitation, the effect 2 to cementite precipitation and the effect 3 is ascribed to secondary hardening carbide formation.

For correlating the results of the dilatometer analyses, hardness measurements were carried out after various tempering treatments. The results are shown in Fig. 6. It is revealed that the hardness of the CT samples is higher compared to the non-CT samples after tempering up to temperatures of 500°C. After tempering beyond temperatures of 500°C the hardness of CT and non-CT conditions is the same. A decrease of hardness values for both conditions can be observed between 200°C and 400°C (process 2, shown in Fig. 6), which corresponds, according to dilatometer analyses, to the region of cementite precipitation. Additionally, the raise of hardness, which occurs for CT and non-CT samples in excess of 400°C can be attributed to secondary hardening carbide formation, which supports the results from dilatometer analyses.

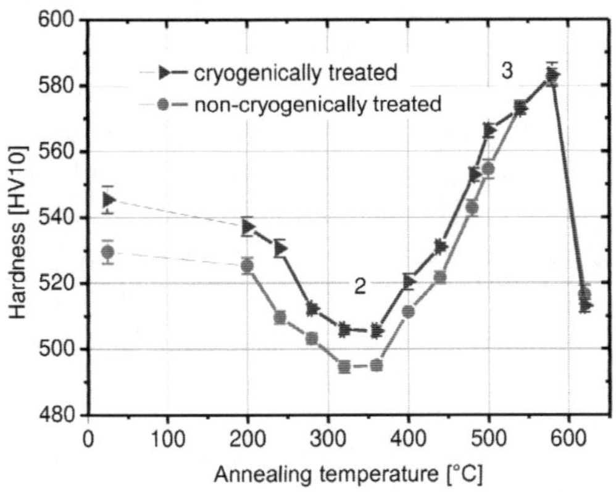


**Figure 6.** Hardness measurements of the CT and non-CT samples after heating (20°C/ min) to various temperatures up to 620°C and cooling to room temperature (100°C/ min). The reduction of hardness can be related to cementite formation (2). The large increase of hardness (3) is caused by the formation of secondary hardening carbides.

## 4. Discussion

The influence of cryogenic treatments on the microstructural evolution during heating was analyzed. It is identified that CT compared to non-CT samples exhibit a more intense cementite formation between 250°C and 400°C and more pronounced secondary hardening carbide precipitation above temperatures of 450°C - 500°C.

By XRD measurements, the austenite phase fraction of the cryogenically treated sample without tempering was determined with approximately 3.5 %. Due to tempering, the austenite content of the CT samples does not change significantly. Hence, it is suggested that for the CT samples neither austenite formation nor austenite transition occurs during tempering. This is also in accordance with the dilatometer heating and cooling (not shown here) curves, as they do not show any indication for austenite transformation. Conversely, the austenite phase fraction of the non-CT samples, which is about 3 % higher compared to the CT samples after tempering up to temperatures of 400°C, decreases due to tempering at higher temperatures. This decrease of the austenite phase fraction is identified to occur during cooling by dilatometer investigations. As the transformation of austenite is accompanied with an increase of volume, it is expected that austenite transforms into a bcc phase, e.g. martensite.

The lower hardness values of the non-CT compared to the CT conditions after tempering up to temperatures of 500°C can be ascribed to the higher austenite phase fraction of non-CT samples. Since the austenite of non-CT samples partially transforms upon cooling after tempering above temperatures of 500°C into martensite, the hardness values of the non-CT and CT conditions are comparable, as also the austenite phase fraction is in the same range.

The decrease of hardness between 200°C and 400°C can be attributed to a reduction of carbon content in martensite. This reduction of carbon content is a result of carbon segregation to lattice defects, e.g. dislocations, which is followed by the formation of carbides such as cementite. However, the reduction of carbon in martensite causes a decrease of tetragonality, which is accompanied with a stress relief in martensite. This process causes a reduction of hardness, whereas the formation of cementite principally produces an increase of hardness. As the increase of hardness produced by cementite formation is to neglect compared to the hardness reduction due to the relief of stresses in martensite, the hardness decreases in the temperature range between 200°C and 400°C.

Furthermore, it is observed that the decrease of hardness in the temperature range between 200°C and 400°C is as high for CT samples as for non-CT samples. As carbon diffusion out of martensite causes this hardness decrease, it is assumed that for CT and non-CT conditions an equal reduction of carbon content in martensite occurs. In contrast to that, it is observed by dilatometer analyses that the cementite precipitation reaction for CT samples is more intense, indicating that more carbon is available for cementite formation in CT samples as in non-CT samples. Since it is suggested that a significant amount of carbon may also diffuse into austenite, the lower austenite phase fraction of CT samples compared to non-CT samples (in the temperature range between 200°C and 400°C) is probably the reason for the higher availability of carbon for cementite precipitation in CT samples. Generally, austenite has a higher solubility for carbon than martensite. Therefore, carbon is eager to diffuse from the supersaturated martensite into austenite. As non-CT samples exhibit two times as much austenite as CT samples in this temperature region, also the amount of carbon which can dissolve in austenite is twice as much. By assuming a maximum carbon enrichment in austenite of 2 wt.% [18, 19, 20], the non-CT samples can dissolve 0.12 wt.%, whereas the CT samples are only able to dissolve 0.06 wt.% of carbon in austenite. Consequently, less carbon is available for forming cementite in the non-CT samples, resulting in a smaller reaction peak for cementite formation.

The increase of hardness beyond temperatures of 400°C is assigned to be a result of secondary hardening carbide precipitation, which may also be influenced by the preceding precipitation processes. Thus, the carbide precipitation reaction in the CT samples is more intense, which was also determined by dilatometer analyses. Alterations in the carbide structure due to cryogenic treatment are also described by Das and Huang et al. [1, 4].

Secondary hardening carbide precipitation near or at the interface between retained austenite and martensite would cause a reduction of carbon content in austenite. Generally, this reduction leads to a decrease of the stability of austenite [21]. Conversely, cryogenic treatment would cause an enhancement of the stability of austenite, as it is explained in the following. It is expected that the thickness of retained austenite particles varies, as estimated by Fig. 3. Waitz et al. [22] reported that larger austenitic particles exhibit a lower stability. As cryogenic treatment causes a partially transformation of retained austenite, it is estimated that due to the cryogenic treatment a transformation of the larger, less stable austenitic regions takes place. As a result, for CT samples the less stable austenitic particles have already transformed during the cryogenic treatment. Hence, no transformation of austenite occurs during cooling of the CT samples after heating to temperatures beyond 400°C. Conversely, during cooling of the non-CT samples some austenite transforms, proving its lower stability. This transformation of austenite into martensite, produces an increase of hardness for the non-CT samples, which overlaps with the increase of hardness due to secondary hardening carbide precipitation. Hence, no comparison between CT and non-CT samples regarding the hardness increase due to formation of secondary carbides can be carried out.

## 5. Conclusion

The investigations on the effect of cryogenic treatment on the microstructural evolution during tempering of a high Co-Ni steel leads to the following conclusions:

- Cryogenic treatment causes a reduction of austenite phase fraction. This reduction is probably accompanied with a stabilization of austenite, as the less stable austenitic regions transform during the cryogenic treatment. Conversely, tempering causes a destabilization of austenite because of carbon reduction in austenite due to secondary hardening carbide formation. As a result of these opposing processes, the austenite of the CT samples does not transform due to the stabilization of austenite by the cryogenic treatment. In contrast, the austenite of the non-CT samples transforms upon cooling when tempering is conducted beyond 400°C, i.e. in the temperature range of secondary hardening carbide precipitation.
- Due to the higher austenite content of the non-CT samples during tempering in the temperature range between 200°C and 400°C, a higher amount of carbon can be dissolved in austenite. Thus, less carbon is present in martensite, causing a reduction of the intensity of the cementite precipitation reaction compared to the CT samples.
- The precipitation of secondary hardening carbides may also be influenced by the preceding precipitation processes. As a result, the CT samples also exhibit a more pronounced secondary hardening carbide precipitation.

## References

[1] Das D, Sarkar R, Dutta A K and Ray K K 2010 *Mat. Sci. Eng. A* **528** 589–603
[2] Zhirafar S, Rezaeian A and Pugh M 2007 *J. Mater. Process. Tech.* **186** 298–303
[3] Molinari A, Pellizzari M, Gialanella S, Straffelini G and Stiasny K H 2001 *J. Mater. Process. Tech.* **118** 350–355
[4] Huang J, Zhu Y, Liao X, Beyerlein I, Bourke M and Mitchell T 2003 *Mat. Sci. Eng. A* **339** 241–244
[5] Stratton P F 2007 *Mat. Sci. Eng. A* **449-451** 809–812
[6] Ayer R and Machmeier P M 1993 *Metall. Trans. A* **24** 1943–1955
[7] Sato K 2002 Ph.D. thesis Universitiy of California, Berkeley
[8] Haidemenopoulos G N 1988 Ph.D. thesis Massachusetts Institute of Technology
[9] Young A 2002 *The Rietveld Method* (Oxford Universitiy Press)
[10] Speich G R and Leslie W C 1972 *Metall. Trans.* **3** 1043–1054
[11] Morra P V, Böttger A J and Mittemeijer E J 2001 *J. Therm. Anal. Calorim.* **64** 905–914
[12] Waterschoot T, Verbeken K and De Cooman, B C 2006 *ISIJ International* **46** 138–146
[13] Bala P, Pacyna J and Krawczyk J 2007 *Archives of Materials Science and Engineering* **28** 517–524
[14] Bala P, Pacyna J and Krawczyk J 2007 *Journal of Achievements in Materials and Manufacturing Engineering* **22** 15–18
[15] Thelning K E 1978 *Steel and its heat treatment* (Butterworth Co Publishers Ltd)
[16] Pacyna J 2011 *Journal of Achievements in Materials and Manufacturing Engineering* **46** 7–17
[17] Cheng L, Brakman C M, Korevaar B M and Mittemeijer E J 1988 *Metall. Trans. A* **19** 2415–2426
[18] Garcia-Mateo C, Caballero F G, Miller M K and Jimenez J A 2012 *J. Mater. Sci.* **47** 1004–1010
[19] Speer J, Matlock D K, De Cooman, B C and Schroth J G 2003 *Acta. Mater.* **51** 2611–2622
[20] Wang J and Van Der Zwaag, Sybrand 2001 *Metall. Trans. A* **32** 1527–1539
[21] Lerchbacher C, Zinner S and Leitner H 2012 *Metall. Mater. Trans. A* **43A** 4989–4998
[22] Waitz T, Antretter T, Fischer F D and Karnthaler H P 2008 *Mater. Sci. Technol.* **24**(8) 934–940

# Induction hardening: Differences to a conventional heat treatment process and optimization of its parameters

**Annika Vieweg[1], Gerald Ressel[1], Petri Prevedel[1], Peter Raninger[1], Michael Panzenböck[2], Stefan Marsoner[1] and Reinhold Ebner[1]**

[1] Materials Center Leoben Forschung GmbH, Roseggerstraße 12, A-8700 Leoben

[2] Department of Physical Metallurgy and Materials Testing, Montanuniversität Leoben, Franz Josef Straße 18, A-8700 Leoben

E-mail: annika.vieweg@mcl.at

**Abstract.** The possibility of obtaining similar mechanical properties with faster heating processes than the conventional ones has been of interest for several years. In the present study, investigations were performed in terms of the influences of such fast heat-treatments on the microstructure and mechanical properties of the material. This investigation compares an inductive with a conventional furnace heat treating process of a 50CrMo4 steel, however only the austenitizing treatment was changed and subsequent quenching and tempering was done in the same way. To this end experiments with a middle frequency generator, using different heating rates and austenitizing temperatures, were conducted and followed by oil quenching of the workpieces. The resulting structures were characterized regarding their microstructures and mechanical properties in order to gather a better understanding of the differences between the inductive and the conventional heat treating process. As a main result it was found, that the fast austenitized samples exhibited worse ductility than the conventional treated material.

## 1. Introduction

The field of induction hardening has become increasingly important in industrial heat treatment processes due to short production times, in line heat treatments and the fact, that they are ecologically more favorable than conventional furnace based heat treatments [1, 2]. Consequently, it is important to obtain mechanical properties, which are comparable to the ones from a longer process but with a much shorter production cycle. Nevertheless, differences in mechanical properties, especially toughness, have been noticed, when induction hardened material is compared to conventionally (oven-) hardened material [3, 4]. These differences may result from the different heating rates and dwell times at austenitizing temperature, as well as different cooling rates through spray quenching [5]. In conventional heat treatments the material is slowly heated to austenitizing temperature and kept within the furnace for a certain time. During the dwell time the austenite can develop a homogeneous composition with a low density of crystal defects. Conversely, induction heating uses high heating rates $\dot{T}$ (up to $\dot{T} = 100\,\mathrm{K/s}$) and short - sometimes even no - dwell times. This may lead to chemical inhomogeneities and mechanical properties have been found to not hold up to the standards of conventionally treated material [3]. Furthermore, the effect, that the heat input within the surface, the so-called

penetration depth, localizes 83 % of the induced power, leads to faster heating of the surface of the workpiece, whereas the core temperature lags behind the surface temperature. This effect is highly dependent on the frequency and the magnetic, as well as the electric properties of the material.
In this study the influence of the austenitizing treatment on the microstructure and the mechanical properties is studied. Therefore, the inductive, fast heat treatment is compared to a longer conventional one. Following recommended literature [6], the challenge of incomplete homogenization was solved by choosing higher temperatures for the inductive heat treatment, to obtain an almost homogeneous structure even in a short period of time. The other characteristics of the heat treatment (e.g. quenching, tempering) were kept constant in order to be able to solely study influences of fast austenitizing treatments.

## 2. Experimental

The steel investigated in this study was a 50 CrMo4 steel with the chemical composition as listed in Table 1.

**Table 1.** Chemical composition of 50CrMo4 as used in this study.

| C | Si | Mn | Cr | Mo | P | S |
|---|---|---|---|---|---|---|
| 0.49 | 0.27 | 0.71 | 1.05 | 0.18 | 0.016 | 0.010 |

A well defined initial ferritic-pearlitic microstructure was adjusted using a heat treating process stated in the literature [7]. Cylindrical specimens with a diameter of 22 mm were austenitized at 850 °C for 20 min, cooled down to 650 °C, held for 1500 s and afterwards air-cooled to room temperature. Samples with this ferritic-pearlitic microstructure were put into the hot oven at 850 °C, held at temperature for 20 min and quenched in oil in order to obtain a conventionally quenched structure. Quenched and tempered (QT) samples were obtained by subsequent tempering within the oven at 540 °C for 40 min and afterwards air cooled.
For induction hardening a middle frequency generator (20 kHz, 30 kW) was used for the through heating of the workpieces of a diameter of 22 mm. The temperature evolution during heating was calibrated using a type S thermo-couple located in the center of the workpiece and which was positioned at the center of the induction coil. By applying different power values the samples were heated to 950 °C and 1000 °C for both temperatures within a processing time of 24 and 100 s. The heating power was around 15 kW for 50 K/s and 2 kW for the heating rate of 10 K/s. For the chosen power values a surface temperature evolution $T_S$ was measured 2 mm underneath the surface with a type S thermo-couple in order to get a rough estimate for the temperature profile over the cross-section. Temperatures at the surface exceeded the core temperature, depending on the heating rate. For 50 K/s the temperature 2 mm underneath the surface was 150 °C higher than the core temperature, whereas the samples heated with 10 K/s received an overheating of 100 °C. It can be assumed, that the temperature at the surface was somewhat higher.
The detailed parameter-set and the nomenclature of the samples is listed in Table 2. The so-called processing time is defined as the time for the whole austenitization treatment, including heating and dwell time, but without quenching and tempering time, which was the same for every process. The workpiece was statically held within the induction coil and dropped into an oil-bath at the end of the heating process. Subsequent tempering was done in the same manner as for the conventional QT-samples (oven 540 °C for 40 min, air cooling).
Microstructural analysis was conducted at the part which was positioned at the center of the induction coil. Optical light microsopic (OLM) investigations were done on samples etched with 3 % nital and prior austenite grain etching was achieved using a solution of picric acid

**Table 2.** Parameter-set and nomenclature of the samples. Processing time solely corresponds to the time for the austenitizing treatment.

| Name | Processing time [s] | Austenitizing Temperature [°C] | Dwell time [s] | Heating rate $\dot{T}$ [K/s] |
|---|---|---|---|---|
| CONV. | 1800 | 850 | 1200 | approx. 15 |
| 10-1 | 100 | 950 | 0 | 10 |
| 10-2 | 100 | 1000 | 0 | 10 |
| 50-1 | 24 | 950 | 0 | 50 |
| 50-2 | 24 | 1000 | 0 | 50 |

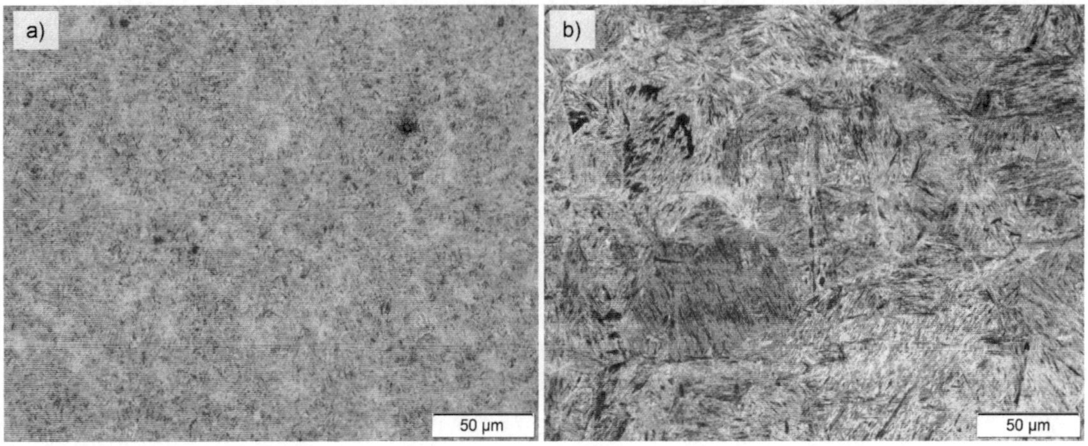


**Figure 1.** OLM images of the conventional (CONV.) (a) and the 50-2 (b) state after quenching.

with benzenesulfonic acid as wetting agent. The mean grain size was evaluated using the line intercept method according to ASTM E 112-12. The near surface grain size was done around 500 $\mu$m below the surface.

Samples for mechanical testing were manufactured within the uniformly heated part of the workpiece, which was located in the center of the induction coil. Vickers hardness measurements were conducted at a Qness Q10A+. Mean hardness was evaluated for 8 points at half radius measured every 45 ° and an additional hardness profile was measured over the cross-section of the workpiece. Phase fraction analysis of retained austenite was carried out using X-ray diffraction (XRD) on a Xstress 3000 G2 diffractometer by Stresstech. Charpy impact energy was investigated at three samples for each condition at a Zwick RKP 450 at room temperature. Scanning electron microscopic (SEM) investigations were conducted using a Zeiss EVO MA25.

## 3. Results and Discussion

### *3.1. Differences of the as-quenched state*

In a first step the microstructural characteristics at the center of the sample in the as-quenched condition were evaluated. OLM pictures, as presented in Figure 1, show the differences between the sample CONV. and 50-2. The conventional heat treated sample exhibits a fine microstructure. The inductively heated sample shows a significantly coarser microstructure, in which the martensitic laths can be observed very clearly. The higher temperature for the inductively heat treated sample (1000 °C) leads to a coarser microstructure, despite the short processing time. To evaluate dimensions of the microstructural characteristics, prior austenite grain sizes were analyzed.

Figure 2 depicts that the grain sizes of the samples are different for the different processing times and temperatures. The higher the austenitizing temperature the larger the austenite grain size, despite the shorter processing time. A difference in the austenite grain size can be observed between the surface and the core (open and closed symbols) of the workpiece. This occurs especially for high $\dot{T}$, since the over heating for fast heating rates is more pronounced than for slower heating. A higher generator power is needed to achieve the mentioned heating rates and because of the skin effect a higher power density at the surface of the workpiece was obtained. The result is a fast or even an over-heated surface (up to 150 °C), represented through larger prior austenite grains for higher $\dot{T}$. This effect occurs as well for a higher austenitizing temperature and a longer processing time (10-2), also due to over-heating and longer holding times. This leads to a pronounced variation of the prior austenite grain size for the sample 10-2 over the cross-section. It can be stated, that, the slower the heating and the austenitizing temperature is, the more uniform the through heating of the workpiece and the smaller the difference between prior austenite grain sizes over the cross-section. It has to be noticed that even for short processing times and lower temperatures (Figure 2, 50-1 and 50-2) the austenite grain size is higher than for the conventional heat treatment, which might be due to increased overheating for the high heating rates.

In order to elucidate the influence of the processing on hardness, the mean hardness of the as-quenched state is depicted in Figure 3. The mean hardness is higher at lower processing times processing times and maximum temperatures. Although the hardness increase is not even 10 HV10, this might be the result of the higher processing temperatures, which promote carbon solution within the matrix and thus enabling harder structures [8]. The amount of retained austenite (depicted through open symbols in Figure 3) is decreasing over the processing time. For the CONV. samples 7 % retained austenite were measured, whereas the specimen 50-2 comprises 14 %. The higher percentage of retained austenite at lower processing times leads to the conclusion, that higher amounts of carbon are dissolved in the matrix, which might also be a result of the high processing temperatures and the increasing over-heating phenomenon for the fast austenitized samples.

### *3.2. Differences of the tempered state*

Figure 4 depicts the microstructures of the samples CONV and 10-2 after the tempering process. The fast austenitized sample (10-2) exhibits a coarser structure. Furthermore, the carbide distribution does show slight differences in the tempered samples. The conventional state shows few large globular carbides (indicated by the arrows in Figure 4a), which are expected to be undissolved carbides stemming from normalizing treatment. Apparently, not all carbides from the normalized structure did dissolve during the conventional austenitizing treatment. Therefore, it can be assumed, that the dwell time for the conventional process should be longer. The precipitating carbides, which form during tempering, occur finely dispersed, globular as well as in a plate-like shape (indicated by ellipsoids). Analyses of the plates in more detail reveals, that most of the plates show a tendency of becoming globular and the plate breaks apart.

Specimen 10-2 (Figure 4b) does not show larger globular carbides from the normalizing treatment, which can be attributed to the higher austenitizing temperature, which obviously causes full dissolution of all carbides. The precipitating carbides occur mostly in a plate-like shape and shows a tendency to precipitate strikingly at the former austenite grain boundaries (indicated by the arrow in Figure 4b) and at martensite lath boundaries. The inductive sample showed the effect of beginning spherodization of the precipitating carbides in less distinct than the conventional sample. Precipitation of carbides at dislocations, at martensite laths and at prior austenite grain boundaries, as well as the plate-like shape has been observed in other studies [9]. The decomposition of retained austenite is also an important consequence during

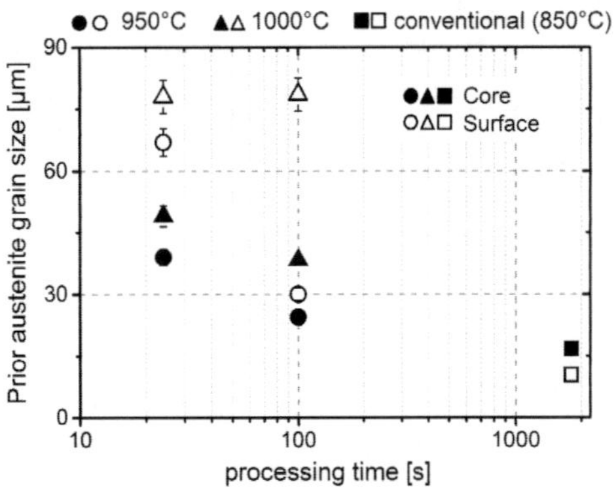


**Figure 2.** Austenite grain size for the different the processing time. The open symbols represent the grain size at the surface of the workpiece, whereas closed symbols refer to the core. Circles correspond to austenitizing temperature of 950 °C, triangles to 1000 °C and the square refers to the conventional heat treatment at 850 °C within the oven.

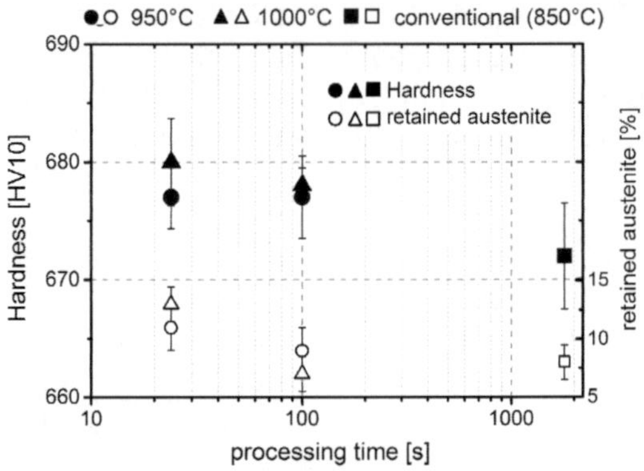


**Figure 3.** Mean hardness, measured at half radius, and percentage of retained austenite of the as-quenched samples. Hardness is depicted by closed symbols and percentage of retained austenite by open symbols. Symbols are in accordance with Figure 2.

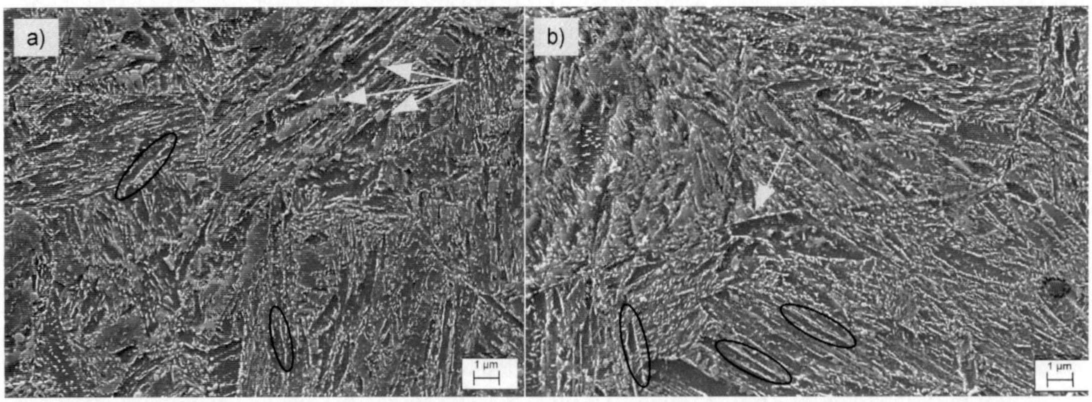


**Figure 4.** (a) SEM image of the conventionally heat treated material (CONV), arrows are pointing towards larger carbides, ovals are indicating plate-like carbide shapes; (b) SEM image of 10-2, arrow is pointing at carbides on a prior austenite grain, ovals are indicating plate-like carbides.

and after the tempering process. All samples comprise no retained austenite in the tempered state, therefore, it is clear that the retained austenite did decompose during tempering or during cooling after tempering. The kinetics of this decomposition were not analyzed, therefore it is not clear if the austenite decomposed only during heating or as well during the dwell time at tempering temperature of 540 °C.

All samples were tempered in the same manner and Figure 5 compares the hardness for the inductively and conventionally heat treated samples. Hardness profiles were measured along the diameter of the workpiece and shows no significant differences between core and surface. The hardness of the tempered samples is somewhat lower at longer processing times (approx.

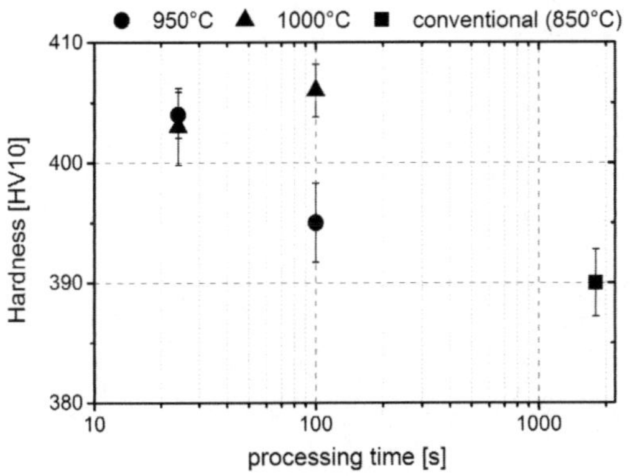


**Figure 5.** The mean hardness of the tempered samples at half radius for the different processing times. Symbols are in accordance with Figure 2.

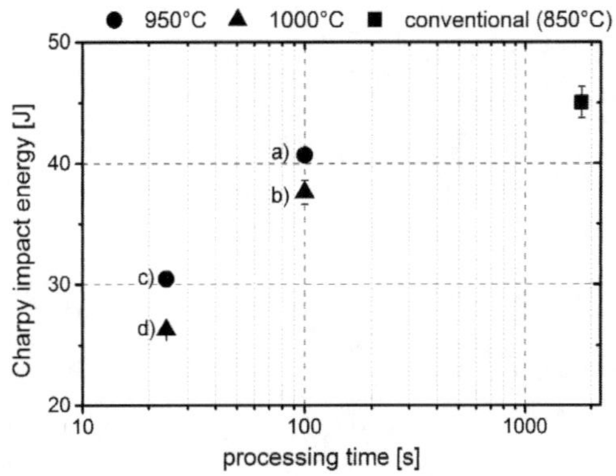


**Figure 6.** Charpy impact energy in Joule for the different processing times. Symbols are in accordance with Figure 2 and the letters correspond to the nomenclature of fracture surfaces in Figure 7.

15 HV10) and with lower austenitizing temperatures. Differences in hardness appear through the slightly different tempering behaviour depending on the prior austenitizing treatment. The different amounts of carbon dissolved within the matrix and retained austenite might be an explanation and might therefore be more pronounced with higher austenitizing temperatures and higher amounts of retained austenite (e.g. specimen 50-2).

The Charpy impact values (see Figure 7) follow a reversed trend compared to the hardness evolution, as they higher at longer processing times and lower austenitizing temperature. The conventionally heat treated sample obtains an impact value of 45 J, whereas the Charpy impact energy of the sample heated inductively to 1000 °C within 24 s (50-2) drops down to 26 J. The inductively heated samples show a higher hardness than the conventionally heat treated one, which might be one reason for the lower Charpy impact energy. A decrease of a Charpy impact value of almost 20 J seems to be large for an increase in hardness of only 15 HV10, hence, it is assumed that other effects might influence the toughness of the inductive heated specimens.

In order to clarify this assumption, detailed analyses of the Charpy impact fracture surfaces have been carried out (Figure 7). They show increasing percentage of intergranular and some fractions of cleavage fracture for shorter processing time and higher austenitizing temperature. Sample 50-2 shows high amounts of intergranular failure, which corresponds to a low impact energy. At higher impact energies the fraction of intergranular and cleavage failure is lower, and seems to be non existent for the sample 10-1. The CONV. samples show as well solely ductile fracture surface (not depicted in the Figure). The increasing percentage of intergranular fracture leads to the conclusion, that the low impact values are caused by this type of microscopic fracture mechanism. The increasing intergranular fracture sites indicate, that certain prior austenite grain boundaries are weakened by the fast austenitizing treatment. As seen in Figure 4(b) the tempering carbides precipitation is favored at former austenite grain boundaries, hence their position and shape (plate-like) can be assumed as a weakening mechanism [10]. Furthermore, impurities (e.g. P, Sb) can not be excluded as a reason for intergranular failure. Segregation during austenitiziation and pile ups ahead of the tempering carbides during tempering has been reported, especially for P, in the literature [11, 12].

Nevertheless, it is not yet clear why intergranular failure occurs for the inductively heated samples and SEM investigations were not sufficient to analyze the fracture surfaces in regard to this problem. This is subject of further studies.

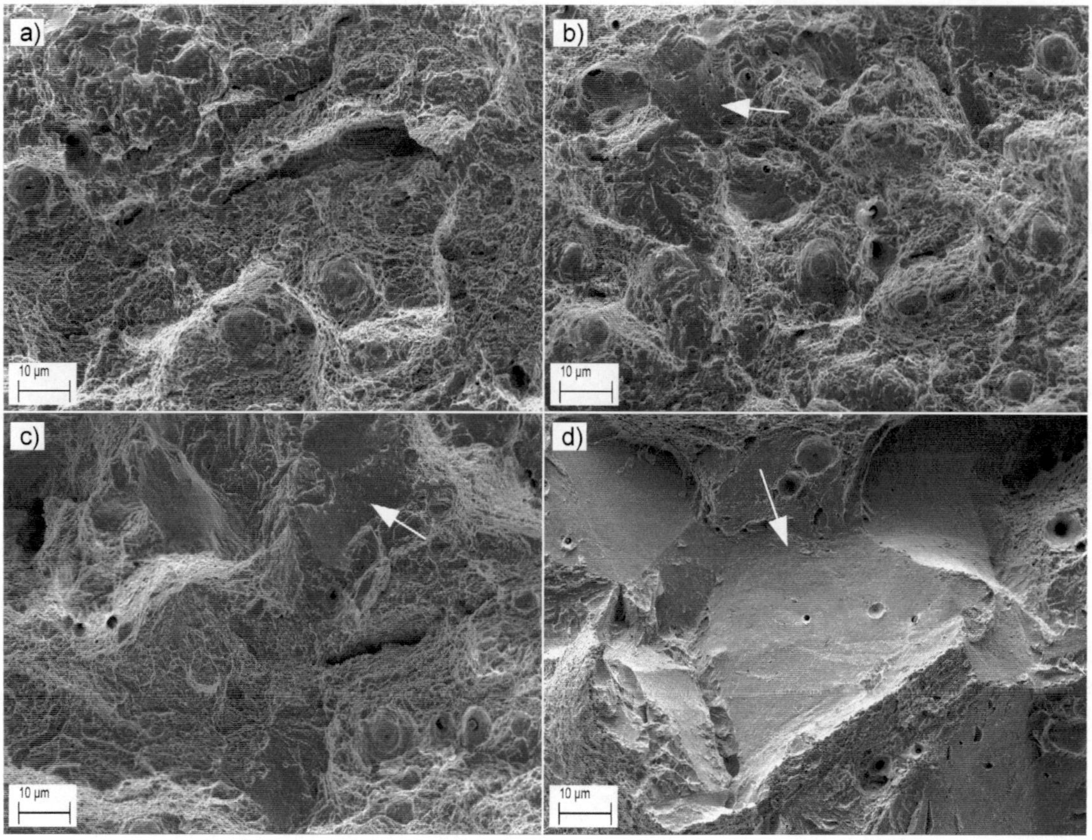


**Figure 7.** SEM-images of the fracture surfaces of the samples 10-1(a), 10-2 (b), 50-1 (c) and 50-2 (d). Arrows are pointing to intergranular fraction sites.

## 4. Summary

This study points out differences of microstructure and mechanical properties between a conventional and an inductive heat treatment of a 50CrMo4 steel.

An important difference between the inductively and conventionally heat treatment, is that during the inductive treatment temperature gradients occur over the cross-section, which is due to the skin effect. The gradient between surface and core is larger for higher heating rates, since the power intensities have to be increased. Temperature calibration at the core of the specimen lead o an overheated surface of the workpiece (up to 150 °C for a heating rate of 50 K/s). The inductively heat treated samples were austenitized at higher temperatures (950 °C and 1000 °C) compared to the conventional one (850 °C), which lead to increasing prior austenite grain sizes. Apparently, at higher temperatures a larger volume fraction of carbides dissolves during the heat treatment and the prior austenite grain size is increased. This agrees with the fact, that in the same specimens an increase in hardness and amount of retained austenite has been observed, which is a result of a higher carbon content in the matrix.

Subsequent tempering shows differences in carbide precipitation, due to different amounts of carbon dissolved as well as varying percentages of retained austenite. Further investigations might be helpful to analyse the kinetics of the tempering process in dependance on the prior austenitizing treatment, as well as detailed analysis of the precipitating carbides via e.g. TEM. A decrease in toughness for the inductively austenitized samples, represented by Charpy impact

energy, can be traced back to a combination of three phenomena: increase in hardness, increase in grain size and the change of the fracture behavior - from ductile to partly intergranular. The reason for the latter one requires further investigations, especially of the condition of the prior austenite grain boundaries regarding precipitations and impurities.
Nevertheless, samples which were inductively heated with $\dot{T} = 10\,\mathrm{K/s}$ to both austenitizing temperatures show a good combination of hardness and Charpy impact energy. Despite the decrease of impact energy compared to the chosen conventional process, it would still be sufficiently high for the industrial specification of this steel (30 J). Furthermore, the samples show a sufficient hardness and the processing time is over 10 times shorter than for the conventional treatment. Heating rates as high as $\dot{T} = 50\,\mathrm{K/s}$ are difficult to control and lead to unsatisfactory toughness properties and are therefore not recommendable according to this study.

## 5. Acknowledgement

Financial support by the Austrian Federal Government (in particular from Bundesministerium für Verkehr, Innovation und Technologie and Bundesministerium für Wissenschaft, Forschung und Wirtschaft) represented by Österreichische Forschungsförderungsgesellschaft mbH and the Styrian and the Tyrolean Provincial Government, represented by Steirische Wirtschaftsförderungsgesellschaft mbH and Standortagentur Tirol, within the framework of the COMET Funding Programme is gratefully acknowledged.

## 6. Literature

[1] Zinn S and Semiatin S 1988 *Elements of induction heating: Design, control, and applications* (EPRI)
[2] Rudnev V 2003 *Handbook of induction heating* (Dekker)
[3] Sackl S, Leitner H, Zuber M, Clemens H and Primig S 2014 *Metall Trans A* **45** pp. 5657–5666
[4] Clarke A, MKMiller, Field R, Coughlin D, Gibbs P, Clarke K, Alexander D, Powers K, Papin P and Krauss G 2014 *Acta Mat* **77** pp. 17–27
[5] Liščić B, Luty W and Tensi H 1991 *Theory and technology of quenching* (Berlin and New York: Springer-Verlag)
[6] Gebril M, Aldlemey M and Kablan A 2014 Effect of austenization temperatures and times on hardness, microstructure and corrosion rate of high carbon steel *Design and Computation of Modern Engineering Materials* (*Advanced Structured Materials* vol 54) ed Ächsner A and Altenbach H (Springer International Publishing) pp 421–428
[7] Orlich J 1973 *Atlas zur Wärmebehandlung der Stähle* (Max-Planck-Institut für Eisenforschung)
[8] Bleck W 2004 *Werkstoffkunde Stahl für Studium und Praxis* 2nd ed (Aachen: Mainz)
[9] Speich G and Leslie W 1972 *Metall Trans* **3** pp. 1043–1054
[10] Briant C L and Banerji S K 1978 *Intern Metals Rev* **23** pp. 164–199
[11] Rellick J and McMahon C 1974 *Metall Trans* **5** pp. 2439–2450
[12] Grabke H J 1989 *ISIJ Int* **29** pp. 529–538

# Estimation of Material Parameters in Nonlinear Hardening Plasticity Models and Strain Life Curves for CuAg Alloy

**J Srnec Novak[1], D Benasciutti[1], F De Bona[1], A Stanojević[2], A De Luca[3], Y Raffaglio[3]**

[1]DIEGM, University of Udine, Via delle Scienze 206, Udine 33100, Italy
[2]Chair of Mechanical Engineering, Montanuniversität Leoben, Franz Josef-Straße 18, Leoben 8700, Austria
[3]Centro Ricerche Danieli, Via Nazionale 41, Buttrio (UD) 33042, Italy

E-mail: jelena.srnec@uniud.it

**Abstract**. This work investigates the cyclic response and low-cycle fatigue behaviour of a CuAg alloy used in crystallizer for continuous casting lines. Therefore isothermal strain-based fatigue tests are first performed on CuAg specimens at different temperature levels (20 °C, 250 °C, 300 °C). The evolution of stress-strain loops recorded during the cyclic tests is used for the parameter identification of several nonlinear hardening models (nonlinear kinematic, nonlinear isotropic). Cyclic stress-strain data from experiments are compared with results from numerical simulations with the identified material parameters, showing a satisfying agreement. Critical examination of numerical results from different models is also performed. Finally, the strain-life fatigue curves estimated from experimental data are compared with approximate strain-life equations (Universal Slopes Equation, 10% Rule) which are obtained from simple tensile tests. The material parameters determined in this work can conveniently be used as inputs in a elasto-plastic finite element simulations of a crystallizer.

## 1. Introduction

Mechanical components in steel-making plants are often exposed to cyclic thermo-mechanical loadings and then exhibit a cyclic elasto-plastic behaviour and fatigue damage. In continuous casting lines, a typical example is the crystallizer, which is a long hollow component where the molten steel starts to solidify. Thermo-mechanical finite element (FE) analysis requires suitable models to properly simulate the cyclic elasto-plastic material response of the crystallizer as well as other components under thermo-mechanical loading (e.g. anode of electric arc furnace). The cyclic stresses and strains calculated by FE simulations are then compared to the experimentally obtained fatigue lives in order to estimate the component service life.

Over the last fifty years several theories to describe the elasto-plastic and viscoplastic material behaviour (plasticity, creep, relaxation) have been developed and further improved [1-3]. Some of them have become readily available in commercial finite element software and are used for every-day industrial design.

The capability of a material model to correctly represent a material behaviour observed in experiments is the first criterion for model selection. Material models generally depend on several parameters, which have to be calibrated on experimental results. By increasing the model complexity the number of material parameters usually increases as well. Complex numerical algorithms are often used to identify multiple parameters simultaneously and optimisation routines are recommended [4-6].

Sometimes it may be difficult, especially for non-experienced engineers, to understand which model is most suitable for their application or which parameters do really affect the material response. The choice of the material model for FE simulations, especially in industrial applications, often is the result of a trade-off among various needs, such as model complexity, computation time and experimental data available for parameter identification.

As a contribution to material modelling and parameter identification this work presents the results of a research activity aimed at experimental testing and parameter identification of a CuAg alloy used for crystallizers of continuous casting lines. Isothermal strain-based low-cycle fatigue (LCF) tests are performed at different temperatures (20 °C, 250 °C, 300 °C). Experimental stress-strain loops are used for parameter identification of nonlinear kinematic (Armstrong-Frederick, Chaboche) and nonlinear isotropic hardening models. Numerical simulations with the identified material parameters are then compared with experimental data, showing that the considered material models are adequate to represent the elasto-plastic behaviour of the CuAg alloy. Strain-life fatigue lines estimated from experimental data are finally compared with some approximate analytical equations (Universal Slopes Equation, 10% Rule), which are estimated from simple tensile tests.

## 2. Nonlinear hardening models: theoretical background

The combined material model (nonlinear kinematic + nonlinear isotropic) is able to capture elasto-plastic behaviour of a material under cyclic loading. In the case of combined hardening, the yield surface can both translate and expand as shown in Figure 1. The von Mises yield criterion is given by:

$$f = \sqrt{\frac{3}{2}(\sigma' - \alpha'):(\sigma' - \alpha')} - R - \sigma_0 = 0 \tag{1}$$

where $\boldsymbol{\sigma}'$ and $\boldsymbol{\alpha}'$ are the deviatoric parts of the stress and the back stress tensor, respectively, $\sigma_0$ is the initial yield stress and $R$ is the drag stress. Kinematic part is controlled by $\boldsymbol{\alpha}$ (translation of the yield surface), while the isotropic part is related to $R$ (expansion of the yield surface). The simplest kinematic model is the linear kinematic hardening model developed by Prager (1949), which assumes that the evolution of $\boldsymbol{\alpha}$ is collinear with the plastic strain tensor:

$$\mathrm{d}\boldsymbol{\alpha} = \frac{2}{3} C \mathrm{d}\varepsilon_{\mathrm{pl}} \tag{2}$$

where $C$ is the initial hardening modulus. Armstrong and Frederick (1966) modified Prager's model by adding a recall term, which introduces a fading memory effect to the strain path. The recall term is called dynamic recovery. As a result, the nonlinear evolution of $\boldsymbol{\alpha}$ is obtained [1]:

$$\mathrm{d}\boldsymbol{\alpha} = \frac{2}{3} C \mathrm{d}\varepsilon_{\mathrm{pl}} - \gamma \boldsymbol{\alpha} \mathrm{d}\varepsilon_{\mathrm{pl,acc}} \tag{3}$$

where $\gamma$ defines the rate at which hardening modulus starts to decrease as the plastic strain develops. Chaboche further extended the Armstrong-Frederick model by superimposing two or more nonlinear kinematic hardening models:

$$\boldsymbol{\alpha} = \sum_i \boldsymbol{\alpha}_i \;\; ; \;\; \mathrm{d}\boldsymbol{\alpha}_i = \frac{2}{3} C_i \mathrm{d}\varepsilon_{\mathrm{pl}} - \gamma_i \boldsymbol{\alpha}_i \mathrm{d}\varepsilon_{\mathrm{pl,acc}} \tag{4}$$

The integration of equation (3) with respect to $\varepsilon_{pl}$, for uniaxial loading, leads to:

$$\alpha = \psi \frac{C}{\gamma} + \left( \alpha_0 - \psi \frac{C}{\gamma} \right) \exp\left[ -\psi\gamma\left(\varepsilon_{pl} - \varepsilon_{pl,0}\right)\right] \tag{5}$$

where $\psi=\pm1$ indicates the flow direction, $\varepsilon_{pl,0}$ and $\alpha_0$ are the initial values of plastic strain and back stress, respectively, at the beginning of the considered loading branch. Assuming tension ($\psi=1$) and zero initial plastic strain and back stress, equation (5) becomes:

$$\sigma = \sigma_0 + \frac{C}{\gamma}\left[1 - \exp\left(-\gamma\varepsilon_{pl}\right)\right] \tag{6}$$

Equation (6) is proposed to estimate material parameters $C$ and $\gamma$ from a single stabilized stress-strain loop [1, 7]. Analytical integration of (5), for tension and compression, gives the relation between the stress amplitude and the plastic strain amplitude for the stabilized cycle [1, 7]:

$$\sigma_a = \sigma_0 + \frac{C}{\gamma}\tanh\left(\gamma\varepsilon_{pl,a}\right) \tag{7}$$

or in the form of Chaboche model:

$$\sigma_a = \sigma_0 + \sum_{i=1} \frac{C_i}{\gamma_i}\tanh\left(\gamma_i\varepsilon_{pl,a}\right) \tag{8}$$

where $\sigma_a$ is the stress amplitude and $\varepsilon_{pl,a}$ is the plastic strain amplitude. Equation (7) and (8) are suitable to estimate parameters from several stabilized stress-strain curves [1].

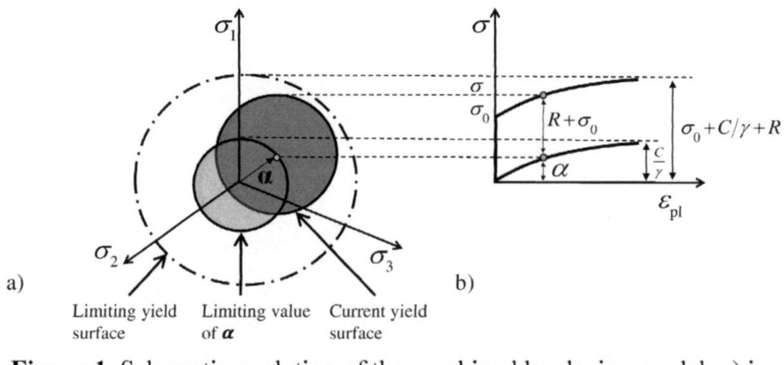


**Figure 1.** Schematic evolution of the combined hardening model: a) in stress space and b) in uniaxial tension.

The nonlinear isotropic hardening is expressed by the following equation [1]:

$$\mathrm{d}R = b\left(R_\infty - R\right)\mathrm{d}\varepsilon_{pl,acc} \tag{9}$$

where $R_\infty$ is the saturation value of the yield surface, $b$ is the speed of stabilization and $\varepsilon_{pl,acc}$ is the accumulated plastic strain. $R_\infty$ can be either positive or negative, giving rise to cyclic hardening or softening, respectively. The relation between $R$ and $\varepsilon_{pl,acc}$ is obtained after integration:

$$R = R_\infty\left[1 - \exp\left(-b\varepsilon_{pl,acc}\right)\right] \tag{10}$$

## 3. Experimental testing

Isothermal low cycle fatigue (LCF) tests were performed to characterize the cyclic stress-strain behaviour and fatigue life of CuAg alloy at three temperature levels (20 °C, 250 °C, 300 °C). Several specimens were tested at different strain ranges for each temperature. All isothermal LCF tests were carried out in strain controlled mode with a triangular loading waveform and with a fully reversed strain ratio $R_\varepsilon$=-1. The examined strain rate was 0.01 $s^{-1}$. Tests were interrupted before specimen failure when the maximal stress decreased by 80%. LCF testing at 20 °C were performed on the servo-hydraulic Instron-Schenck test rig with a nominal force ±250 kN, while the Instron extensometer with a gauge length of 12.5 mm and a range of ±5 mm was used to measure elongation during testing. The specimens were clamped by mechanical clamping grips. LCF testing at 250 °C and 300 °C were performed on the Instron test rig with a nominal force of ±100 kN. The temperature was applied by the induction heating system with a 10 kW medium frequency generator, Hüttinger TIG 10/300. The temperature was measured within the gauge length with a pre-stressed type K loop thermocouple. To measure an elongation at high temperatures an MTS extensometer, model 632.53F-14 with a gauge length of 12.6 mm and a range of ±1.8 mm was used. Test specimens were clamped by water cooled hydraulic clamping grips. Results of experimental tests are used for model calibration and estimation of the strain-life equations, as shown in the next paragraphs.

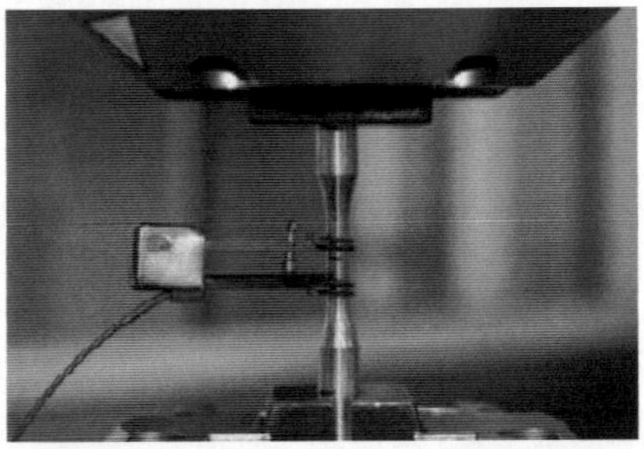

**Figure 2.** Mechanical clamping jaws with extensometer for room temperature.

**Figure 3.** Hydraulic clamping jaws, HT extensometer and heating apparatus.

## 4. Identification of material parameters

Firstly, the Young's modulus and the yield stress were estimated as they define the elastic region. The Young's modulus was determined by using both the tensile portion of the first hysteresis loop ($E_1$) and the stabilized stress-strain loop ($E_s$). As shown in Table 1, the Young's modulus is a function of the temperature and it also seems to depend on the applied strain amplitude ($\varepsilon_a$). Furthermore, the Young's modulus seems to slightly decrease during the applied cyclic loading.

**Table 1.** Estimated values of the Young's modulus.

| Temp. (°C) | $\varepsilon_a$=0.3% | | $\varepsilon_a$=0.4% | | $\varepsilon_a$=0.5% | | $\varepsilon_a$=0.7% | | Average values | |
|---|---|---|---|---|---|---|---|---|---|---|
| | $E_1$ (MPa) | $E_s$ (MPa) | $E_1$ (MPa) | $E_s$ (MPa) | $E_1$ (MPa) | $E_s$ (MPa) | $E_1$ (MPa) | $E_s$ (MPa) | $E_1$ (MPa) | $E_s$ (MPa) |
| 20 | 119700 | 116900 | 119900 | 115900 | 118200 | 113500 | 118800 | 114900 | 119080 | 110900 |
| 250 | 108500 | 93760 | 108600[a] | 98530[a] | 105400 | 90140 | 103900 | 85100 | 106600 | 94758 |
| 300 | 105600 | 97930 | 104300 | 98820 | 101900 | 95770 | 103400 | 87690 | 103800 | 94792 |

[a] $\varepsilon_a$=0.35%

The initial yield stress ($\sigma_0$) was identified as a point on the tensile portion of the first hysteresis loop where plastic strain occurs; the actual yield stress ($\sigma_{0*}$) was measured by using the stabilized stress-strain loop (at half number of cycles to failure). The evolution of the yield stress with increasing number of cycles enabled to determine the hardening or softening characteristics of the material, as can be seen in Table 2. In all evaluated cases $\sigma_{0*} < \sigma_0$, confirming a softening behaviour of the material.

**Table 2.** Estimated values of the yield stress.

| Temp. (°C) | $\varepsilon_a$=0.3% | | $\varepsilon_a$=0.4% | | $\varepsilon_a$=0.5% | | $\varepsilon_a$=0.7% | | Average values | |
|---|---|---|---|---|---|---|---|---|---|---|
| | $\sigma_0$ (MPa) | $\sigma_{0*}$ (MPa) | $\sigma_0$ (MPa) | $\sigma_{0*}$ (MPa) | $\sigma_0$ (MPa) | $\sigma_{0*}$ (MPa) | $\sigma_0$ (MPa) | $\sigma_{0*}$ (MPa) | $\sigma_0$ (MPa) | $\sigma_{0*}$ (MPa) |
| 20 | 121.64 | 87.95 | 138.76 | 91.44 | 117.97 | 84.25 | 154.94 | 114 | 130 | 86 |
| 250 | 111.34[a] | 51.71[a] | 103.46 | 50.86 | 140.31 | 50.98 | 80.68 | 53.47 | 113 | 50 |
| 300 | 124 | 44.7 | 103.3 | 48.23 | 116.37 | 40.49 | 122.5 | 43.48 | 110 | 45 |

[a] $\varepsilon_a$=0.35%

The estimation procedure of nonlinear kinematic parameters ($C_i$, $\gamma_i$) and nonlinear isotropic parameters ($R_\infty$, $b$) could be performed separately. In fact, for fully-reversed symmetrical stress cycles the kinematic model stabilizes after a single cycle. As the contribution of the isotropic model in the first cycle is actually small, it could be neglected [1]. Similarly the kinematic model does not influence significantly the subsequent cycles.

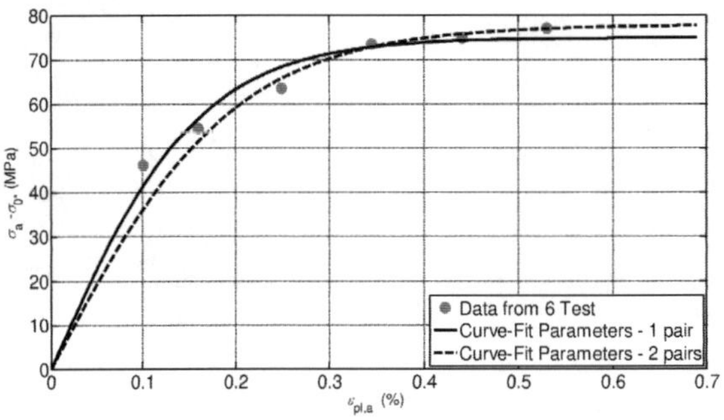


**Figure 4.** Curve fitting by the least squares method, using data of 6 hysteresis loops ($T$=20 °C).

Nonlinear kinematic hardening material parameters, $C_i$ and $\gamma_i$ were estimated after the determination of Young's modulus and yield stress. One and two pairs of material parameters were estimated using only the plastic regions of each stabilized cycle obtained under imposed $\varepsilon_a$. For a given stress-strain loop, the stress amplitude ($\sigma_a$), the plastic strain amplitude ($\varepsilon_{pl,a}$) and the actual yield stress ($\sigma_{0*}$) were measured at a stabilized cycle (half-life cycle) and plotted as in Figure 4. The procedure was repeated at each strain amplitude. Equations (7) and (8) were fitted to the measured points to obtain one and two pairs of material parameters $C_i$, $\gamma_i$, respectively. The material parameters obtained by the method here described are suitable to be used for different $\varepsilon_a$ values.

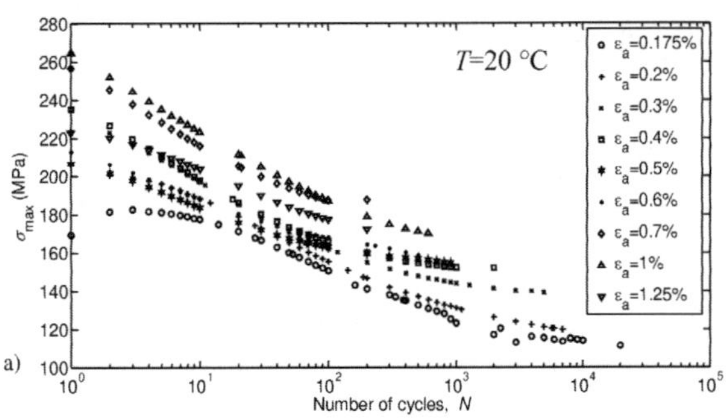


**Figure 5.** Stress amplitude vs. number of cycles.

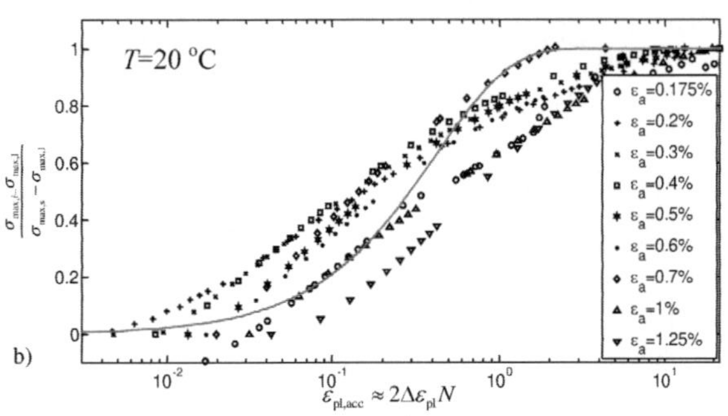


**Figure 6.** Fitting of Eq. (13) to find parameter $b$.

Nonlinear isotropic parameters were finally estimated. The maximum stress ($\sigma_{max}$) was measured for each stress-strain cycle and plotted against the number of applied cycles ($N$), see Figure 5, showing a softening behaviour. The procedure was repeated for each imposed strain amplitude and temperature. The saturation stress ($R_\infty$) was determined as the difference in the maximum stress of the first cycle ($\sigma_{max,1}$) and the stabilized one ($\sigma_{max,s}$). Figure 5 shows that the saturation stress ($R_\infty$) depends also on the applied strain amplitude; this behaviour was also observed in [4] for a nickel base superalloy. For each temperature an average $R_\infty$ is calculated, see Table 3. The speed of stabilization ($b$) is estimated by fitting to experimental data (see Figure 6) the following expression proposed by [1]:

$$\frac{\sigma_{max,i}-\sigma_{max,1}}{\sigma_{max,s}-\sigma_{max,1}} \cong \frac{R}{R_\infty} = 1-\exp\left(-b\varepsilon_{pl,acc}\right) = 1-\exp\left(-2b\Delta\varepsilon_{pl}N\right) \qquad (11)$$

where $\sigma_{max,i}$ is the current maximum stress for the $N^{th}$ cycle. Equation (11) is shown in Figure 6. A reasonable correlation is obtained, although a modification of the evolution rule of $R$ seems necessary. Estimated material parameters used in numerical simulations are shown in Table 3.

**Table 3.** Estimated material parameters used in numerical simulations.

| | | | | | One pair | | Two pairs | | | |
|---|---|---|---|---|---|---|---|---|---|---|
| Temp. (°C) | $E$ (MPa) | $\sigma_0$ (MPa) | $R_\infty$ (MPa) | $b$ | $C_1$ (MPa) | $\gamma_1$ | $C_1$ (MPa) | $\gamma_1$ | $C_2$ (MPa) | $\gamma_2$ |
| 20 | 119080 | 130 | -75.7 | 2.352 | 46250 | 617.2 | 38160 | 505.7 | 679.5 | 274 |
| 250 | 106600 | 113 | -80.2 | 3.894 | 45340 | 820.9 | 290600 | 8699 | 8772 | 349.5 |
| 300 | 103800 | 110 | -76.6 | 5.293 | 40080 | 832.8 | 27530 | 894.9 | 12760 | 731.1 |

## 5. Numerical simulation

Figure 7 a)-c) shows a comparison between experimental data and results of numerical simulations for a uniaxial cyclic loading. Simulations are performed with the material parameters identified in the previous Section. The nonlinear kinematic model uses both one and two pairs of ($C_i$, $\gamma_i$) parameters, estimated at 20 °C, 250 °C and 300 °C. A better agreement between experimental and simulated data is obtained with only one pair of parameter at 20 °C, 250 °C; while at 300 °C one and two pairs give quite the same shape of stress-stain loops.

The comparison between experimental and simulated stress-strain loops is also performed at different strain amplitudes ($\varepsilon_a$) to confirm that the estimated material parameters are suitable to be used over a wider interval of strain ranges. As can be seen in Figure 8, all three simulations are in good accordance with the experimental data. Based on the obtained results, a kinematic model with only one pair is used in the following simulations.

Finally, the combined kinematic and isotropic material model is used to simulate 50 cycles at strain amplitude $\varepsilon_a$=0.5%. Material parameters used in the simulation are taken from Table 3. Comparison between experimental stress-strain loops (1$^{st}$ and stabilized cycle) and simulated loops (every 5$^{th}$ cycle is plotted for a better overview) is shown in Figure 9. It has to be noted, however, that the small value of $b$ does not allow the material to reach the stabilized state within the simulated 50 cycles.

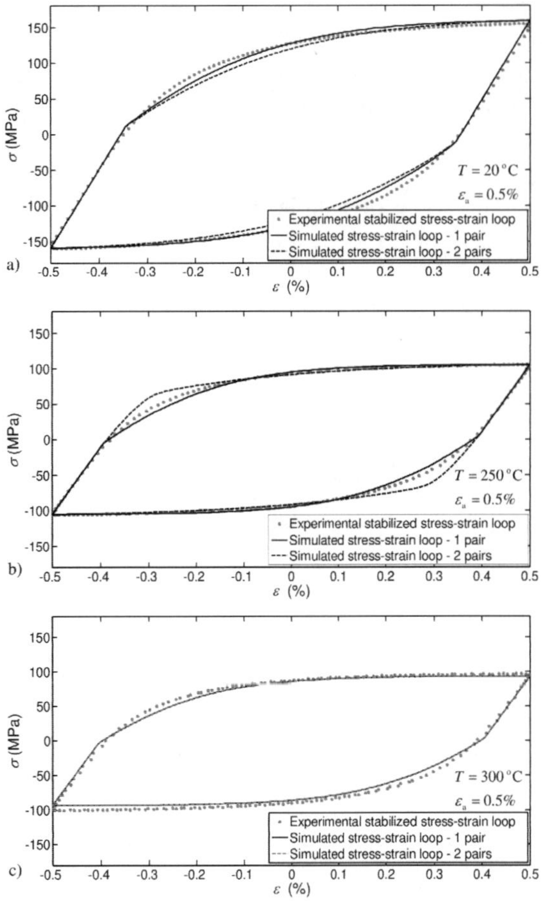


**Figure 7.** Comparison between experimental and simulated stress-strain loops obtained with one and two ($C_i$, $\gamma_i$) pairs at different temperatures.

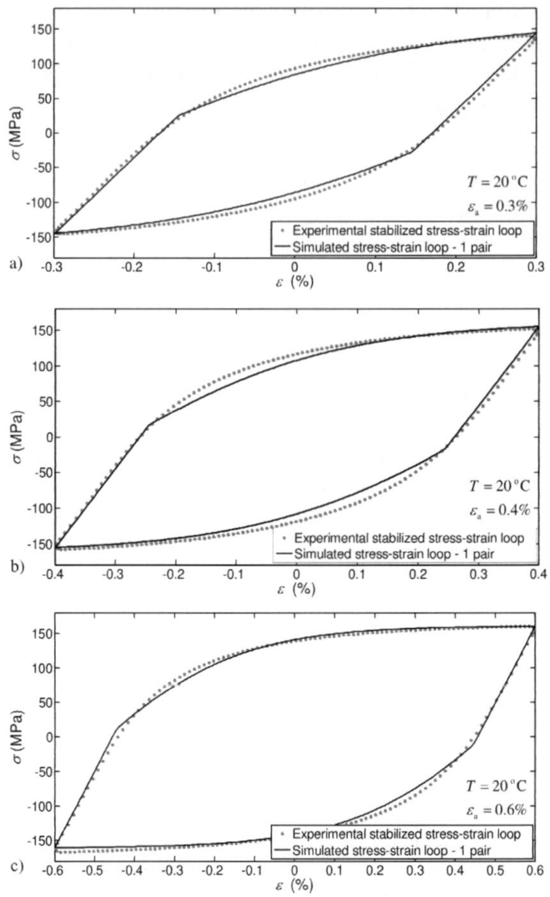


**Figure 8.** Comparison between experimental and simulated stress-strain loops for different values of $\varepsilon_a$.

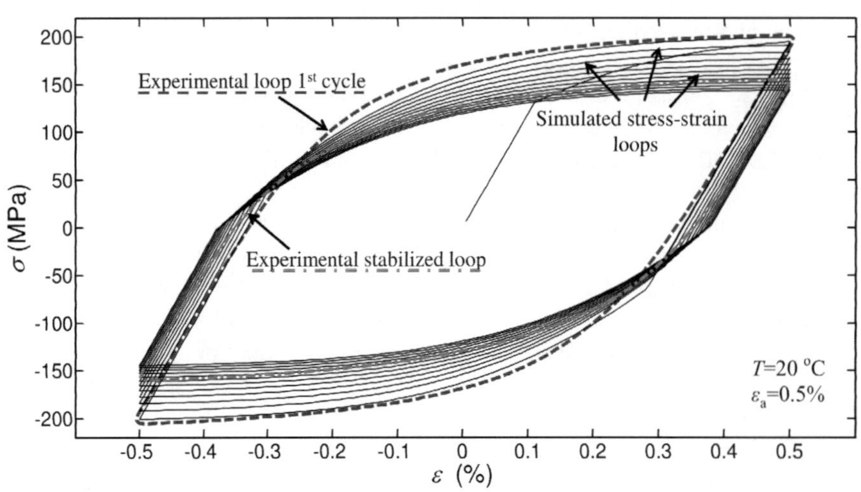


**Figure 9.** Comparison between experimental and simulated stress strain loops for combined material model.

## 6. Strain-life fatigue curves

The Manson–Coffin-Basquin equation is used to relate the cycles to failure $N_f$ to the strain amplitude [8]:

$$\frac{\Delta\varepsilon}{2} = \frac{\Delta\varepsilon_{el}}{2} + \frac{\Delta\varepsilon_{pl}}{2} = \frac{\sigma'_f}{E}(2N_f)^e + \varepsilon'_f(2N_f)^c \tag{12}$$

where $\Delta\varepsilon$, $\Delta\varepsilon_{el}$, $\Delta\varepsilon_{pl}$ are total, elastic and plastic strain ranges, respectively. The fatigue strength coefficient ($\sigma'_f$), fatigue strength exponent ($e$), fatigue ductility coefficient ($\varepsilon'_f$) and fatigue ductility exponent ($c$) are estimated from isothermal LCF data at different temperature, see Table 4.

**Table 4.** Estimated material parameters used in numerical simulations.

| Temp. (°C) | $\sigma'_f$ (MPa) | $e$ | $\varepsilon'_f$ | $c$ |
|---|---|---|---|---|
| 20 | 359.1 | -0.1031 | 0.07689 | -0.3942 |
| 250 | 253.6 | -0.1018 | 0.2942 | -0.5311 |
| 300 | 240.4 | -0.1068 | 0.4258 | -0.5708 |

Approximate methods are often used to estimate the strain-life curve from static strength data. An example is the Universal Slopes (US) equation, which assumes that, for all materials, elastic and plastic lines have unique slopes 0.12 and 0.6, respectively:

$$\Delta\varepsilon = \Delta\varepsilon_{el} + \Delta\varepsilon_{pl} = 3.5\frac{\sigma_{uts}}{E}N_f^{-0.12} + D^{0.6}N_f^{-0.6} \tag{13}$$

where $\sigma_{uts}$ is the ultimate tensile strength and $D$ is the ductility, which is related to the cross-area reduction in a tensile test. Although the US equation was originally proposed for steel at room temperature [9], an attempt is made here to apply it to CuAg alloy at 20 °C and 250 °C.
Influence of high temperatures and creep can reduce fatigue life by up to 90%. Therefore, 10% Rule assumes that at high temperature, only 10% of the life estimated by the US equation will actually be achieved. The upper bound of time life is then given by the US equation, while the lower bound of time life is given by the 10% Rule.

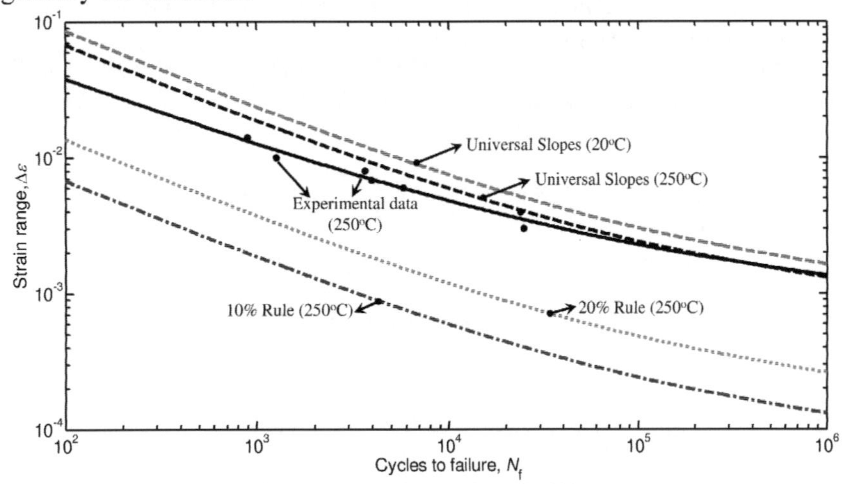


**Figure 10.** Comparison of strain-life curves.

Figure 10 shows a comparison of strain-life curves, estimated by different methods. The US equation (20 °C - upper bound) seems to be under conservative compared to the US (250 °C) curve. Therefore, to be on a safe side, the 10 % Rule (lower bound) and the 20% Rule (average) were calculated based on US (250 °C). It is worth noting how the 10% Rule seems to be over conservative, at least for this alloy at this temperature.

## 7. Conclusions

This work investigates the cyclic response and the fatigue life of a CuAg alloy. Isothermal low-cycle fatigue tests have been performed at different temperatures (20 °C, 250 °C, 300 °C) to determine the stress-strain response and the experimental fatigue life. Several material models (nonlinear kinematic, nonlinear isotropic) have been then calibrated with experimental results. In particular, parameters for the nonlinear kinematic model (one and two pairs of $C_i$, $\gamma_i$) have been estimated from LCF experimental data, as well as parameters for nonlinear isotropic model ($R_\infty$, $b$). Numerical simulations for uniaxial loading with the identified material parameters have been compared with experimental results, showing a quite good agreement. Finally, the strain-life curves estimated from experimental data have been also compared with approximate methods (Universal Slopes Equation, 10% Rule), which estimate the strain-life curve from static strength data of a tensile test. A satisfactory agreement has been observed.

The results presented in this work would thus permit/allow one to perform an effective service life assessment of steel-making components made of CuAg alloy, when a design approach based on Finite Element modelling and elasto-plastic analysis have to be followed.

## References

[1] Lemaitre J and Chaboche J L 1990 *Mechanic of Solid Materials* (Cambridge: Cambridge University Press)

[2] Chaboche J L 2008 A review of some plasticity and viscoplasticity constitutive theories *Int. J. Plasticity* **24** 1642–1693

[3] Chaboche J L and Rousselier G 1983 On the plastic and viscoplastic constitutive equations – Part I: Rules developed with internal variable concept *J. of Pressure Vessel Technol.* **105** 153-158

[4] Zhao L G, Tong B, Vermeulen B and Byrne J 2001 On the uniaxial mechanical behaviour of an advanced nickel base superalloy at high temperature *Mech. Mater.* **33** 593-600

[5] Broggiato G B, Campana F and Cortese L 2008 The Chaboche nonlinear kinematic hardening model: calibration methodology and validation *Meccanica* **43** 115–124

[6] Franulovic M, Basan R and Prebil I 2009 Genetic algorithm in material model parameters' identification for low-cycle fatigue *Comput. Mater. Sci.* **45** 505–510

[7] Halama R, Sedlak J and Šofer M 2012 Phenomenological modelling of cyclic plasticity Numerical Modelling edited by Peep Miidla *Intech* 329-351

[8] Manson S S and Halford G R 2006 *Fatigue and Durability of Structural Materials* ASM International

[9] Manson S S 1968 A simple procedure for estimating high-temperature low-cycle fatigue, *Exp. Mechanics* **8** 349-355

# Analysis of the thermo-mechanical deformations in a hot forging tool by numerical simulation

**R.L-Cancelos[1], F. Varas[2], E. Martín[3], I. Viéitez[2]**

[1]Departamento de Ingeniería de los Materiales, Mecánica Aplicada y construcción. E. de Ing. Industrial, Universidad de Vigo, Campus Marcosende, E-36310 Vigo
[2] Departamento de Matemática Aplicada II. E. de Ing. de Telecomunicación, Universidad de Vigo, Campus Marcosende, E-36310 Vigo
[3] Departamento de Ingeniería Mecánica, Máquinas y Motores Térmicos y Fluidos. E. de Ing. Industrial, Universidad de Vigo, Campus Marcosende, E-36310 Vigo

E-mail: rlopezcancelos@uvigo.es

**Abstract**. Although programs have been developed for the design of tools for hot forging, its design is still largely based on the experience of the tool maker. This obliges to build some test matrices and correct their errors to minimize distortions in the forged piece. This phase prior to mass production consumes time and material resources, which makes the final product more expensive.

The forging tools are usually constituted by various parts made of different grades of steel, which in turn have different mechanical properties and therefore suffer different degrees of strain. Furthermore, the tools used in the hot forging are exposed to a thermal field that also induces strain or stress based on the degree of confinement of the piece. Therefore, the mechanical behaviour of the assembly is determined by the contact between the different pieces.

The numerical simulation allows to analyse different configurations and anticipate possible defects before tool making, thus, reducing the costs of this preliminary phase.

In order to improve the dimensional quality of the manufactured parts, the work presented here focuses on the application of a numerical model to a hot forging manufacturing process in order to predict the areas of the forging die subjected to large deformations.

The thermo-mechanical model developed and implemented with free software (Code-Aster) includes the strains of thermal origin, strains during forge impact and contact effects. The numerical results are validated with experimental measurements in a tooling set that produces forged crankshafts for the automotive industry. The numerical results show good agreement with the experimental tests. Thereby, a very useful tool for the design of tooling sets for hot forging is achieved.

## 1. Introduction

Nowadays forging process remains a widely used technology [1]. Since the development of computers and their diffusion in companies of metal forming, there have been many researchers and developers who have designed models and software capable of simulating processes of forging [2].

Within the broad spectrum covered by forging problems, some researchers have focused on evaluating the elastic deformation occurring in forging tools by the method of finite element. One of the early work in this field is done by Lange et al. [3], where an analysis of the causes of failure of forging dies due to fatigue phenomena, with certain simplifications assumed (as an axisymmetric problem is carried out). A further work is published by Nagao et al. [4], they perform an analysis of the states of stress that occur in the inserts during the forging operation for automotive industry, also solved for an axisymmetric problem. Qin and Balendra [5] study the influence of the elasticity of the die relative to the final dimensions of the part in the case of extrusion forging.

The work presented by Joun et al. [6] focuses on the analysis of sequence control for hot forging considering the behavior of the die as infinitely rigid. With this simplification Joun et al. [7] studied the elastic behavior of the die in cold forging applications. First, they calculated the forging pressure under the assumption of die infinitely rigid. Then, the calculated pressures are imposed on the die modeling its elastic behavior. They also considered the various parts of the tool and applied a penalization algorithm to solve the contact.

Ou and Armstrong [8] study the important influence of the elasticity of the forging die and the thermal distortion in the process of manufacturing a titanium turbine blade for the aeronautic industry. But they only focus on the die, they do not include all the set tool and the geometry of the piece is relatively simple.

Jun et al. [9] present a work with a practical approach to systematically estimate the geometric dimensions of pieces manufactured by cold forging. They make the assumption that the matrices are rigid to calculate the forging stress. The characteristics of the studied process allow them to use an axisymmetric model.

In the present work a valid model for the evaluation of induced stress and strain in hot forging dies is solved, considering the elastic behavior of the tool and the influence of the thermal field. The model is validated by acquiring data during the manufacture of crankshafts for the automotive industry

## 2. Description of the process

The shaping of crankshafts is performed in a 4000T mechanical press in two stages. Therefore two pairs of matrices are used (called prior and final die). The control operation is semiautomatic and it is performed by an operator. Between each forged crankshaft two operations are performed, the cooling and the lubrication of the surfaces of the dies.

The forging tool is formed of two parts, the upper and the lower, being both symmetric. This study focuses on the behavior of the lower part because it is the one exposed to a larger thermal field.

In figure 1 a simplified scheme of the lower part used by CIE Galfor S.A. company is shown. Geometrical details are not shown for confidentiality reasons. This tool is formed by 12 pieces built in different steel qualities. The set broadly contains a holder (P1), a support slab (P2), the dies of the first and second press stroke (D1 and D2 respectively) and a third lateral block (P3) that fills in the empty space. Most of the pieces simply rest lying ones on the others and are fixed by means of elastic joints named disc springs that are responsible for the compression of the set D1, D3 and P3. These joints are located between the wedges (P6-1, P6-2, P6-3 and P7) and the dies and the piece P3.

Following European codification (EN 10027/2:1992) the dies D1 and D2 are made of steel W Nr 1.2344 (table 1 shows the physical properties and their dependency with the temperature), the piece P1 of W Nr 1.27381 (table 1) and the rest of W Nr 1.27141 (table 1).

**Table 1.** Dependency with the temperature of the physical properties of steels.

| *Steel* | $T\ (°C)$ | $E(GPa)$ | $\rho\left(\frac{kg}{m^3}\right)$ | $\lambda\left(\frac{W}{m°C}\right)$ | $c\left(\frac{J}{kg\ K}\right)$ | $\alpha\left(\frac{mm}{mm\ °C}\right)$ |
|---|---|---|---|---|---|---|
| 1.2344 | 20 | 210 | 7800 | 25 | 460 | 11.0 |
| | 300 | 188 | 7726 | 27.9 | 526 | 11.8 |
| 1.2738 | 20 | 210 | 7800 | 25 | 460 | 12.8 |
| | 300 | 193 | 7730 | 33.4 | 557 | 13.8 |
| 1.2714 | 20 | 210 | 7800 | 25 | 460 | 12.5 |

| | | | | | |
|---|---|---|---|---|---|
| 300 | 194 | 7710 | 36.7 | 513 | 13.5 |

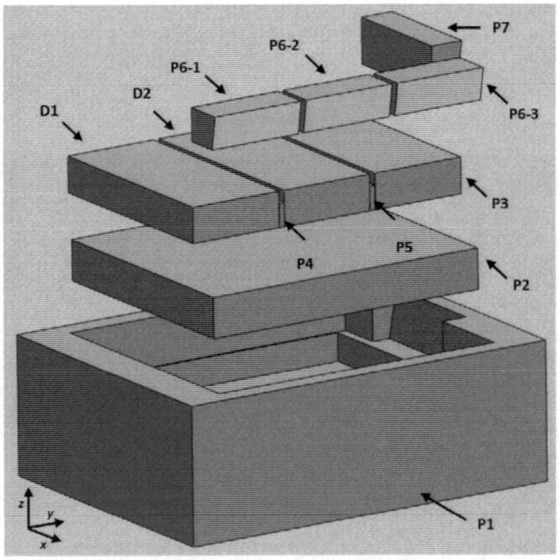


**Figure 1.** Schematic representation of the lower tool. D1 and D2 indicate the position of the dies of the first and second press stroke respectively.

A picture of the lower part tool in its work position is shown in figure 2.

**Figure 2.** Lower part tool located in its work position.

## 3. Implemented model

The nature of the physical problem requires solving two models: a thermal and a mechanical model. Therefore the model presented solves the coupling between the thermal problem and thermo-mechanical as a one-way problem. This simplification is done because it is considered that there is no heat generation on the contact surfaces and the opening of the contact due to the press strokes does not influence the heat transfer between the different pieces because the characteristic stroke time is very small (0.1s).

### *3.1. Thermal Model*

Although the problem is essentially a transitory one, the resolution of the thermal problem arises as a steady state (eq 1) because the forging characteristic time is much shorter than the thermal inertia.

$$0 = div\ (k\nabla T)\ \text{in}\ \Omega \tag{1}$$

The heat source for the tool set are the lower die surfaces (that remain some time in contact with the hot billet). To model its effect thermographies (figure 3 and 4) are used. Therefore the temperature distribution of the die surfaces are imposed as a Dirichlet boundary condition (eq 2).

$$T = T_0\ \text{on}\ \Gamma_D \tag{2}$$

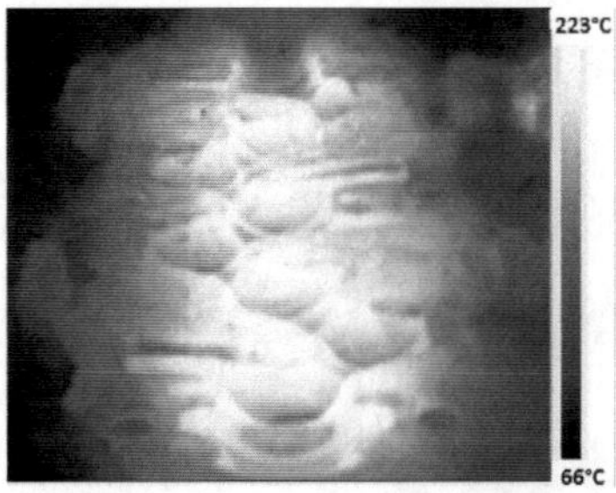


**Figure 3.** Thermography of prior die.

**Figure 4.** Thermography of final die.

The internal surfaces that are not in contact with others are considered as insulated (eq 3).

$$k\,\frac{\partial T}{\partial n} = 0\ \text{on}\ \Gamma_i \tag{3}$$

As observed in figure 2, the front of the tool is exposed to the environment. Therefore the heat flux through the boundary is solved by a model of convection (eq 4) to the environment. The value of the coefficient $h$ is 10 $\frac{W}{m^2\,k}$ (see Incropera [10] and Chao et al. [11]) and the temperature of the environment is $20°C$.

$$k\,\frac{\partial T}{\partial n} = h\,(T_\infty - T)\ \text{on}\ \Gamma_f \tag{4}$$

The rest of the external boundaries are in contact with the press. The strategy for modeling the heat transfer is performed by using an equivalent heat transfer coefficient (eq 5) Cooper et al. [12]. The value used is 40 $\frac{W}{m^2k}$ [13] while the selected temperature of the heat sink for the press foundation is $T_\infty = 15°C$.

$$k\,\frac{\partial T}{\partial n} = h_{eq}\,(T_\infty - T)\ \text{on}\ \Gamma_f \tag{5}$$

*3.2. Mechanical Model*

As the nature of the deformation is elastic it is assumed that the behavior of different materials is elastic and isotropic (as Jun et al [9]), being the material properties variable with temperature (see table 1).

As the tool is constituted with several pieces, the surfaces that come in to contact must fulfill equation 6. To model the contact between the different parts a penalization algorithm is used as in the work of Joun et al [7]. The total number of contacts are 35, 16 in $x$ direction, 10 in $y$ direction and 9 in $z$ direction.

$$\bar{\bar{\sigma}}^i(\vec{u})\,\vec{n}^i = p_N\,\vec{n}^i + p_T\,\vec{t}^i\ in\ \Gamma_c^i \tag{6}$$

where $\Gamma_c^i$ are the surface boundaries of each piece that comes into contact with the surface of a different one. $p_N$ stands for the normal stress in the contact surfaces while $p_T$ represents the tangential stresses.

The final form of the forging piece is conditioned by the shaping of the footprint at the moment of maximum load. Thus, the mechanical problem is solved for the instant under the maximum load. In addition, it is also necessary to decouple the problem of the forged piece from the elastic deformation of the tool due to the size of the problem and the computational cost. This procedure has been used in previous works like Jun et al. [9].

The calculation of forging stress (see eq. 7) under the assumption of rigid die was made by the company through the use of the commercial software FORGE3© (see results on figure 5). The imposed stress on the final die is described by the equation 7 as a boundary condition of Neumann type.

$$\bar{\bar{\sigma}}^i(\vec{u})\,\vec{n}^i = \vec{f}_\Gamma^{FOR} \; in \; \Gamma_s^D \tag{7}$$

where $\Gamma_s^D$ is the footprint surface of the die and $\vec{f}_\Gamma^{FOR}$ is the known forge surface stress.

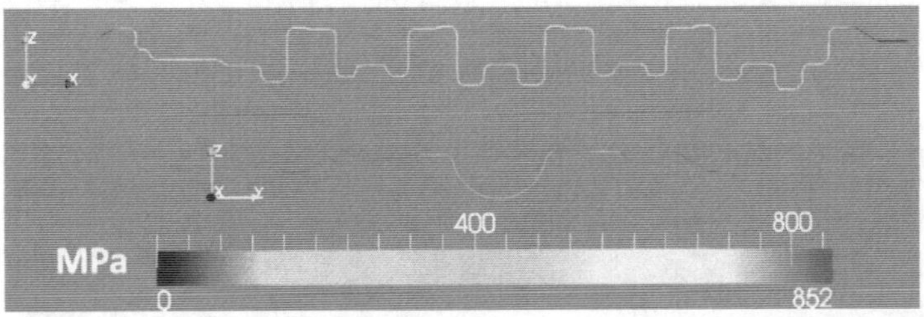


**Figure 5.** Stresses field over the longitudinal section of die D2 obtained with FORGE3®.

The lower boundary of the tool is placed, with no restriction, over a steel surface. Thus, this surface can expand in x and y directions (in order to manipulate the tool there is a gap in y direction). To mimic this behavior the movement is only blocked in z direction (eq. 8). In order to prevent an ill-posed problem it is necessary to include certain reactions in x and y directions in the base of the tool that permit the thermal expansion but that do not alter the mechanical problem. This is achieved by including in the model very low stiffness springs in the nodes of the tool base. These springs are modeled as nodal discrete elements without dimensions (0D).

$$u_z^{P1} = 0 \; in \; \Gamma_u^{P1} \tag{8}$$

Because of the type of fixing between the back part of the tool and the outer piece, this part is modeled assuming no movement in the x direction, and then, simulating the contact problem between the tool and the exterior.

$$\vec{u}^{P1} = 0 \; in \; \Gamma_u^{P1(-x)} \tag{9}$$

In all the wedges used in the tool, certain disc springs are placed for the elastic coupling between the wedges and the dies D1 and D2 and the supplementary piece P3, being in charge of absorbing the different expansions between the pieces while keeping them at work position. Due to the difficulty that a detailed numerical model for the disc springs entails, a simplification was done, modeling each disc spring as a cylindrical ring with an equivalent elastic module (dependent on the temperature) that reproduce the same rigidity than the real disc springs, whose values are indicated in table 2. The equivalent elastic module is obtained using the real force exerted on the set and its displacement.

**Table 2.** Equivalent elastic module (as a function of temperature) of the disc springs located between the wedges and the dies.

| $T\,(°C)$ | $E_{equi}\left(\frac{kN}{mm^2}\right)$ |
|---|---|
| 20 | 82 |
| 300 | 75 |

Each wedge is fixed to the group through two bolts. Due to the reduced dimensions of the bolts in relation to the complete tool, the elements are not taken into account in the geometrical model but the forces induced by them are. The pressing force in each bolt is of 287N (estimated taking into account the contact surface of the head of the bolts)

## 4. Experiment description

To evaluate the mechanical behavior of the final die is decided to monitor its base due to the impossibility of placing sensors on the surface of the footprint. The physical quantities evaluated are the unitary strains and the temperature. For this purpose two types of sensors are used, 8 self-compensated strain gauges (G1 to G8) and 7 thermocouples (T1 to T7) type k. To accommodate the sensors is necessary to perform a machining of the base die. In figure 6 it is shown the position of the sensors on the basis of the final die (D2).

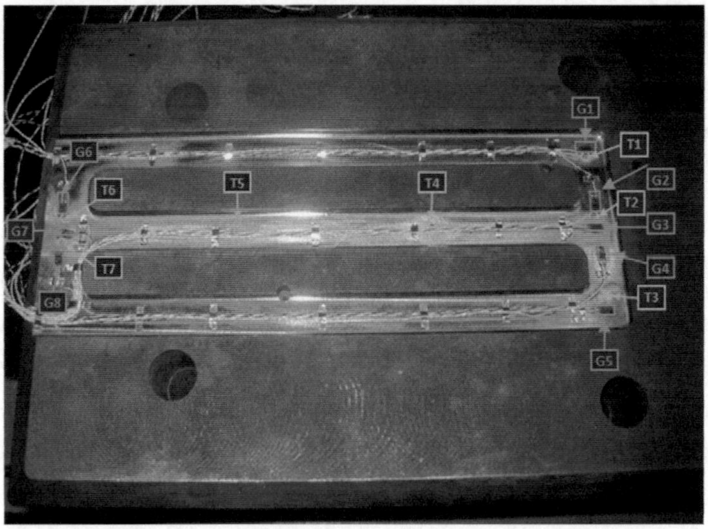


**Figure 6.** Sensors position on the base of the final die (D2).

The temperature values are used to validate the thermal problem and also to correct the strain signals registered as explained in Watson [14] due to the thermal output.

The even gauges recorded the longitudinal strain while the odd gauges the transversal one. The gauges G1 to G5 are positioned at the central zone of the base and the G6 to G8 at an extreme. The signal of the gauge G5 was lost at the beginning of the experiment.

Data acquisition was performed during the manufacturing of a series of crankshafts, and a total of forty forging strokes for the final die were recorded.

## 5. Experimental and numerical results for the thermal problem

Figure 7 shows the temperature value for each thermocouple and each final forging stroke. It can be observed the temperature variation and the absence of a defined pattern.

The strong variability of the process makes necessary to determine the temperature field by a representative mean value, so a statistical treatment of the data is performed. The results of this statistical analysis are shown in Figure 8 using a box plot for each thermocouple. The maximum temperatures were recorded for T4 and T5 thermocouples (188°C). The measures for the thermocouple T5 had the highest variability. The thermocouples of the center zone (T1-T3) recorded a temperature around 165°C, while the thermocouples of the extreme zone recorded the minimum average temperature, around 157°C.

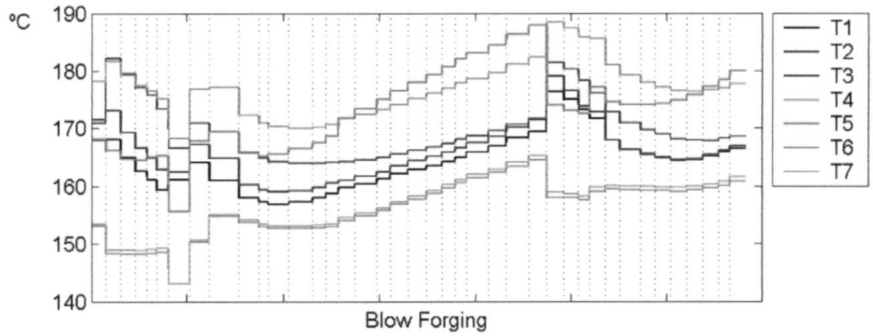


**Figure 7.** Signal recorded by thermocouples for each final forging stroke.

The numerical result of the temperature field inside the forging tool is shown in figure 9. The highest temperature values are located in the dies area and decrease toward the external boundaries.

Figure 8 shows also the numerical values extracted from the probe points located on the same coordinates as the thermocouples, being the values represented by blue blades. Although there are some differences between numerical results and the mean temperatur, all numerical values are within the range of temperatures recorded for each thermocouple. Therefore it is considered that the proposed thermal model reproduces the recorded average field and is used to evaluate the thermal contribution to the mechanical problem.

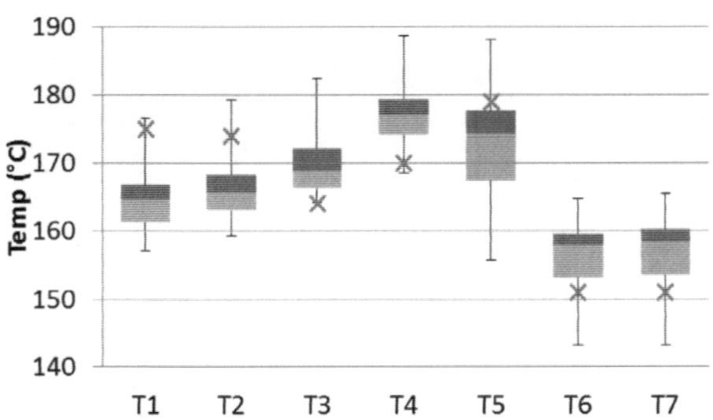


**Figure 8.** Comparison between temperature measurements registered (box plot) and numerical values (blue blades).

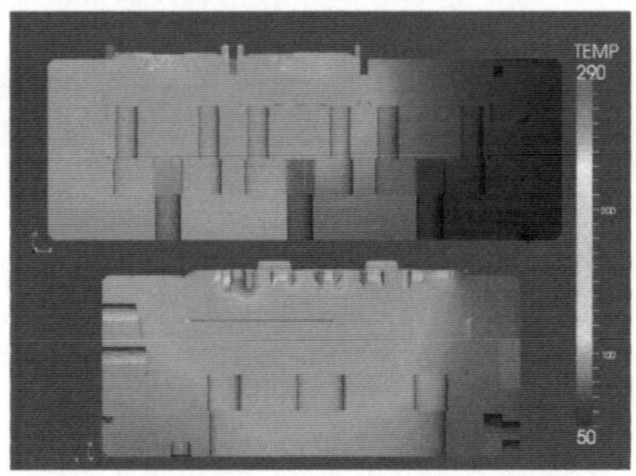


**Figure 9.** Thermal field simulation result for the thermal field.

### 6. Experimental and numerical results for the mechanical problem

The data recorded and corrected by the strain gauges are shown in figure 10 for the forty press strokes. The corrections are made based on the thermal output and the free expansion [15]. The last one is estimated based on the minimum temperature value extracted from the result of the thermal problem (62.70°C) and the coefficient of thermal expansion of the steel. The free expansion estimated value is 470με.

For each signal recorded from each strain gauge, two types of strain can be extract. The basis signal corresponds to the strain of thermal origin and the peak signal corresponds to the maximum strain at the time of maximum load. The recorded data is processed statistically just as in the case of temperature data. Then, two average fields were calculated, the strain corresponding to the thermal field and the strain corresponding to the thermal field and the press stroke. These results are represented by a box plot, in figure 11, that includes the results for the thermal strain, and in figure 12 that shows the results for the thermal strain and the strain produced by the press stroke.

Mean values of the thermal deformation for each gauge have a similar value, around 1650με and it is similar in the central zone and in the extreme. As for the values corresponding to blow forging are also similar between zones, except for the recorded values of the gauge G7. These decrease their values with respect to strain of thermal origin. On the other hand, the recorded values of the even gauges are greater than the values of the odd gauges. Therefore, transverse strain of the die is greater than the longitudinal. The value of the thermal deformation represents a third of total deformation.

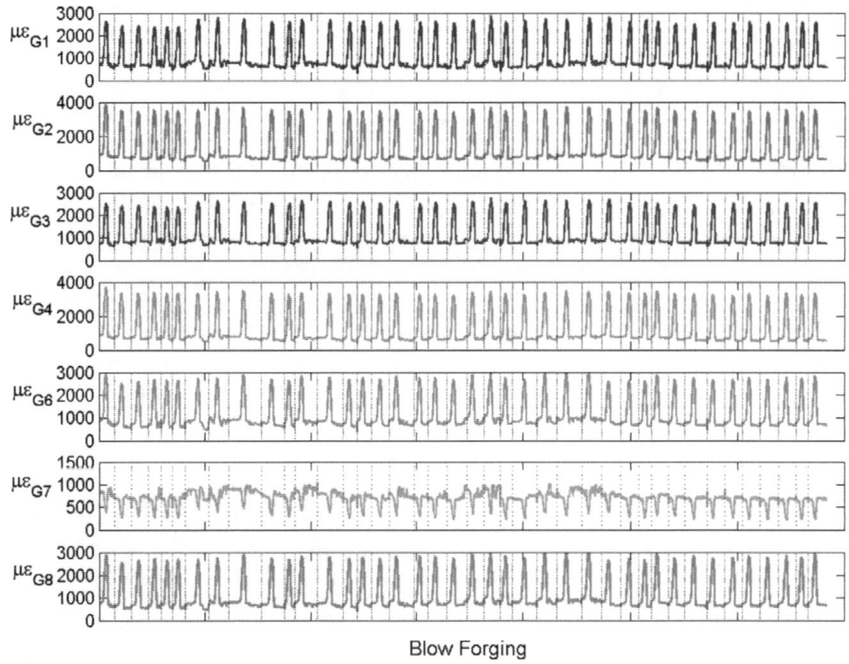


**Figure 10.** Register of the corrected measures of the strain gauges for the experimental data.

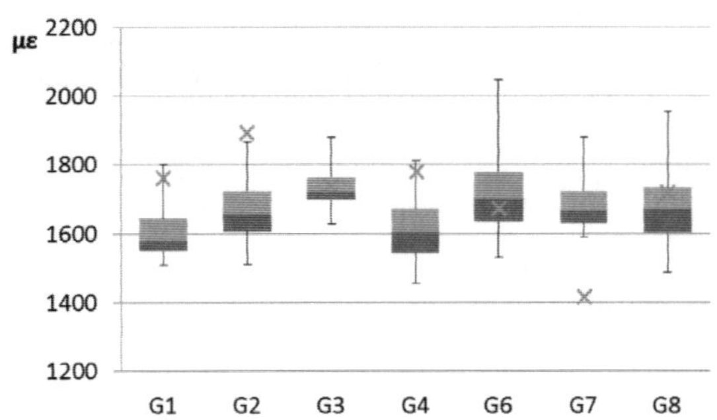


**Figure 11.** Strain of thermal origin for the experimental data and simulated results.

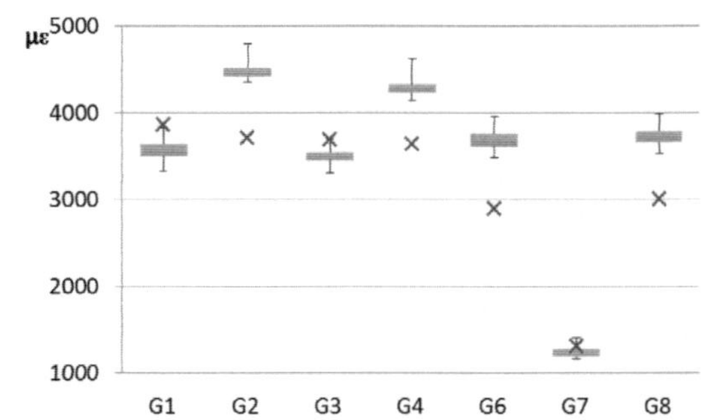


**Figure 12.** Strain at the instant of maximum load for the experimental data and simulation results.

Figures 11 and 12 also includes the numerical result of the simulation (represented by blades). The results for the strain of thermal origin in the probe points are similar to the values recorded. The values of G1, G3, G4, G6 and G8 are within the range of measured values. The largest relative errors correspond to the calculated values for G2 and G7 (14% y 17% respectively).

As for the results at the instant of maximum load, the calculated values for the gauges G1 and G3 are those falling within the measured range. The biggest relative errors correspond to the results for the G6 and G8 probe points (26% y 24% respectively), followed by G2 and G4 (20% y 18% respectively). The numerical values corresponding to the central zone are similar. This situation is inherited from forging stress calculations on the assumption of rigid die resulting in a homogeneous efforts on the central die.

Despite the difference between the measured and calculated values for both cases (thermal and thermo-mechanical) it is considered that the numerical result is optimal, because the relative errors calculated are less than 26%. This validates the proposed model.

Focusing now on the numerical results of the validated model, figure 13 shows the calculated displacement for the set tool along two transverse cuts. The positive displacement of the prior die is due to the thermal expansion. The final die is placed at the top center of the assembly and it is shown individually in figure 14.

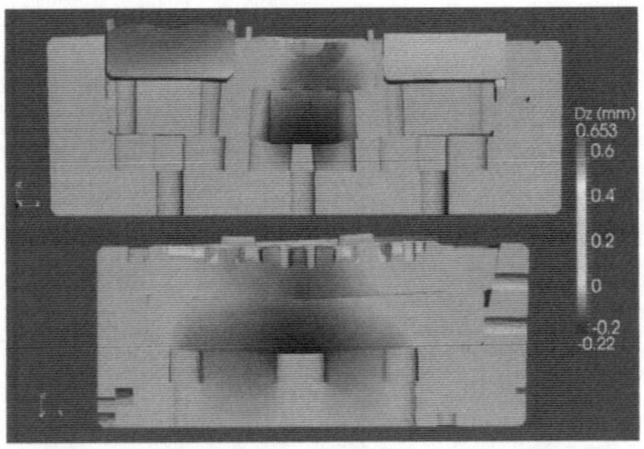


**Figure 13.** Displacement simulation result of the set tool (x50).

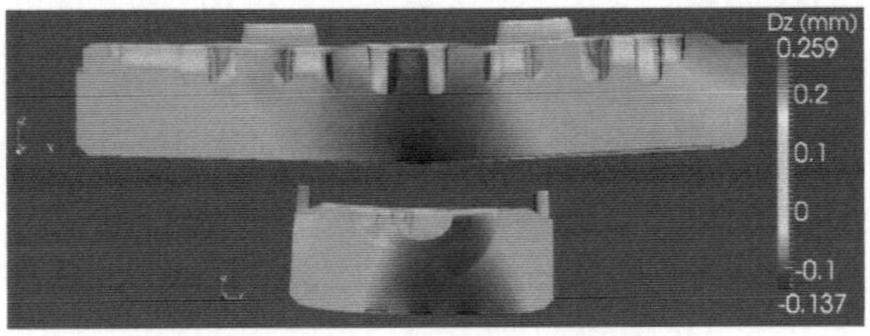


**Figure 14.** Displacement simulation result of the final die (x50).

Figure 15 compares thermal and mechanical deformations represented by a longitudinal section of the surface of the final footprint. Thus, based on the results, the contribution of the thermal field to the total strain field is significant and it must be taken into account when evaluating the mechanical behavior of the tools used in hot forging.

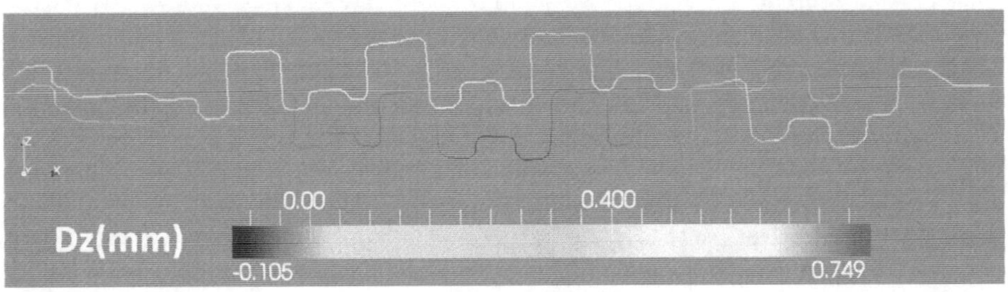


**Figure 15.** Comparison between displacement of thermal origin and thermo-mechanical (x50).

## 7. Conclusions

This paper presents a model to estimate the strain that occur in hot forging tools and it is validated with experimental measures obtained in a hot forging manufacturing process.

From the experimental results, it can be concluded that the thermal field does not show a fully defined pattern, there are fluctuations throughout the production process. This indicates that the thermal problem is very complex but making a series of careful simplifications it can be calculated an averaged thermal field representative of the real problem. The proposed thermal model reproduces the recorded average field and, in consequence, is used to evaluate the thermal contribution to the mechanical problem.

There are also fluctuations in the strain field but these fluctuations cannot be directly related to the fluctuations of the thermal field. There are many dependent factors besides temperature such as the temperature of the steel billet, the degree of lubrication of the die, positioning parts, etc. That are not taken into account due to their complexity.

Currently the design process of forging dies is a costly process that consists of several test phases. The model proposed in this paper intends to reduce the design phase and therefore its costs.

## 8. Acknowledgement

This work has been carried out in collaboration with company CIE Galfor S.A. (Ourense, Spain) and was partially supported by funds of Xunta de Galicia (Galicia, Spain) and CDTI (Ministerio de Economía e Industria, Spain) under the project from the "Design and Industial Productions" program, code 10DPI126E and MTM2013-47800-C2-2P.

## References

[1] T. Takami, Production engineering strategies and metalworking at Toyota motor corporation, Procedia Engineering 81 (2014), 5-17.

[2] P. Hartley, I. Pillinger, Numerical simulation of the forging process, Computer Methods in Applied Mechanics and Engineering 195 (48) (2006), 6676-6690.

[3] K. Lange, A. Hettig, M. Knoerr, Increasing tool life in cold forging through advanced design and tool manufacturing techniques, Journal of Materials Processing Technology 35 (3) (1992), 495-513.

[4] Y. Nagao, M. Knoerr, T. Altan, Improvement of tool life in cold forging of complex automotive parts. Journal of Materials Processing Technology 46 (1) (1994), 73-85.

[5] Y. Qin, R. Balendra, FEM simulation of the influence of die-elasticity on component dimensions in forward extrusion. International Journal of Machine Tools and Manufacture 37 (2) (1997), 183-192.

[6] M. Joun, H. Moon, R. Shivpuri, Automatic simulation of a sequence of hot-former forging processes by a rigid-thermoviscoplastic finite element method. Journal of Engineering Materials and Technology 120 (4) (1998), 291-296.

[7] M. Joun, M. Lee, J. Park, Finite element analysis of prestressed die set in cold forging. International Journal of Machine Tools and Manufacture 42 (11) (2002), 1213-1222.

[8] H. Ou, C. Armstrong, Die shape compensation in hot forging of titanium aerofoil sections. Journal of Materials Processing Technology 125 (2002), 347-352.

[9] B. Jun, S. Kang, M. Lee, R. Park, M. Joun, Prediction of geometric dimensions for cold forgings using the _nite element method. Journal of Materials Processing Technology 189 (1), (2007) 459-465.

[10] F. P. Incropera, Introduction to heat transfer. John Wiley & Sons. (2011)

[11] Y. J. Chao, X. Qi, W. Tang, Heat transfer in friction stir welding-experimental and numerical studies. Journal of Manufacturing Science and engineering 125 (1) (2003), 138-145.

[12] M. Cooper, B. Mikic, M. Yovanovich, Thermal contact conductance. International Journal of Heat and Mass Transfer 12 (3) (1969), 279-300.

[13] E. J. Caron, K. J. Daun, M. A. Wells, Experimental heat transfer coefficient measurements during hot forming die quenching of boron steel at high temperatures. International Journal of Heat and Mass Transfer 71 (2014), 396-404.

[14] R. Watson, in: W. Sharpe (Ed.) Springer Handbook of Experimental Solid Mechanics, 1st Edition. Springer, Ch. Bonded Electrical Resistance Strain Gages, Springer, 2008, pp. 297-302.

[15] R. B. Hetnarski, M.R. Eslami, 2009. Thermal stresses: advanced theory and applications, first ed., Springer, (2009).

# Laser-assisted metal spinning for an efficient and flexible processing of challenging materials

**C Brummer[1], S Eck[2], S Marsoner[2], K Arntz[1], F Klocke[1]**

[1]Fraunhofer Institute for Production Technology IPT, Steinbachstr. 17, 52074 Aachen, Germany
[2]Materials Center Leoben Forschung GmbH, Roseggerstr. 12, 8700 Leoben, Austria

[1]E-mail: christoph.brummer@ipt.fraunhofer.de

**Abstract.** The demand for components made from high performance materials like titanium or nickel-based alloys as well as strain-hardening stainless steel is steadily increasing. However, conventional forming operations conducted on these materials are generally very laborious and time-consuming. This is where the limitations of metal spinning also become apparent. Using a laser to apply heat localized to the forming zone during metal spinning facilitates to enhance the formability of a material. In order to analyse the potential of the new manufacturing process, experimental investigations on laser-assisted shear forming and multi-pass metal spinning have been performed with austenitic stainless steel X5CrNi18-10, nickel-based alloy Inconel 718 and titanium grade 2. It could be demonstrated that the formability of these materials can be enhanced by laser-assistance. Besides the resulting enhancement of forming limits for metal spinning of challenging materials, the forming forces were reduced and the product quality was improved significantly.

## 1. Introduction

### *1.1. Motivation and state-of-the-art of metal spinning*

The demand for highly stressed components for the use in several industrial sectors is forcing manufacturers to use more innovative materials. The call for reduction in fuel consumption in automotive industry led to an increased demand for thin sectioned parts with reduced weight and improved corrosion resistance. Titanium with a density of 60% of steel while maintaining a high specific strength (i.e. ultimate tensile strength UTS divided by density) has gained in importance in production [1]. Consequently, commercially pure titanium grade 2 (3.7034) was used as a representative for this class of materials for the investigations presented in this work and will be denominated as "Ti grade 2" throughout the rest of this work. For high-temperature applications nickel-based alloys are frequently used. Nickel-based alloys provide a high corrosion resistance and maintain a high material strength at elevated temperatures [2]. As prominent representative of Ni-based alloys, Inconel 718 (2.4668) has been investigated in this work. Inconel 718 is used in gas turbines, aircraft engines, spacecraft, nuclear reactors, pumps, and tooling [3]. Because of the high costs of titanium and nickel-based alloys, stainless steel still is the most common material if the

[1] Corresponding author

demands of the application can be met with such a material. A prominent representative is the widely used austenitic steel X5CrNi18-10 (1.4301) and has thus been investigated in this work.

Due to their outstanding mechanical properties, materials like titanium, nickel-based alloys and stainless steel are generally regarded as being difficult to machine. Product developers and their suppliers are therefore faced with the challenge of realizing complex parts economically using these materials. In contrast to other forming technologies metal spinning as incremental forming technology offers advantages such as high geometrical flexibility, short set-up times and low tooling costs as well as the option of performing several process steps in one single clamping [4]. Metal spinning processes thus represent a technologically and economically efficient alternative to other manufacturing techniques [5] and offer the possibility for near net shape manufacturing of thin sectioned lightweight parts [4]. They are also particularly suitable for machining high-strength materials since the tools affect only a small localized area and therefore require significantly lower forming forces than those needed for deep drawing, for example [6]. However, even metal spinning technology is pushed to its limits when processing difficult to machine materials.

In industrial practice, mainly two different approaches for heat treatment have been adopted to expand the forming limits, even though both show considerable drawbacks in terms of production. One of these approaches is recrystallization annealing usually performed in separate furnaces alternating with several cold forming steps. As consequence long throughput times incur that usually involve high efforts for set-up, transport and storage because the geometrical flexibility per forming step is limited. Using gas burners instead to reduce the material strength during processing (see figure 1), the energy introduced into the work piece can only hardly be controlled [7]. This can cause detrimental impact on the microstructure of the finished component. An innovative approach to avoid the aforementioned drawbacks is the use of a high power laser for the selective heating of the zone directly ahead of the forming roller. Like that deformation degrees can be enhanced significantly while the applied thermal energy to the work piece can be controlled reproducibly and reduced to a minimum because the heating is locally and temporally limited (see figure 2).

**Figure 1.** Gas burner assisted metal spinning of stainless steel [source: Radkersburger Metallwarenfabrik GmbH, Austria].

**Figure 2.** Laser-assisted metal spinning of stainless steel

*1.2. Laser-assisted metal spinning*

In general plastic deformation of metallic materials is based on dislocation movement. By further deformation the density of the dislocations increases, this leads to higher strength (strain hardening) on the one hand and to a decrease of the fracture strain (ductility) on the other hand. Elevating the temperature of the metal increases the mobility of the dislocations or activates other mechanisms for dislocation movement. Therefore the material's strength drops (thermal softening) and in most cases the ductility of the material increases with increasing temperature [1]. Quantitative measurements of the temperature dependence of the material's strength and ductility are given in section 3.1 for all three metals presented in this work.

The use of a laser as heat source in metal spinning operations is aiming at the generation of an adequate temperature field for the material being processed in order to reduce both the yield stress and any strain hardening occurring during the forming process. Therefore a laser beam is used to heat the area of the work piece just ahead of the forming roller contact zone as illustrated in figure 3. Thus the main benefits of laser-assisted metal spinning operations are:

- Improved formability of challenging materials such as titanium and nickel-based alloys
- Improved component quality due to temporally and locally limited heating
- Improved reproducibility compared to manual heating by means of gas burners
- Reduced forming forces

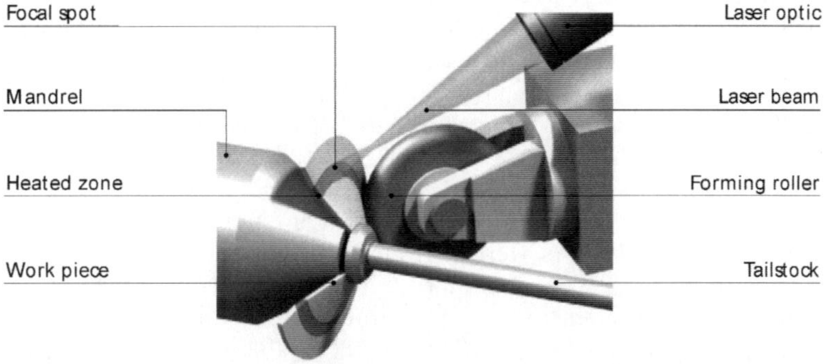


**Figure 3.** Process principle of laser-assisted multi-pass spinning

Former developments of laser integrations into spinning lathes, e.g. published in [8] or a provisional experimental set-up published by [9] facilitated to proof the general feasibility and the advantages of laser-assisted shear forming and multi-pass spinning operations but suffered several limitations. Especially with regard to multi-pass spinning limited flexibility and high programming efforts have been deficits. Thus a recently developed laser-integration into a spinning lathe published in [10] has been used for this work facilitating to realize laser-assisted multi-pass spinning operations without limitations. Scope of this publication is:

- To demonstrate the functionality of the newly developed laser integration by means of increased formability achieved on stainless steel, commercially pure titanium and a nickel-based alloy via laser-assisted metal spinning operations,
- To demonstrate the amount of force reduction achieved through laser heating during metal spinning operations and
- To discuss part quality achieved via laser-assisted forming in comparison with cold forming.

## 2. Experimental

The experimental set-up in a spinning lathe is shown in figure 4. For the investigations a high power diode laser with maximum output power of 5 kW has been used in combination with a 8 x 3 mm² focussing optic with homogenized laser intensity distribution. The experimental set-up included two two-colour-pyrometers for temperature measurement respectively control of laser power and a dynamometer mounted behind the forming roller in order to record forming forces in three directions.

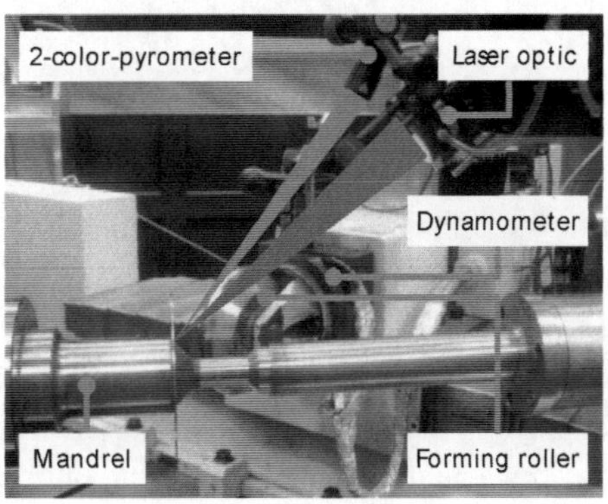

**Figure 4.** Experimental set-up for laser-assisted multi-pass metal spinning

For all investigations presented in this work, the initial sheet metal thickness was 2 mm. The rotational speed was between 250 and 300 $min^{-1}$ and the feed rate varied between 50 and 300 mm $min^{-1}$ depending on the material and forming operation. For the investigations on multi-pass spinning the same forming paths have been used for the forming operation and the feed rate has been adjusted in order to generate a suitable temperature field in the forming zone. Forming operations have been performed with controlled temperature on the work piece surface in front of the forming roller by means of a pyrometer closed-loop control of the laser output power.

## 3. Results

### *3.1. Material characterization prior to forming*

Tensile tests were performed to characterize the material properties at room temperature (RT) and at elevated temperatures. The tensile tests were performed with flat specimens taken from the same 2 mm thick sheet metal blanks that were used for the forming tests. As a measure for the materials strength and toughness at the different forming temperatures the ultimate tensile strength (*UTS*) and fracture strain *A* were evaluated according to DIN EN 6892-2. The results are summarized in table 1.

**Table 1.** Ultimate tensile strength and fracture strain in relevent temperature range for hot forming

| | Temperature range [°C] | Ultimate tensile strength [MPa] | | Fracture strain [%] | |
|---|---|---|---|---|---|
| | | Room temperature | Elevated temperature | Room Temperature | Elevated temperature |
| Ti grade 2 | 400…600 | 430 | 130…60 | 31 | 40…>100 |
| Inconel 718 | 800…900 | 890 | 636…180 | 55 | 55…80 |
| X5CrNi18-10 | 600…800 | 700 | 360…150 | 53 | 40…80 |

As all three materials are common in metal production, detailed investigations on the evolution of the local microstructure and the corresponding material properties during forming operations can be found in literature, e.g. in [11] for Inconel 718.

### *3.2. Analysis of formability by laser-assisted shear forming*

A first indicator for the formability of a blank material with the chosen process parameter set is the appearance of cracks in the final product. The processing of Inconel 718 and stainless steel have been possible in cold condition without crack formation. But compared to hot forming a clearly noticeable spring-back of the material occurred. Furthermore, the images in figure 5 show that in the case of Ti grade 2 forming at room temperature led to failure by circumferential cracking along the whole perimeter at a very early stage of the shear forming process.

**Figure 5.** Left: Image of Ti grade 2 formed at room temperature with the shear forming setup. A large crack around the whole perimeter appeared at the early stage of deformation. Right: Application of laser heating enabled a successful forming operation.

By using a laser to heat the Ti blanks locally to temperatures between 450 °C and 600 °C in the forming zone the shear forming process could be successfully finished. Figure 6 shows images of Ti cones that were formed at 450 °C and 600 °C, respectively. Laser heating to 450 °C resulted in a successful completion of the forming process (figure 6 a), however the final product showed cracks on the inside of the cones (figure 6 b). Figure 6 c) shows a representative depth profile measured with a NanoFocus µsurf light-optical profilometer in the region of those cracks. The profile shows that the cracks were between 100 and 150 µm deep. Using the laser to heat the Ti blanks to 600 °C (figure 6 d) resulted in Ti cones without these cracks on the inside, as demonstrated by the images (figure 6 e) and the corresponding profile (figure 6 f).

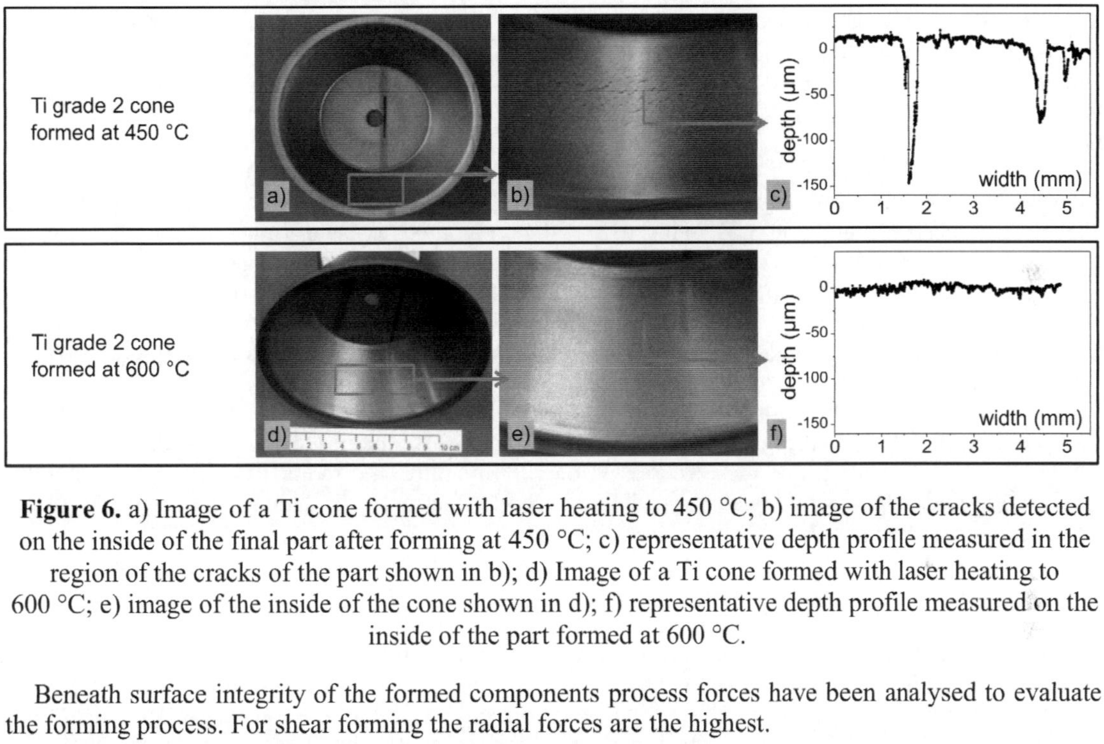


**Figure 6.** a) Image of a Ti cone formed with laser heating to 450 °C; b) image of the cracks detected on the inside of the final part after forming at 450 °C; c) representative depth profile measured in the region of the cracks of the part shown in b); d) Image of a Ti cone formed with laser heating to 600 °C; e) image of the inside of the cone shown in d); f) representative depth profile measured on the inside of the part formed at 600 °C.

Beneath surface integrity of the formed components process forces have been analysed to evaluate the forming process. For shear forming the radial forces are the highest.

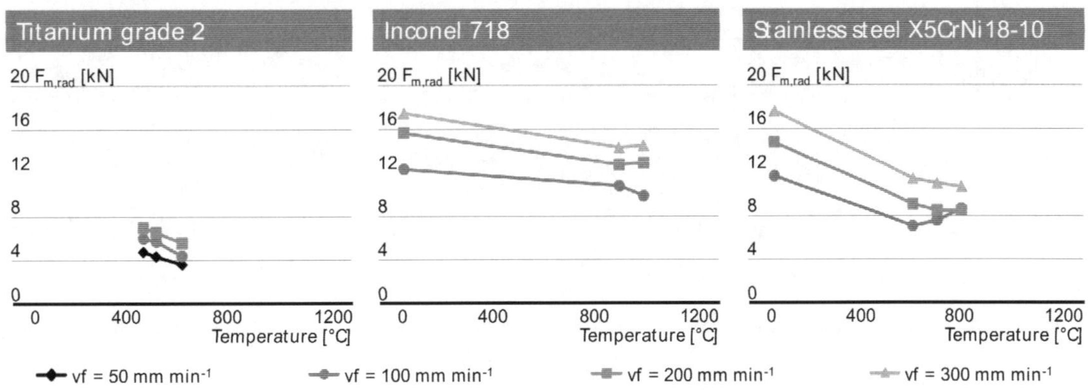


**Figure 7.** Mean radial forming forces for conventional and laser-assisted shear forming of titanium grade 2, Inconel 718 and stainless steel X5CrNi18-10

Thus in order to evaluate the laser-assisted process compared to cold forming, the mean radial forces during processing have been used as characteristic factor in this work. Figure 7 shows the mean radial forming forces for the investigated materials in dependence of forming temperature and feed rates.

As described above the cold shear forming of the used titanium grade 2 was not possible without failure within the investigated parameter range. Using a laser to heat the forming zone locally to temperatures between 450 °C and 600 °C the mean radial forces decreased nearly linearly in this range for the applied feed rates $v_f$ of 50, 100 and 200 mm $min^{-1}$. Comparing the values between 450 °C and 600 °C the mean radial force has been reduced by up to 27%. In order to achieve the defined temperature in the forming zone moderate laser power (< 1.75 kW) was required. With feed rates $v_f$ of 100 and 200 mm $min^{-1}$ small surface cracks occurred on the inside of the components which resulted most likely from non-sufficient heat transfer through the entire material thickness due to the low thermal conductivity of titanium. Thus feed rates below 100 mm $min^{-1}$ should be applied for the laser-assisted shear forming of Ti grade 2. Alternatively a bigger laser spot can be applied in order to increase the heated area.

For laser-assisted processing of the nickel-based alloy Inconel 718 the forming temperatures 900 and 1000 °C have been investigated. With a feed rate of 100 mm $min^{-1}$ and forming temperature of 1000 °C the mean radial force was reduced by 20% compared to the forming operation at room temperature. Applying the feed rates $v_f$ of 200 and 300 mm $min^{-1}$ a mean laser power of 4 to 4.5 kW was required for the forming operation but the maximum available laser power of 5 kW has not been sufficient to ensure a forming temperature of 900 respectively 1000 °C during the whole forming process. The slightly increasing trend of mean radial forces between the values for 900 and 1000 °C reflects this fact.

For the laser-assisted shear forming of stainless steel X5CrNi18-10 the mean radial force decreased most intensively with increasing forming temperature. At 800 °C the radial force reduced up to 40% compared to cold forming. For the feed rate $v_f$ of 100 mm $min^{-1}$ the radial force increased slightly between 600 and 800 °C which was most likely caused by thermal expansion of the tools due to the comparatively slow feed rate.

### *3.3. Laser-assisted multi-pass spinning*

Figure 8 shows the laser-assisted multi-pass spinning operation of Inconel 718. At the beginning of the process the sheet metal blank has been pre-heated (a) in order to reduce thermal gradients in the work piece resulting in heat removal out of the forming zone. During the forming operation (b) the required forming temperature is generated in the forming zone directly in front of the forming roller. At the end of the process a flow forming step has been performed for achieving a homogenous wall thickness.

**Figure 8.** Laser-assisted multi-pass spinning of Inconel 718: a) Pre-heating, b) intermediate forming step, c) flow forming operation

During the forming operation the temperature has been kept constant by means of a pyrometer closed-loop-control. In the case of Inconel 718 it has not been possible to form the component at forming temperatures below 900 °C without failure (see figure 9, left). By using a controlled forming temperature of 900 and 1000 °C the cylindrical component with spinning ratio 2.0 could successfully be manufactured in one single clamping by laser-assisted multi-pass spinning (figure 9, right).

**Figure 9.** Non-sufficient heating during forming of Inconel 718 resulted in component failure by circumferential cracking (left). Laser-assisted multi-pass spinning of Inconel 718 facilitated the successful manufacture of a cylindrical component with spinning ratio 2.0 (right).

The manufacture of the defined spinning part made of Ti grade 2 was basically possible at room temperature but resulted in crack formation at the rim of the work piece and poor dimensional accuracy due to spring back. By laser-assisted hot forming between 400 and 600 °C the component could successfully be manufactured without any cracks and enhanced dimensional accuracy. For 600 °C forming temperature the deviation in roundness has been reduced by 77% compared with forming at room temperature. At the same time the surface roughness Ra was reduced from 0.8 μm to 0.4 μm.

Stainless steel X5CrNi18-10 could also be processed at room temperature but with very poor dimensional accuracy due to spring back. Compared to the cold formed component the forming forces could be reduced by 30%. Comparing the results of forming at room temperature and 700 °C the dimensional accuracy could be improved in terms of deviation in roundness by 43% and the surface roughness Ra was reduced from 0.5 μm to 0.2 μm.

Laser-assisted multi-pass spinning facilitated for all three materials to eliminate conventionally needed intermediate annealing operations in separate furnaces. Like that the manufacture in one single clamping becomes possible reducing the throughput times drastically. Additionally, parts with complex shapes can be formed without the negative impact caused by conventionally used gas burners for simultaneous heating of the work piece during manufacturing.

*3.4. Mechanical properties after laser-assisted spinning*

Vickers hardness measurements have been chosen as an indicator for the local material properties after the shear forming operations shown in section 3.2. Table 2 shows a comparison of hardness measurements performed after RT and laser assisted shear forming of stainless steel X5CrNi18-10, Ti grade 2 and Inconel 718 with 200 mm $min^{-1}$ feed rate. HV1 hardness measurements across the walls of the cones showed a uniform hardness across the walls within the measurement accuracy limit of ± 1.5 HV1. All three materials showed a comparable reduction of the wall thickness that is typical for the shear forming process, as discussed in section 2 and reference [7].

For all three materials the position of the minimum wall thickness in the final part correlated with the position of the maximum hardness, as given in the $4^{th}$ column in table 2. This indicates a correlation between the local degree of deformation and the local hardness, i.e. strain hardening. The relative hardness increase and the corresponding wall thickness reduction are indicated in the last two columns in table 2.

**Table 2.** Influence of forming parameters on the local hardness after laser assisted shear forming

| | Forming temperature | Initial blank hardness | Hardness at position of min. wall thickness | Hardness increase at min. wall thickness | Wall thickness reduction |
|---|---|---|---|---|---|
| | [°C] | HV1 | HV1 | [%] | [%] |
| | 25 | 160 | 370 | 131 | 38 |
| X5CrNi18-10 | 600 | | 300 | 88 | 44 |
| | 800 | | 300 | 88 | 44 |
| | 25 | 144 | - | - | - |
| Ti Grade 2 | 450 | | 150 | 5 | 38 |
| | 600 | | 180 | 25 | 45 |
| | 25 | 209 | 440 | 111 | 38 |
| Inconel 718 | 900 | | 440 | 111 | 39 |
| | 1000 | | 400 | 91 | 46 |

For stainless steel X5CrNi18-10 the wall thickness reduction and the strain hardening were the same for 600 °C and 800 °C forming temperature despite the ongoing reduction of the forming forces indicated in figure 7. For Ti formed at 450 °C and 600 °C the measured higher hardness at 600 °C seems to be in contradiction with the reduced forming forces shown in figure 7. The low hardness measured on the cones formed at 450 °C can be explained by the micro-cracks detected on the Ti cones formed at 450 °C and shown in figure 6. The presence of the micro cracks reduces the local strength of the material and thus leads to the lower Vickers hardness. For Inconel 718 the higher forming temperature of 1000 °C resulted in a reduction of the maximum hardness and the forming forces but less reduction of the wall thickness indicating a reduced strain hardening at the higher forming temperature.

## 4. Summary and outlook

In this paper the effects of laser heating simultaneously to shear forming and multi-pass metal spinning on forming forces, formability and the preliminary investigations of the resulting material properties have been demonstrated for Ti grade 2, Inconel 718 and stainless steel X5CrNi18-10. For all three materials the formability could be enhanced significantly while improving the component quality at the same time. Laser-assisted multi-pass spinning facilitated to eliminate conventionally needed intermediate annealing operations in separate furnaces thus to manufacture complex geometries in one single clamping. Especially the successful manufacture of components with a high spinning ratio made of Inconel 718 is remarkable, because cold forming resulted in failure of the work piece at an early stage of the forming process. Like that the potential of the technology has been proven to reduce the throughput times drastically. Because the heat is applied only briefly to the area in which it is needed the thermal load of the work piece can be reduced to a minimum and controlled accurately by the use of a laser.

In order to benefit from the advantages of laser-assisted shear forming and multi-pass spinning operations in an industrial environment the main objectives for further developments are to develop a suitable machine tool and CAM-module. This will facilitate to make the new process available and manageable for industrial companies in order to process high-strength materials flexibly with increased reproducibility.

## 5. Acknowledgement

This research has been funded under the Seventh Framework Programme (FP7) of the European Union under grant agreement EasyForm (315111).

## 6. References

[1] Weißbach W 2010 *Werkstoffkunde* Wiesbaden, D Vieweg Teubner.
[2] Maderbacher H, Oberwinkler B, Gänser H P, Tan W, Rollet M and Stoschka M 2013 The influence of microstructure and operating temperature on the fatigue endurance of hot forged Inconel® 718 components *Materials Science and Engineering A* **585** 123–131
[3] Davis J 2000 *ASM Speciality Handbook: Nickel, Cobalt, And Their Alloys* ASM International Matrials Park, OH
[4] Bewlay B P and Furrer D U 2006 Spinning. In: Semiatin S L, Marquard E, Lampman H, Karcher C and Musgrove B (Eds.), *Metalworking: Sheet Forming*. ASM International, Matrials Park, OH 367-374
[5] Groche P, Fritsche D, Tekkaya E A, Allwood J M, Hirt G and Neugebauer R, 2007. Incremental Bulk Metal Forming. In: *Annals of the CIRP* **56** 635-656
[6] Wong C C, Dean T A and Lin J 2003 A review of spinning, shear forming and flow forming processes *International Journal of Machine Tools & Manufacture* **43** 1419-1435
[7] Music O, Allwood J. and Kawai K 2010 A review of the mechanics of metal spinning. *Journal of Materials Processing Technology* **210** 3-23
[8] Klocke F and Wehrmeister T 2004 Laser-Assisted Metal Spinning of Advanced Materials. In: Geiger M and Otto A (Eds.) *Proceedings of the 4th Laser Assisted Net Shape Engineering conference LANE* 1183-1192
[9] Romero P, Otero N, Cabrera J M and Masagué D 2010 Laser Assisted Conical Spin Forming of Dual Phase Automotive Steel. Experimental demonstration of Work Hardening Reduction and Forming Limit Extension *Physics Procedia* **5** 215-225
[10] Klocke F and Brummer C 2014 Laser-assisted metal spinning of challenging materials. *Procedia Engineering* **81** 2385-2390
[11] Drexler A, Maderbacher H, Povoden-Karadeniz E, Gänser H P, Ecker W, Oberwinkler B and Fischersworring-Bunk A 2015 Yield stress evolution in Inconel 718 samples under standard heat treatment process conditions of turbine disks In: *Proceedings of the European Conference on Heat Treatment 2015 & 22nd IFHTSE Congress*

# State parameter-based modelling of microstructure evolution in micro-alloyed steel during hot forming

**H Buken[1] and E Kozeschnik[1, 2]**

[1] Institute of Materials Science and Technology, Vienna University of Technology, Getreidemarkt 9, 1060 Vienna, Austria
[2] Materials Center Leoben Forschung GmbH, Roseggerstraße 12, 8700 Leoben, Austria

E-mail: heinrich.buken@tuwien.ac.at

In steel production, thermo-mechanical treatment at elevated temperatures is an inevitable step for controlling the microstructure and, thus, the mechanical-technological properties of the final product. One of the main goals in modelling microstructure evolution is the prediction of progress and interaction of hardening and softening mechanism at temperatures, where reheating, hot rolling, finish rolling and coiling are typically carried out. The main mechanisms that need to be accounted for are precipitation, grain growth, solute drag, recovery, recrystallization and phase transformation, which are to be described as functions of temperature, external loading and chemical composition of the material. In the present work, we present a new approach for dealing with these problems and apply it to the thermal and mechanical loading of microalloyed steel. Within this model, we quantitatively predict, for instance, the phenomenon of recrystallization stop in the presence of precipitation. The computational treatment is verified against experimental data from literature, where good agreement is achieved.

## 1. Introduction

For the case of single phase alloys that is, e.g., C- Mn- steel in the austenitic range, many constitutive laws are available that are capable of providing more or less sufficient knowledge about the static recrystallization behaviour [1]. In multiphase alloys, such as V-microalloyed steel, recrystallization kinetics can become more complicated and constitutive laws are inappropriate to describe recrystallization kinetics since a further dimension enters and influences the system: Particles cause a pinning pressure on grain boundaries. The magnitude of this Zener Drag effect is mainly determined by the precipitated phase fraction and the precipitate size but has its physical roots in the saving of energetically unfavourable total grain interface area.

Deformation provides a higher dislocation density in the material and affects the precipitation process by providing a higher nucleation site density and a higher diffusion rate in the matrix. As a consequence of this interplay of mechanisms, a description of microstructure kinetics with a single Avrami- curve [2] is not possible. Amongst many others, Medina et al. [3] show recrystallization plateaus in the presence of the common microelements Ti, Nb and V.

Modelling approaches are available in literature describing the simultaneous precipitation and recrystallization kinetics either phenomenological or physically based. Medina et al. [3] use a framework of interrelated Avrami based equations. Developing for each composition a grain size and strain dependent formula to calculate a critical maximum temperature where barely a plateau can be measured, they classify recrystallization kinetics into two regimes: a fast one with a low activation energy and a slow one where precipitates pin the grain boundaries. Nonetheless, this approach is not capable of illustrating the experimentally observed curve shapes but shows an assessment of transformation times in dependence of various material conditions and thermo-mechanical processing routes.

Zurob et al. [4] suggest a physical approach for modelling those plateaus. By simulating both, strain-induced precipitation and recrystallization kinetics, they interconnect the microstructure and precipitate evolution by the integral value of driving and retarding pressures. Once the Zener force exceeds the driving force, grain growth and recrystallization progresses come to a stop. Due to coarsening, (Ostwald ripening), the number density of precipitates decreases, they exert less influence on grain boundary movement and the Zener pressure decreases. At this point, the sum of driving pressure and retarding pressure are positive again and grains continue to grow. In this approach, Zurob et al. deliver simulation results, which show the experimentally observed curve progression.

Although some models describing coupled precipitation kinetics and microstructure evolution are available, in most approaches, either the precipitation and recrystallization kinetics are modelled implicitly, such that one set of parameters is only valid for one composition, or that the modelling of precipitation kinetics is fitted with phenomenological parameters (usually interface energies), such that the coupled model delivers microstructure results (e.g. recrystallized fraction) being in good agreement with experimental results.

In the present work and in contrast to the conventional approaches, we utilize a precipitate kinetics framework that is based on independent thermo-physical quantities (thermodynamic and kinetic databases). From these, temperature, composition and size dependent system parameters, such as interfacial energies [5–7], are computed. The state parameters are then evolved as a function of chemical composition, heat treatment and deformation regime as well as microstructure (grain size, dislocation density, etc.). For simulation of microstructure evolution, we utilize a recently developed single class recrystallization model, in which the driving forces and mobilities are computed by means of the approaches introduced subsequently. In this framework, we describe the precipitation and recrystallization interaction of low carbon steel, where the recrystallization kinetics are influenced by precipitate phases.

## 2. Modelling approach

*2.1 Model development*

The thermo-kinetic simulations of precipitation with coupled microstructure evolution are carried out with the software package MatCalc (version 6.00.007). The physical concept and functionality of the precipitate evolution, as delivered by MatCalc, is explained for the case of micro alloying elements in steel elsewhere [8,9]. To the precipitation kinetics routines of MatCalc, we couple a microstructure evolution model, which is introduced next.

The nucleation rate of recrystallized grains is modelled as

$$\dot{N}_{\mathrm{rx}} = \begin{cases} C_{\mathrm{rx}}\left(\frac{\pi}{6}\delta^2 D\right)^{-1} \exp(\frac{-Q_{\mathrm{rx}}}{RT})(1-X_{\mathrm{rx}}) & ,\delta \geq \delta_{\mathrm{crit}} \\ 0 & ,\delta < \delta_{\mathrm{crit}} \end{cases} , \tag{1}$$

where $\delta$ is the subgrain diameter, $D$ is the mean un-recrystallized grain diameter, $C_{rx}$ is a calibration coefficient, $Q_{rx}$ is an activation energy equal to the value for substitutional self-diffusion along grain boundaries, $X_{rx}$ is the recrystallized fraction, $R$ is the universal gas constant and $T$ is temperature. The inverse term including the grain size and the subgrain size captures the condition that only subgrains (LAGB), which are in contact with a high-angle grain boundary can become nuclei for recrystallization as experimentally observed in ref. [10].

A second condition for nucleation is that the subgrain exceeds a critical size, $\delta_{crit}$. This is given by the ratio of the interfacial energy of the LAGB, $\gamma_{LB}$ and the driving pressure, $P_D$, with

$$\delta_{crit} = \frac{3\gamma_{LB}}{P_D} = \frac{3\gamma_{LB}}{0.5\mu b^2 \rho}. \tag{2}$$

The driving pressure is introduced by deformation and calculated by means of a total dislocation density, $\rho$, the shear modulus, $\mu$, and the Burgers vector, $b$. Before nucleation takes place, the subgrain has to grow in order to overcome this critical size. On the one hand, the subgrain size is reduced by the deformation-induced dislocation that are stored in the subgrain boundaries. Nes [11] and Estrin [12] introduced the principle of similitude, where the dislocation density is correlated to a subgrainsize, $\delta_0$, by means of a material parameter K, which is taken as starting subgrain size directly after deformation. On the other hand, the subgrain grows due to the driving pressure of capillary, where $M_{eff, LB}$ is the effective LAGB mobility.

$$\delta = \delta_0 + \dot{\delta} \cdot t = \left(\frac{K}{\sqrt{\rho}}\right) + \left(M_{eff, LB} \frac{3\gamma_{LB}}{\delta}\right) \cdot t \tag{3}$$

The growth velocity of stable nuclei, $\dot{D}_{rx}$, is expressed by the product of an effective HAGB mobility, $M_{eff, HB}$, and the driving pressure as

$$\dot{D}_{rx} = M_{eff, HB} P_D (1 - X_{rx}). \tag{4}$$

With increasing degree of recrystallization, less deformed volume is available in which the recrystallized grains can grow into. To account for that, the growth velocity is balanced with the unrecrystallyzed volume fraction, $1\text{-}X_{rx}$.

Deformation induces an increased dislocation density in the material and originates a driving pressure for recrystallization. The deformation-induced dislocation density is calculated by means of a Kocks-Mecking-type approach [13], which has been extended for static recovery by Sherstnev et al. [14].

$$\dot{\rho} = \frac{M\sqrt{\rho}}{Ab}\dot{\varphi} - 2B\frac{d_{ann}}{b}\rho M\dot{\varphi} - 2CD_{Dis}\frac{\mu b^3}{k_B T}\rho \tag{5}$$

with the Taylor factor, $M$, the critical dislocation annihilation distance, $d_{ann}$, the substitutional self-diffusion coefficient at dislocations, $D_{Dis}$, the Boltzmann constant, $k_B$, the strain rate $\varphi$, and material parameters $A$, $B$, $C$.

The influence of precipitates on recrystallization kinetics is incorporated via the Zener pressure, $P_Z$, which is given for a spatial distribution of spherical precipitates by Nes et al. [15] as

$$P_Z = \frac{3f\gamma_{HB}}{2r}, \tag{6}$$

where $\gamma_{HB}$ is the HAGB energy, $f$ is the phase fraction of precipitates and $r$ is the precipitate radius. In contrast to most other approaches [16–18], we do not incorporate the Zener pressure into the driving force term, which may lead to a complete growth stop of the recrystallization front if the retarding force exceeds the driving force. In our treatment, we assume that the precipitates, which are arranged at the pinned grain boundary, are interconnected along a high velocity diffusion path, which allows for fast local coarsening of precipitates. Hence, the Zener pressure locally decreases and allows for a further boundary movement till a new front of precipitates is encountered, where the process of local coarsening is repeated. We aggregate this mechanism in the form of an effective mobility, which is written as

$$M_{\mathrm{eff}} = \begin{cases} \left(\dfrac{P_{\mathrm{D}} - P_{\mathrm{Z}}}{P_{\mathrm{D}}}\right) M_{\mathrm{free}} + \left(1 - \dfrac{P_{\mathrm{D}} - P_{\mathrm{Z}}}{P_{\mathrm{D}}}\right) M_{\mathrm{pinned}} & , P_{\mathrm{D}} > P_{\mathrm{Z}} \\ M_{\mathrm{pinned}} & , P_{\mathrm{D}} \le P_{\mathrm{Z}} \end{cases} \quad (7)$$

where the effective boundary mobility is calculated as superposition between a free mobility, $M_{\mathrm{free}}$, in the absence of pinning, and a minimum mobility $M_{\mathrm{pinned}}$, caused by precipitate pinning. This new approach provides a possibility for boundary movement even if the mean Zener pressure of the system exceeds the driving pressure for recrystallization.

*2.2 Model Parameters*

To verify the new model on experimental data, an evaluation of important parameters is performed first, which is summarized in table 1. Since the grain boundary mobility is a most important input parameter in our recrystallization model, we express this quantity in relation to well-assessed diffusion coefficients. Turnbull [19] suggests an upper bound for the boundary mobility, $M_{\mathrm{TB}}$, with

$$M_{\mathrm{free}} = \eta_{\mathrm{free,HB}} \cdot M_{\mathrm{TB}} = \eta_{\mathrm{free,HB}} \cdot \frac{\omega D_{\mathrm{GB}} V_{\mathrm{m}}}{b^2 RT}, \quad (8)$$

with the substitutional diffusion coefficient at grain boundaries, $D_{\mathrm{GB}}$, the grain boundary width, $\omega$ and the molar volume of the material, $V_{\mathrm{m}}$. Since atomic attachment kinetics are not accounted for within this approach, we introduce a dimensionless linear factor, $\eta_{\mathrm{free,HB}}$, which conserves the temperature dependency of the free mobility as the ratio of the grain boundary diffusion coefficient and temperature. We adopt $D_{\mathrm{GB}}$ in austenite from a recent diffusion assessment of Stechauner and Kozeschnik [20]. The linear factor is adjusted to the mobility measurements of Zhou et al. [21] of plain C-Mn steel in austenite, where good agreement is achieved with a value of 1.5%.

The temperature dependency of the pinned mobility should be equal to the free mobility. Kirchner [22] suggests for precipitate coarsening at grain boundaries the same temperature dependency as Turnbull for the grain boundary mobility. Thus, we model the pinned mobility as fraction of the free mobility, which delivers with Eq. 4 a reduced growth rate in the presence of precipitates as

$$M_{\mathrm{pinned}} = \eta_{\mathrm{pinnedHB}} \cdot M_{\mathrm{free}}. \quad (9)$$

Application of this concept (section 3) shows that the mobility decreases by two orders of magnitude to 3% due to the retarding pressure exerted by the precipitates. The driving pressure for recrystallization is mainly determined by the deformation-induced dislocations and interacts with both nucleation and growth of the new microstructure. Therefore, the parameters (*A*, *B*, *C*) of the extended Kocks- Mecking model (Eq. 5) are adjusted such as to reproduce flow curve results of Hernandez et al. [23]. The observed dislocation densities do not exceed $7 \cdot 10^{14}$ $\mathrm{m}^{-2}$ and deliver appropriate flow stresses in the sense of the Taylor forest hardening law with input parameters similar to those mentioned in ref. [4]. The activation energy for nucleation of recrystallization (Eq. 1) mainly determines the temperature dependency of the nucleation rate. Occurring at the grain boundary, nucleation is assumed to obey the temperature dependency of grain boundary diffusion. Thus, this value is also taken from ref. [20]. The linear factor, $C_{\mathrm{rx}}$, is an adjusting parameter, which balances the nucleation suppressing effect of the activation energy, but it is left constant over the range of considered materials (section 3). A collection of the elaborated parameters is given in table 1.

**Table 1.** Simulation input parameters

| Parameter | Unit | Value | Ref. |
|---|---|---|---|
| $\eta_{\text{free,HB}}$ | - | 0.015 | [19–21] |
| $\eta_{\text{pinned,HB}}$ | - | 0.03 | This work |
| A | - | 50 | This work |
| B | - | 5 | This work |
| C | - | $5\cdot10^{-5}$ | This work |
| $Q_{rx}$ | kJ mol$^{-1}$ | 145 | [20] |
| *Crx* | s$^{-1}$ | 1e6 | This work |
| $\gamma_{\text{HB}}$ | J m$^{-2}$ | 1.311-0.005·T | [4] |
| $\gamma_{\text{LB}}$ | J m$^{-2}$ | $0.5\cdot\gamma_{\text{HB}}$ | This work |

## 3. Validation

Finally, the present microstructure model with combined precipitation kinetics is verified on experimental results from Medina et al. [24], who measured the progress of static recrystallization in diverse V-microalloyed steels. Table 2 summarizes the different compositions, starting grain sizes, $D_0$ and solid solution temperatures, $T_{\text{sol}}$, of the investigated materials.

**Table 2.** Considered materials at $\varepsilon = 0.35$ and $\dot{\varepsilon} = 3.62 s^{-1}$

| Steel | C [wt.-%] | V [wt.-%] | N [wt.-%] | $T_{\text{sol}}$ [°C] | $D_0$ [μm] |
|---|---|---|---|---|---|
| S1 | 0.125 | 0.065 | 0.0123 | 1056 | 167 |
| S2 | 0.113 | 0.095 | 0.0144 | 1106 | 162 |

The two considered materials are experimentally analysed at a constant strain, ε, strain rate, $\dot{\varepsilon}$, and at different temperatures. The different precipitate solution temperatures, which are calculated with the thermokinetic software MatCalc, indicate different driving forces for precipitation. Steel S1 has a lower driving force for precipitation than steel S2 due to lower V and N contents. Therefore, the retardation of recrystallization due to precipitation is more pronounced in the more supersaturated alloy S2. At lower temperatures (below 1000°C), recrystallization starts to interact with precipitation and both steels show in that temperature range a recrystallization plateau. The more supersatured system S2 shows, in comparison to steel S1, recrystallization plateaus at lower levels of recrystallized fractions, because the Zener pressure exceeds the driving pressure for recrystallization earlier. When the retarding pressure exceeds the driving pressure, further kinetics are limited by the pinned mobility and the progress of recrystallization is decelerated. At higher temperatures ($T\geq1000$°C), there is no interaction between precipitation and precipitation observed and both alloys show nearly the same microstructure evolution kinetics. Figure 1 depicts the results of our simulation approach in comparison to the experimental measurements of Medina et al [24].

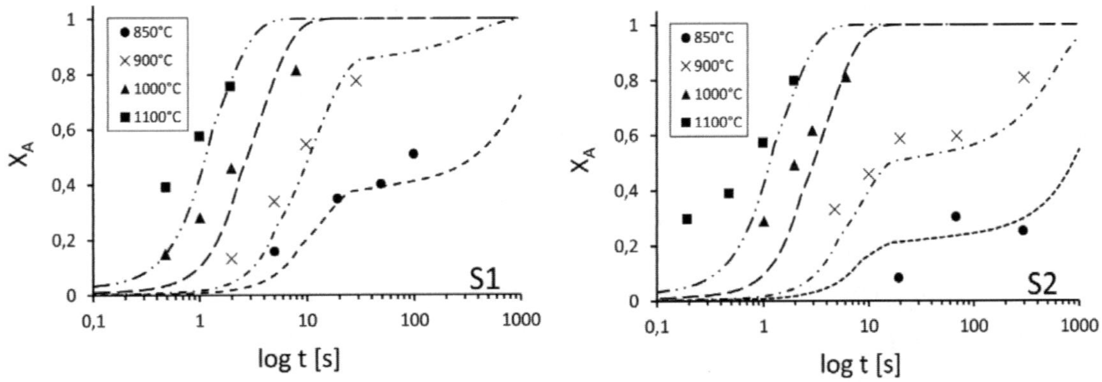


**Figure 1.** Recrystallization experiments vs. simulation for steel S1 and S2 at different temperatures

## 4. Summary

A new microstructure model is presented, where the static recrystallization process is described by the mechanisms of nucleation and growth. The growth rate of recrystallized grains is given by the product of grain boundary mobility and driving pressure. In contrast to several other models, the precipitate influence is incorporated into the mobility term and not the driving pressure term, which allows for a further growth of the recrystallized fraction even if the Zener force exceeds the driving force. The model is verified against experimental data from ref. [24], where good agreement is achieved.

## 5. References


[1] Beynon J and Sellars C M 1992 Modelling Microstructure and its effects during multipass hot rolling *ISIJ Int.* **32** 359–67

[2] Avrami M 1940 Kinetics of Phase Change. II Transformation-Time Relations for Random Distribution of Nuclei *J. Chem. Phys.* **8** 212–24

[3] Medina S F and Mancilla J E 1996 Static recrystallization modelling of hot deformed microalloyed steels at temperatures below the critical temperature *ISIJ Int.* **36** 1077–83

[4] Zurob H S, Brechet Y and Purdy G 2001 A model for the competition of precipitation and recrystallization *Acta Mater.* **49** 4183–90

[5] Sonderegger B and Kozeschnik E 2009 Generalized Nearest-Neighbor Broken-Bond Analysis of Randomly Oriented Coherent Interfaces in Multicomponent Fcc and Bcc Structures *Metall. Mater. Trans. A* **40** 499–510

[6] Sonderegger B and Kozeschnik E 2009 Size dependence of the interfacial energy in the generalized nearest-neighbor broken-bond approach *Scr. Mater.* **60** 635–8

[7] Sonderegger B and Kozeschnik E 2010 Interfacial Energy of Diffuse Phase Boundaries in the Generalized Broken-Bond Approach *Metall. Mater. Trans. A* **41** 3262–9

[8] Radis R and Kozeschnik E 2010 Concurrent Precipitation of AlN and VN in Microalloyed Steel *Steel Res. Int.* **81** 681–5

[9] Radis R and Kozeschnik E 2012 Numerical simulation of NbC precipitation in microalloyed steel *Model. Simul. Mater. Sci. Eng.* **20** 055010

[10] Hansen S S, Vander Sande J B and Cohen M 1980 Niobium carbide precipitation and austenite recrystallization in hot rolled microalloyed steels *Metall. Trans. A* **11A** 387–402

[11] Nes E 1997 Modelling of work hardening and stress saturation in FCC metals *Prog. Mater. Sci.* **41** 129–93
[12] Estrin Y, Tóth L S, Molinari A and Bréchet Y 1998 A dislocation-based model for all hardening stages in large strain deformation *Acta Mater.* **46** 5509–22
[13] Kocks U F and Mecking H 2003 Physics and phenomenology of strain hardening: The FCC case *Prog. Mater. Sci.* **48** 171–273
[14] Sherstnev P, Lang P and Kozeschnik E 2012 Treatment of Simultaneous Deformation and Solid- State Precipitation in Thermo-Kinetic Calculations *Eccomas 2012* 8
[15] Nes E, Ryum N and Hunderi O 1985 On the Zener drag *Acta Metall.* **33** 11–22
[16] Nes E 1976 The effect of a fine particle dispersion on heterogeneous recrystallization *Acta Metall.* **24** 391–8
[17] Humphreys F J 1997 A unified theory of recovery, recrystallization and grain growth, based on the stability and growth of cellular microstructures—II. The effect of second-phase particles *Acta Mater.* **45** 5031–9
[18] Zurob H S, Hutchinson C R, Brechet Y and Purdy G 2002 Modelling recrystallization of microalloyed austenite: effect of coupling recovery, precipitation and recrystallization *Acta Mater.* **50** 3075–92
[19] Turnbull D 1951 Theory of grain boundary migration rates *Trans. AIME* **191** 661–5
[20] Stechauner G and Kozeschnik E 2014 Assessment of substitutional self-diffusion along short-circuit paths in Al, Fe and Ni *Calphad* **47** 92–9
[21] Zhou T, O'Malley R J and Zurob H S 2010 Study of grain-growth kinetics in delta-ferrite and austenite with application to thin-slab cast direct-rolling microalloyed steels *Metall. Mater. Trans. A Phys. Metall. Mater. Sci.* **41** 2112–20
[22] Kirchner H O K 1971 Coarsening of grain-boundary precipitates *Metall. Trans.* **2** 2861–4
[23] Hernandez C A, Medina S F and Ruiz J 1996 Modelling Alloy Flow Curves in Low and Microalloyed Steels *Acta Metall.* **44** 155–63
[24] Medina S F, Mancilla J E and Hernandez C A 1993 Influence of Vanadium on the static recrystallization of austenite in microalloyed steels *J. Mater. Sci.* **28** 5317–24

# The Crack Initiation and Propagation in threshold regime and S-N curves of High Strength Spring Steels

**N Gubeljak[1], J Predan[1], B Senčič[2] and M D Chapetti[3]**
[1] University of Maribor, Faculty of Mech. Eng., Maribor, Slovenia
[2] Štore Steel d.o.o., Steelwork company, Štore, Slovenia
[3] Laboratory of Experimental Mechanics, INTEMA, CONICET – University of Mar del Plata. Juan B. Justo 4302, Argentina

E-mail: nenad.gubeljak@um.si

**Abstract.** An integrated fracture mechanics approach is proposed to account for the estimation of the fatigue resistance of component. Applications, estimations and results showed very good agreements with experimental results. The model is simple to apply, accounts for the main geometrical, mechanical and material parameters that define the fatigue resistance, and allows accurate predictions. It offers a change in design philosophy: It could be used for design, while simultaneously dealing with crack propagation thresholds. Furthermore, it allows quantification of the material defect sensitivity. In the case of the set of fatigue tests carried out by rotational bending of specimens without residual stresses, the estimated results showed good agreement and that an initial crack length of 0.5 mm can conservatively explain experimental data. In the case of fatigue tests carried out on the springs at their final condition with bending at R = 0.1 our data shows the influence of compressive residual stresses on fatigue strength. Results also showed that the procedures allow us to analyze the different combinations of initial crack length and residual stress levels, and how much the fatigue resistance can change by changing that configuration. For this set of tests, the fatigue resistance estimated for an initial crack length equal to 0.35 mm, can explain all testing data observed for the springs.

## 1. Introduction

Spring leaves usually fail by surface or subsurface induced fatigue failure from an inclusion, which size is relatively large in leaf spring. The fatigue crack growth area is often less than $1mm^2$. Usually, spring leaves fail by subsurface induced failure in the vicinity of an inclusion, as is shown in Fig. 1, during fatigue test. The initiation location depends of different parameters associated with the fatigue mechanism that gives rise to a crack, and then it is related with the weakest configuration of material resistance, inclusion size (the inclusion acts as a concentrator), residual stresses and loading. Few investigations of residual stress effect and inclusion size on fatigue failure in very high cycle regime has been performed by Shiozawa and co-workers [1, 2, 3]. They show that in case of rotating bending usually fatigue crack appear from surface inclusion induced failure mode, where compressive stress relaxed and decreased at an early stage of fatigue cycling. Therefore, compress residual stress on the surface are also not stable and uniform for fatigue lifetime prediction, as well. It is known that negative loading R ratio corresponds to negative value of minimum stress intensity factor $K_{min}$ and shift fatigue threshold $\Delta K_{th}$ to higher values. The goal of producer is induced compress residual stresses in layer of leaf exposed to tension applied stress. During the crack initiation period fatigue crack develop during the first loading cycles, but depending on the magnitude of load become quiescent for some time before they eventually succeed in crossing existing barrier as is shown in Fig. 2. The microstructural barrier is the grain boundary, where micro-crack can be retarded or stop. The stress range that leads to a very long or an infinite fatigue lifetime is called fatigue limit $\Delta\sigma_{eR}$. In principle the fatigue limit stress range is measured on smooth specimens and can remain constant even when very short crack (smaller than grain size) are present, as is schematically shown in so-called Kitagawa-Takahashi diagram in Fig. 3 [4]. In this approach the difference between the total applied driving force and the material threshold for crack propagation defines the effective driving force applied to the crack. The initial crack length is given by the position of the strongest microstructural barrier if the material were free of cracks or crack like flaws, or by the greatest defects that acts as a fatigue initiator.

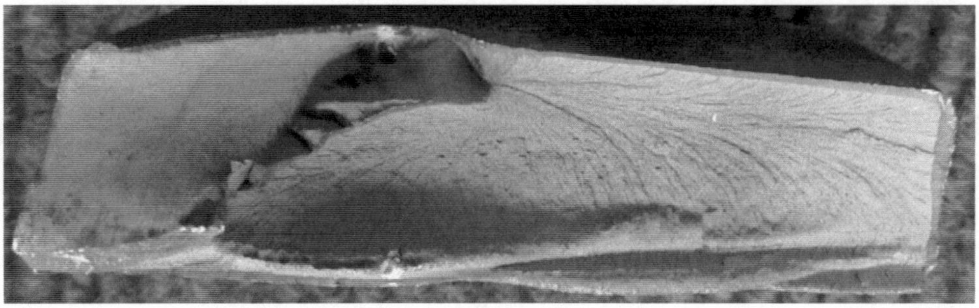

Fig. 1. Leaf spring bar section with inclusion around 0.4 mm under surface at distance 0.8 mm

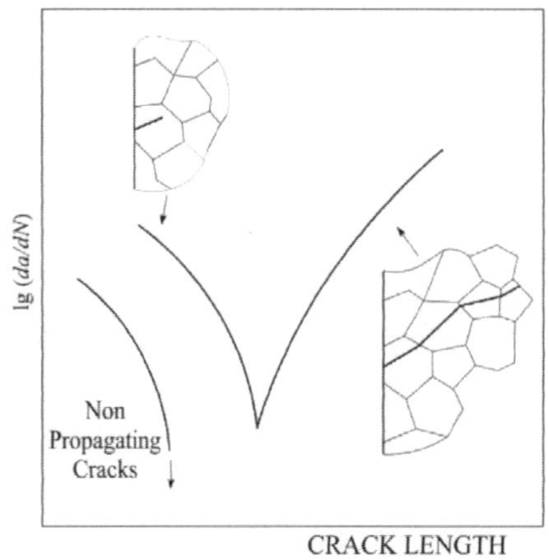


Fig. 2 Schematic view of retarded crack growing

The minimum threshold for fatigue crack propagation is associated to the microstructural barrier that defines the plain fatigue limit and represents the microstructural threshold for crack propagation, as [5]:

$$\Delta K_{dR} = Y \cdot \Delta\sigma_{eR}\sqrt{\pi \cdot d} \quad (1)$$

where $Y$ is the geometrical correction factor. In most cases the nucleated microstructurally short surface cracks are considered semicircular, and the value of $Y$ would then be 0.65[15]. Because the plain fatigue limit depends on the stress ratio $R$, the microstructural threshold also does. The value of $d$ is usually given by the microstructural characteristic dimension, as inclusion or grain size, e.g. for the steel in as-delivered condition $d$ = 10 μm. The grain size in as-delivered condition is consequence of a conventional casting process. The fine grain was obtained by heating to the austenitizing temperature of +870°C and gas quenched a temperature of +60°C, and then tempered for 1 h at +475°C. The inclusion size was reduced by Q+T process of refined spring steel, to size less than 5 μm. However, some inclusions of size more than 0.4 was revealed under surface, representing critical locations for crack initiation.

A total extrinsic threshold to crack propagation, $\Delta K_{CR}$, is then defined by the difference between the crack propagation threshold for long cracks, $\Delta K_{thR}$, and the microstructural threshold, $\Delta K_{dR}$ [5-8]. The development of the extrinsic component is considered to be exponential and a development parameter $k$ is estimated as a function of the same microstructural and mechanical parameter used to define the material threshold for crack propagation. The material threshold for crack propagation as a function of the crack length, $\Delta K_{th}$, is then defined as [5]:

(2) $$\Delta K_{th} = \Delta K_{dR} + \left(\Delta K_{thR} - \Delta K_{dR}\right) \left[1 - e^{-k(a-d)}\right] = Y \, \Delta\sigma_{th} \sqrt{\pi \, a} \qquad a \geq d$$

Where the parameter $k$ is given by:

$$k = \frac{1}{4d} \frac{\Delta K_{dR}}{\left(\Delta K_{thR} - \Delta K_{dR}\right)} \tag{3}$$

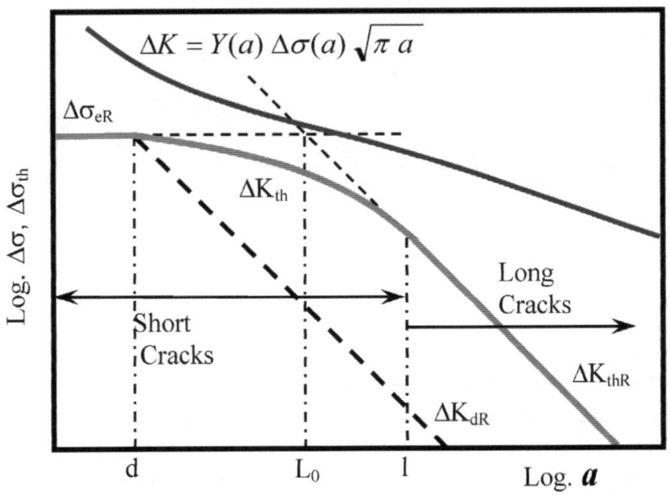


Fig. 3. Threshold curve concept in terms of the stress intensity factor range

Compressive residual stresses are produced by cold reverse bending. These compress residual stresses will act to protect the spring against its tensile service loads. High surface compressive residual stress value is produced by shot-peening. Shot-peening prevent fatigue crack propagation from the surface of bending loaded spring leaf. Properly done shot-peening generates high compress residual stresses and the fatigue crack process starts from the inclusion under the surface. However, producers of spring steel are faced with keeping constant quality of spring. Usually, production of steel has led to a reduction of the inclusion size in order to improve fatigue strength of steel. Springs after heat treatment are subjected to prestressing by cold reverse bending. Figure 4 shows an usual prestressing applied to truck springs. It is then placed in a fixture that loads it exactly as it will be loaded in service but at a level above tensile yield strength. When the load is released, it springs back to a new shape, which is that desired for assembly. However, the elastic recovery has now placed the material that yielded into a residual-stress state, which will be in the opposite (compressive) direction from that of the applied load. Therefore these compress residual stresses will act to protect the spring against its tensile service loads. Properly shot-peening springs can have their fatigue strengths increased to the point that they will fail by yielding instead of failing by fatigue [9]. The two treatments are additive on the upper surface in this case, affording greater protection against tensile stresses in fatigue. Note that if the spring were reversely loaded in service up to yielding, the beneficial compressive stress could be relieved compromising the fatigue life of the spring.

The aim of this study is to clarify the effect of residual stress in spring under surface layers and estimate fatigue life time, as interaction between inclusion size and residual stress under surface. The analysis is carried out by using a fracture mechanic approach and dealing with the resistance curve concept, quantifying the applied driving force and the resistance for different configurations of residual stresses and inclusion sizes.

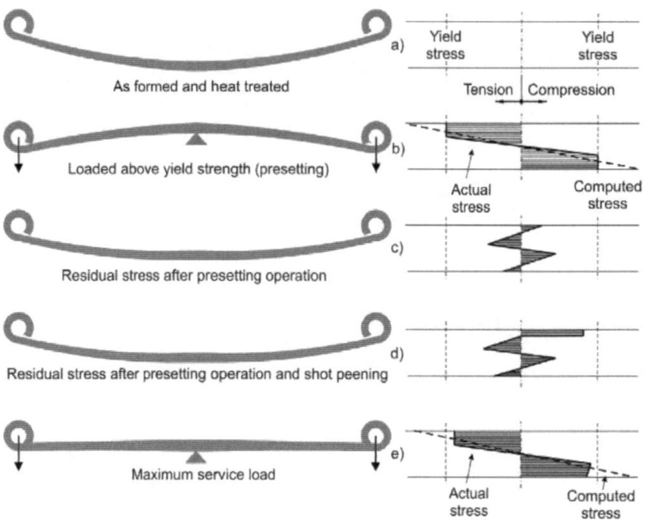


Fig. 4. Residual stresses profile from prestressing and shot-peening a leaf spring [9]

## 2. Material properties and residual stress measurements

Spring steel is delivered to spring producer in hot rolled condition as ferrite-perlite microstructure, with average grain size $d$=10 μm and average hardness 430 $HV$ (42±2 HRc). Its tensile mechanical properties of steel as delivered condition are $R_{p0,2}$=1050 MPa and ultimate tensile strength $\sigma_u$ =1270 MPa in average [10,11]. The spring's manufacturer performs hot rolling, hot bending, eye making and heat treatment. The goal of heat treatment is to achieve fine microstructure of tempered martensite with average grain size $d$=5 μm and average hardness 590 $HV$ (52±2 HRc). Tensile mechanical properties of the steel as delivered conditions are $R_{p0,2}$=1580 MPa and ultimate tensile strength $\sigma_u$=1670 MPa. Inclusions are usually MnS, alumo-silicates and TiN. The size of inclusions varies between 35 μm to 500 μm.

Spring leaf was used for residual stress measurement. Upper part (see Fig. 1) was shot-peening treated and compressive residual stresses are expected, on other side the tensile residual stress should be present. Residual stress measurement was performed at room temperature in ambient conditions by X-ray diffraction method.

Piece of cut spring was used for residual stress measurement. Measurement was performed at shot-peening side of spring in longitudinal and transversal direction. Figure 4 shows the residual stress measurement by Stresstech Rollscan equipment. The measured compressive residual stress on the surface was -493±20 MPa in longitudinal direction. The similar value at the same depth was obtained by Scuracchio and co-workers [12]. The results of residual stresses showed that the maximum residual stress is obtained at a depth of about 0.25 mm and is about 1100-1200 MPa (in compression). It is significantly higher than the one near the surface. For higher depths the residual stress decreases and is in compression about -500 to -800 MPa at a depth of 0.35 mm. Regarding this measurement a residual stress equal to -1100 MPa will be considered for a depth equal to 0.25 mm and -500 MPa for 0.35 mm.

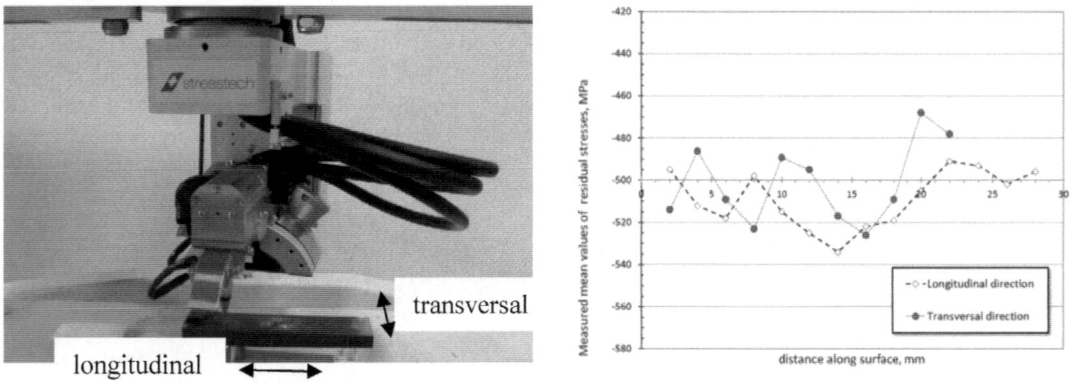


a) measurement by X-ray device   b) Results of residual stress measurement

Fig. 5 Measurement of residual stresses at the polished point 5μm under the surface

Murakami et al showed that the mechanism of fatigue failure starts around the inclusion as "Optical Dark Area" (ODA) as hydrogen embrittlement assisted by fatigue [13,14,15]. Murakami derived the model for the estimation of the threshold $\Delta K_{th}$ as function of $area^{1/2}$:

$$\Delta K_{th} = 3.3 \cdot 10^{-3} \cdot (HV + 120) \cdot (\sqrt{area})^{1/3} \tag{4}$$

where $HV$ is the Vickers hardness, in kgf/mm$^2$, and $area^{1/2}$ is in μm, giving $\Delta K_{th}$ in MPa·m$^{1/2}$. The model works well till the $\Delta K_{th}$ equals the threshold for long cracks. It seems that Murakami expression Eq. (4) works well till a value of $area^{1/2}$ of 1 mm for low strength steel, where fatigue threshold equals the one for long crack. The value of $area^{1/2}$ can be connected with inclusion size, as first iteration for initial fatigue crack length. Equation (4) shows also that the threshold for crack propagation increases with hardness. Therefore, it seems that with increasing of ODA and hardness of high strength material a higher fatigue threshold $\Delta K_{th}$ value can be obtain. Unfortunately, docent of failures of springs steel are caused by different size of inclusions, from 0.1 until 1.5 mm diameter, where spring survived only few or couple of thousand cycles [10, 11].

However, Chapetti shows that opposite trend for fatigue threshold of long crack appear, as is shown in Fig. 6 [6]. For long cracks the threshold decreases with hardness [7], and becomes independent of crack length for a given $R$ ratio. Chapetti shows that fatigue threshold vs. crack length exhibits bimodal behavior, one for short cracks and one for long cracks. Both models provide fatigue limit $\Delta K_{th}$ as a function of crack length, but with Murakami's model it is not possible to estimate fatigue life to failure. The pure fatigue crack propagation threshold for long cracks $\Delta K_{th,R=-1}$ can be estimated by using the following empirical expression [7]:

$$\Delta K_{th,R=-1} = -0.0038 \cdot \sigma_u + 15.5 \tag{5}$$

Where the pure fatigue crack propagation threshold $\Delta K_{th,R=-1}$ (a constant value for given tensile strength or hardness) are in MPa·m$^{1/2}$, the crack length $a$ in mm and the ultimate tensile strength $\sigma_u$ in MPa.

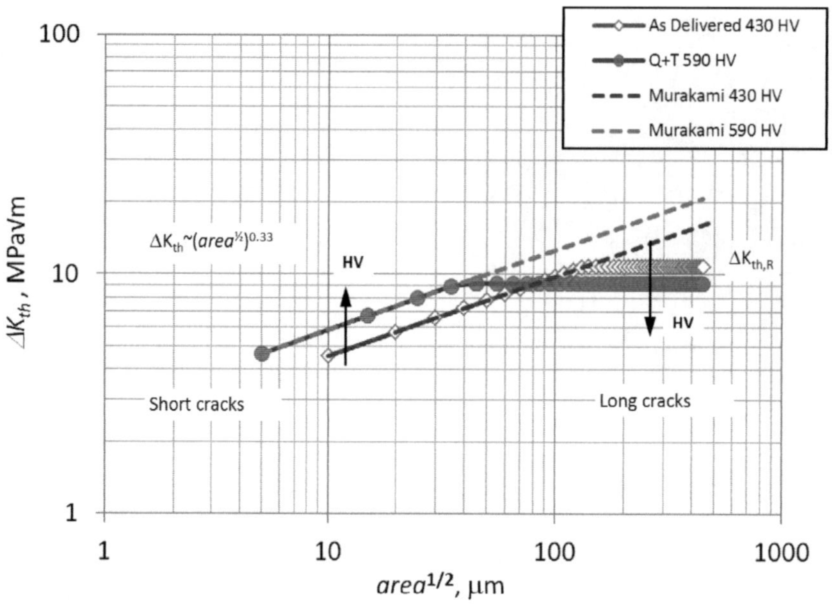


Fig. 6. Fatigue threshold as a function of $area^{1/2}$ parameter for 51CrV4 steels: with hardness 430 $HV$ as delivered and with hardness 590 $HV$ in quenched and tempered condition

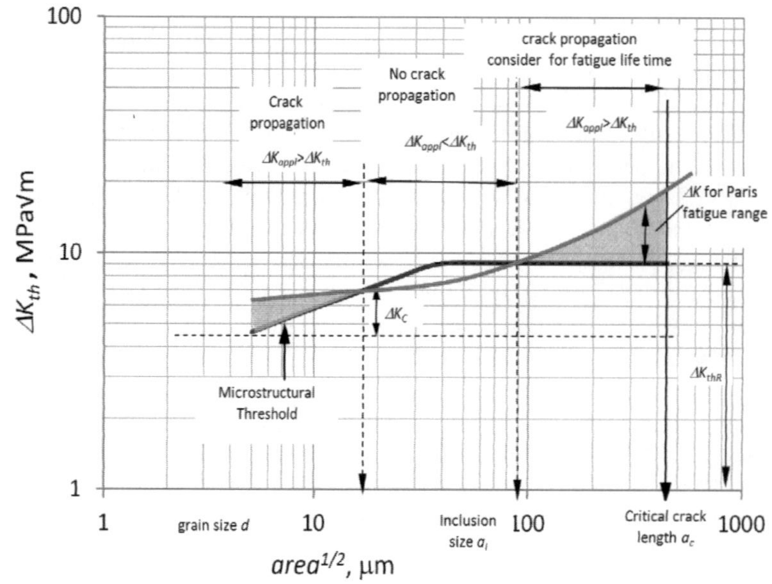


Fig. 7. Fatigue threshold as a function of $area^{1/2}$ vs. fatigue crack driving force for short and long crack propagation

Figure 7 shows schematically by red line the fatigue crack driving force ΔK. In the case of linear fracture mechanics and smooth specimens or spring after shot-peening, the following general expression can be used to estimate the applied driving force as a function of crack length [16]:

$$\Delta K = Y \cdot \Delta\sigma \sqrt{\pi \cdot a} \qquad (6)$$

where $\Delta\sigma$ is the stress range nominally applied at the initiation places.
The crack aspect ratio as a function of crack length has to be defined for the combination of component geometry and loading conditions, which allows definition of the value of the parameter *Y* as a function of crack length, same as in Eq. (1).

Therefore, if the fatigue crack's driving force ΔK is lower than the fatigue threshold $\Delta K_{th}$ respect to loading R ratio the fatigue crack does not grow. $\Delta K_{th}$ depends also about loading R ratio. If the loading R ratio is smaller or even negative the $\Delta K_{th}$ value is higher and fatigue crack will not grow.

A negative loading R ratio is possible to introduce by compressive residual stresses. Philosophy of leaf spring design is based on high strength of steel and induction of high compressive residual stresses in the tension loaded part of the spring. High strength steel grade 51CrV4 in thermo-mechanical treated condition is used for parabolic spring of heavy vehicles.

## 3. Fatigue Testing and Estimation of fatigue crack propagation life

Two groups of experimental results were obtained from representative specimens made from steel with hardness 590 *HV* in quenched and tempered condition. First group of tests are performed on hourglass specimens at room temperature using a four-axis cantilever type rotary bending fatigue machine at a stress ratio R=-1. Hourglass shaped specimens with a grip diameter of 15mm and minimum diameter of 7,5 mm were used. In order to induce minimum compress residual stress a specimens were grinded and electro-polishing in order to remove a layer of the surface about 10 μm thick. In this set of specimens the residual stresses from surface are removed by electro-polishing.

Second group of specimens is single leaf of original spring for tracks. The tests were performed by spring producer as part of regular quality insurance of products. The springs are tested by bending with applied stress amplitude 720±630 MPa. Testing was performed in factory by frequency 1 Hz. This springs had residual stresses and were tested at a stress ratio R=90/1350=0.067.

Quantitative analysis of fatigue crack growth requires a constitutive relationship of general validity be established between the rate of fatigue crack growth, d*a*/d*N*, and some function of the range of the applied stress intensity factor, $\Delta K$ (crack driving force). Besides, it has to take into account the threshold for the whole crack length range, including the short crack regime where the fatigue crack propagation threshold is a function of crack length. Among others, the following relationship meets these requirements [7]

$$\frac{da}{dN} = C \cdot (\Delta K_{appl} - \Delta K_{th,R=-1})^m \qquad (7)$$

where *C* and *m* are Paris range constants obtained from long crack fatigue behavior and $\Delta K_{th,R=-1}$ is the crack growth threshold as lower value of Eq. (4) and (5). The fatigue crack propagation life from crack initiation up to critical crack length $a_c$ can be obtained by integrating expression (5) and using expression (4) for the threshold of the material ($\Delta K_{th,R=-1}$). In the case of smooth specimens or spring after shot-peening, the stress can be considered constant for any crack length, equal to the nominal applied stress $\Delta\sigma_n$. The following general expression can be used to estimate the applied driving force as a function of crack length [8]:

$$\Delta K_{appl} = Y \cdot \Delta\sigma_n \sqrt{\pi \cdot a} \tag{8}$$

where $\Delta\sigma_n$ is the nominal applied stress range.
The crack aspect ratio as a function of crack length has to be defined for the combination of component geometry and loading conditions, which allows definition of the value of the parameter $Y=0.65$ [15] as a function of crack length, same as in Eq. (1).

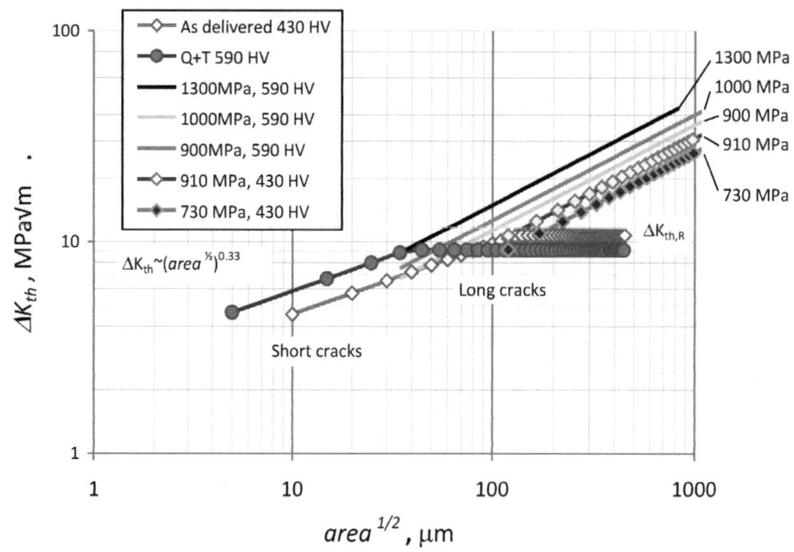


Fig. 8. Fatigue threshold as a function of $area^{1/2}$ parameter for both conditions of steel

Figure 8 shows fatigue crack driving force as difference between applied force $\Delta K_{appl}$ and threshold $\Delta K_{th,R}$ obtained by Chapetti's model by applying Eq. (2). It is possible to determine the number of cycles to failure regarding to different inclusion size and different applied fatigue stress magnitude $\Delta\sigma_n$, by using simple integration of Eq. (7) with experimentally obtained parameters of Paris fatigue crack propagation range ($C=8\cdot10^{-8}$ and $m=3.25$). Fatigue crack propagation occur only if applied crack driving force $\Delta K_{appl}$ is higher than threshold $\Delta K_{th,R}$ and if inclusion size (as $area^{1/2}$ in Fig. 8) is higher than value in intersection between threshold limit curve $\Delta K_{th,R}$ and applied crack driving force line. Fatigue crack is going to propagate until critical crack length $a_c$. The value of critical crack length is determined by fracture toughness of material $K_{IC}$=33.19 MPa·m$^{1/2}$. Figure 8 shows fatigue crack driving force curves vs. initial crack area for constant applied stress $\Delta\sigma_n$. Figure 8 shows that as delivered steel (softer) switch to constant threshold value $\Delta K_{th,R=-1}$=10.769 MPa·m$^{1/2}$ at larger size of inclusion $a_i$=120 μm than quenched and tempered steel, with size of inclusion $a_i$=35 μm and with constant threshold value $\Delta K_{th,R=-1}$=9.154 MPa·m$^{1/2}$.

Figure 9 shows results in form of *S-N* curves obtained by using proposed model for both conditions of 51CrV4 steel. Figure 9 shows the S-N curve obtained from hourglass specimens and points-results of spring leaves testing made of steel in quenched and tempered (Q+T) conditions. The fatigue testing results of hourglass specimens are compared with model where initial crack corresponding to different inclusion's size e.g. $a$=0.15, 0.35 and 0.5 mm. Different stress amplitudes $\Delta\sigma_n$ under constant stress ratio R=-1 and with constant inclusion size $a$, give a different number of cycles to failure. Obtained results as pairs of number of cycles N and applied stress amplitude $\Delta\sigma_n$ at fixed inclusion size are connected to S-N curve. Model provides conservative values at lower number of cycles, lower than 50.000 and quite good agreement at higher number of cycles, at 100.000 cycles or more. Figure 9 shows that applied stress amplitude $\Delta\sigma_n$ at same inclusion size (e.g. $a$=0.25mm) in case of compress residual stresses is few times higher at same number of cycles to failure. The results of spring leaves tests are in low cycle regime and all specimens shows subsurface inclusion-induced failure mode without Granular Bright Facets, as was mentioned in [10], and shown in Fig. 1.

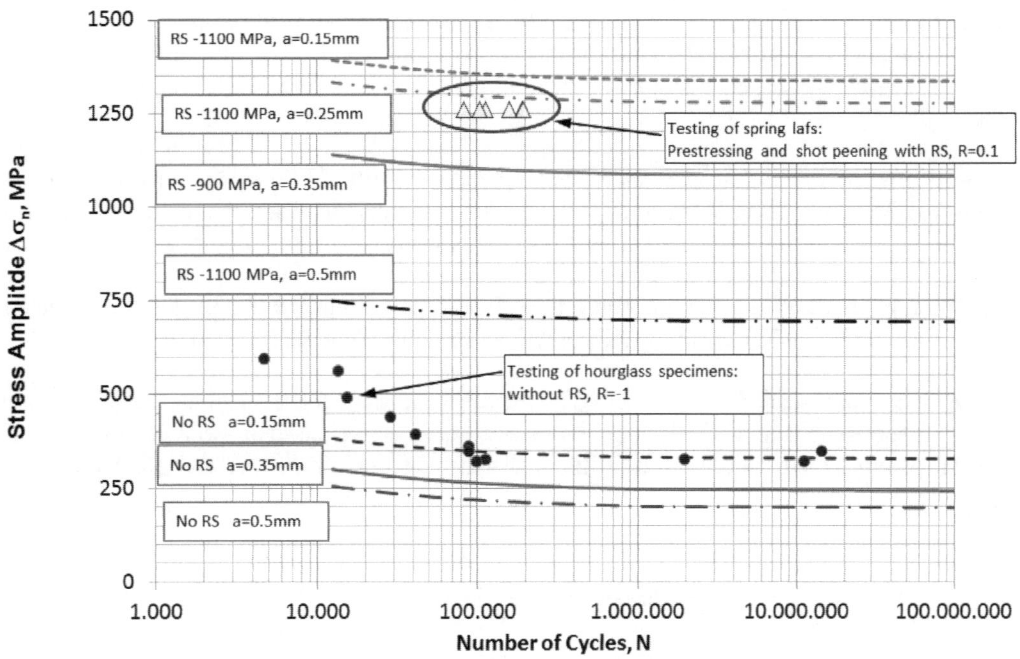


Fig. 9. Prediction of fatigue failure in form of *S-N* curve by using fracture mechanic approach and considering different size of inclusions in Q+T steel

Additional tests were conducted by spring producer on a leaf spring made from same steel with same inclusion size $a_i$ = 0.5 mm [10,11]. As it was mentioned the residual stresses on the surface should be at least $\sigma_{RS}$ = -1100 MPa. Tests were performed on six specimens. All six springs after pre-stressing and shot-peening were subjected the same loading regime (720±630 MPa), therefore, an applied stress amplitude $\Delta\sigma_n$ =1260 MPa and effective ratio $R$ =(-1100+90)/(-1100+1350)=-4. The effective loading ratio R can vary, depending on inclusion distance from surface. The crack initiating inclusion, as is shown in Fig.1, lies 0.4 mm underneath the surface, in this region of crack initiation, the residual stresses are significantly lower!

All leaf springs failed between 80.000 and 195.000 number of cycles as shown in Fig. 8. The experimental results on specimen without residual stresses lie in the same range of cycles, but for an applied stress amplitude $\Delta\sigma_n$=400 MPa and ratio $R$=-1. Therefore, difference between thus two groups of experimental results is consequence of different $R$ ratio, caused by residual stresses.

## 4. Conclusions

Chapetti's model has been used for determination of a threshold $\Delta K_{th}$ from short crack of grain size until critical crack length from less than one or only few millimeters corresponding to applied maximum stress $\Delta\sigma_n/(1\text{-}R)$. Life time of spring material subjected to applied stress amplitude $\Delta\sigma_n$ from microstructural threshold up to critical crack length of high strength steel has been determined, by combining fracture mechanics parameters $K_{IC}$, fatigue propagation Paris range parameters $C$ and $m$, and considering inclusion size $a_i$ as crack initiation area. Results are presented in form of *S-N* curves. Approach shows good agreement with experimentally obtained fatigue results for spring steel in Q+T condition with same inclusion size.

Pre-stressing and shot-peening increase the magnitude of compressive residual stresses on the surface of the leaf spring, but for fatigue the inclusion-induction failure mode is relevant residual stress under the surface. Residual stresses change applied loading ratio R=90/1350=0.067 to negative value R=-4 under the surface.

Results also showed that the procedure allow as to analyze the different combinations of initial crack length and residual stress levels, and how much the fatigue resistance can change by changing that configuration. For this set of tests, the fatigue resistance estimated for an initial crack length equal to 0.35 mm, explain conservatively the high cycle fatigue, but it cannot be applied to estimate low cycles fatigue (high applied nominal stresses). The fatigue lives smaller than 50.000 cycles is associated to low cycle fatigue (LCF) and not to threshold conditions for fatigue crack propagation.

An integrated fracture mechanics approach is proposed to account for the estimation of the fatigue resistance of component. The model is simple to apply, accounts for the main geometrical mechanical and material parameters that define the fatigue resistance of the component and allows acceptable predictions.

It offers a change in design philosophy: It could be used for design, dealing with crack propagation thresholds. Besides, it allows to performed material defect sensitive analysis, quantified by fatigue lives and fatigue limit.

Acknowledgements:
Financial support by the Slovenian Research Agency (ARRS) and the Argentinean Ministry for Sciences in frame of bilateral project is gratefully acknowledged.

**References**

[1] Shiozawa K, Lu L. Internal fatigue failure mechanism of high strength steels in gigacycle regime. *Key Eng Mat* 2008; 378-379:65-80.

[2] Shiozawa K, Hasegawa T, Kashiwagi Y. Lu L. Very high cycle fatigue properties of bearing steel under axial loading condition. *Int J Fatigue* 2009;31:880-8.

[3] Shimatani Y, Shiozawa K, Nakada T., Yoshimoto T., Effect of surface residual stress and inclusion size on fatigue failure mode of matrix HSS in very high cycle regime, *Procedia Engineering* 2010;**2**:873-882.

[4] Tanaka K, Akiniwa Y. Eng Fract Mech 1988;30(6):863–76.

[5] Chapetti MD. Fatigue propagation threshold of short cracks under constant amplitude Loading. Int J Fatigue 2003;25(12):1319–26.

[6] M.D. Chapetti / Engineering Fracture Mechanics 75 (2008) 1854–1863
[7] M.D. Chapetti / International Journal of Fatigue 33 (2011) 833–841
[8] M.D. Chapetti, L.F. Jaureguizahar. Int J Fatigue 43 (2012) 43–53
[9] Black Almen, Residual Stresses and Fatigue in Metals, McGraw-Hill, New York, 1963
[10] Gubeljak N. Predan J. Determination of dynamic strength of spring steel bellow yield stress, Industrial report for Štore Steel d.o.o. (in slovene), University of Maribor, Faculty of Mech. Eng. 2005
[11] Štore Steel: Company report about testing of springs, by spring producer, Nov. 2008
[12] B.G. Scuracchio, N. Batista de Lima, C.G. Schön. Materials & Design:47 (2013), 672-676.
[13] Murakami Y, Nomoto T, Ueda T. Fatigue Fract Eng Mater Struct 1999;22:581–90.
[14] Murakami Y, Nomoto T, Ueda T. Fatigue Fract Eng Mater Struct 2000;23:903–10.
[15] Murakami Y, Nomoto T, Ueda T. Fatigue Fract Eng Mater Struct 2000;23:893–902.
[16] H.O. Fuchs H.O. and Stephens R.I., Metal Fatigue in Engineering, John Wiley & Sons, New York, p. 130, 1980

# Conjoint Forming – Technologies for Simultaneous Forming and Joining

**P Groche, S Wohletz, A Mann, M Krech and V Monnerjahn**

Technische Universität Darmstadt, Institute for Production Engineering and Forming Machines, Otto-Berndt-Str. 2, 64287 Darmstadt, Germany

E-mail: groche@ptu.tu-darmstadt.de

**Abstract.** The market demand for new products optimized for e. g. lightweight applications or smart components leads to new challenges in production engineering. Hybrid structures represent one promising approach. They aim at higher product performance by using a suitable combination of different materials. The developments of hybrid structures stimulate the research on joining of dissimilar materials. Since they allow for joining dissimilar materials without external heating technologies based on joining by plastic deformation seem to be of special attractiveness. The paper at hand discusses the conjoint forming approach. This approach combines forming and joining in one process. Two or more workpieces are joined while at least one workpiece is plastically deformed. After presenting the fundamental joining mechanisms, the conjoint forming approach is discussed comprehensively. Examples of conjoint processes demonstrate the effectiveness and reveal the underlying phenomena.

## 1. Introduction

Forming technologies have been ever since a cost effective way of manufacturing parts. They have established as an important backbone for the human population´s supply with various kinds of products since the industrial revolution. In addition to the sheer generation of the parts' shape, a broad variety of forming technologies can also be utilized to join two or more parts. Within this paper the simultaneous forming and joining is highlighted. Benefits compared to common processing sequences, relevant joining mechanisms and challenges on the way to practical applications are discussed.

In the following the term "conjoint forming" is used when forming and joining operations are carried out in one process step. More precise, conjoint forming can be defined as joining of two or more workpieces by a plastic deformation which is necessary to reach the desired shape of at least one workpiece outside of the joining zone. As summarized by Mori et al. [1] and Groche et al. [2] various manufacturing technologies utilize plastic deformation in order to establish a joint. However, the presented technologies can be classified by being either a sole joining technique e.g. clinching [3] or a forming technique, which also enables the formation of joint e.g. roll bonding [4]. In contrast to joining by plastic deformation, conjoint forming demands for the parts' geometry generation combined with a joint formation, both initiated by plastic deformation. The uses of auxiliary elements like rivets or an integrated welding or brazing do not meet the requirements of the conjoint forming approach.

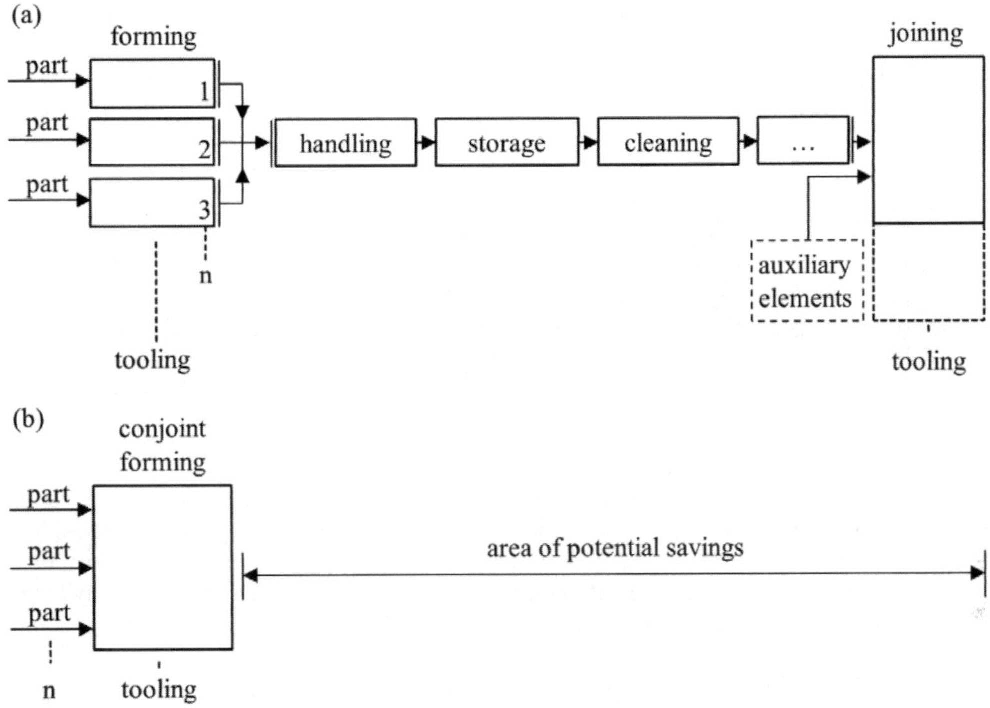


**Figure 1.** Comparison of a sequential (a) and a conjoint forming process layout (b).

Figure 1 illustrates the conjoint forming approach in comparison to a sequential layout, in which forming and joining are performed separately. As it can be seen in the scheme, a sequential process layout demands for multiple tooling and machines, on which the parts are formed and then joined afterwards. In between forming and joining the parts have to be handled, stored and might undergo additional inspection or cleaning prior to the joining. However, additional auxiliary elements, e.g. rivets or screws, can be required for the joining process depending on which joining technology is chosen. Additional joining operations e.g. screwing or friction welding can yield to high cycle times. Thus, conjoint forming offers the shortening of the processing time, because joining and forming are performed during the manufacturing process with a higher productivity.

Conjoint forming processes can be classified by the occurring joint mechanisms. In the following chapter, the classification of joining mechanisms by Mori [1] is used to describe the bond formation.

## 2. Joining mechanisms

The conjoint forming approach combines the joining of at least two parts with the formation of the workpiece's geometry within one process. The approach of conjoint forming is initiated by the plastic deformation of at least one part. Mori divides joining by plastic deformation in metallurgical and mechanical joining (Figure 2) [1]. Mechanisms for both principles are described in the following.

### *2.1. Mechanical joints*

A subdivision of mechanical joints divides joints created by forming into form and force-closed joints (Figure 2). Form-closed connections can be obtained by using fasteners such as bolts, pins or rivets. Form-closed joints without auxiliary elements can be established by forming overlapping parts of the components. The parts in contact with each other offer a positive traction in contact normal direction, which is mainly the applicable load direction. In force-closed connections the relative movement of the joint parts is prevented by friction forces, which result from residual stresses due to a previous deformation [2].

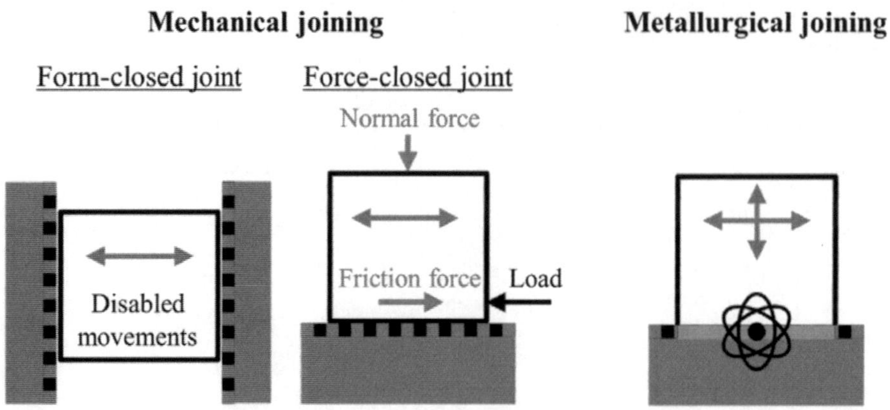


**Figure 2.** Classification of physical joining mechanisms referring to [1], [2], [5].

Main mechanisms of mechanical joints are discussed in [6]. Expansion and compression of tubular joining partners result in a joint (Figure 3). Therefore, two joining partners A and B are positioned to each other (phase 1), which results into an initial gap *g* between them. In the next step the tube (A) is deformed plastically by a radial acting internal pressure *p*. With the occuring contact between the joining partners their deformations proceed until the process is finished (phase 2). In this example, the tube (A) is elastic-plastically deformed, whereas the other joining partner (B) behaves purely elastically. After the elimination of the pressure *p*, springback of the tube is prevented by the other joining partner and leads to the stress $\sigma_{t,rem}$ (phase 3). Consequently, the main prerequisite for a mechanical joint is a remaining stress in the boundary layer of the joining partners.

Based on numerical results presented in [5], the temperature of the joining partners increases locally in the forming zone. Due to different temperatures or coefficients of thermal expansion, different shrinking of the joining partners might occur and lead to modified stresses in the contact zone. Matsumoto et al. develop the indentation of a cold shaft into a hot disc, where the shear bonding stress of the joint increase with higher temperatures due to the thermal shrinkage [7]. Other joining processes require specific temperature conditions. In composite extrusion of reinforcing elements [8] and in thixojoining [9], the formability of the material has to be increased by temperature to allow a processing and create jagged transitions of the joint partners. Further increase of the joint′s strength is generated by form-fits which require the formation of an undercut or intersection of the joint partners.

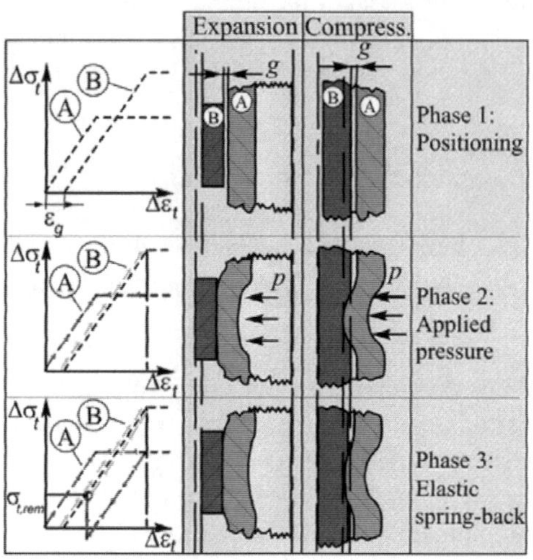


**Figure 3.** Force-closed joint [6]

Mechanical joining processes are influenced by several process parameters. Different process conditions results in different joining strengths due to different remaining stresses in the boundary layer. Force-fit joints without an intersection of the joint partners are dimensioned decisively by its interference pressure, which can be calculated analytically for simple structures. The interference pressure p arises from the yield strength $k_f$, the Young's modulus $E$, the Poisson′s ratios $v$, the ratios of the inner and outer diameter $Q$ and the applied fluid pressure $p_i$ (1). It can be seen, that there is a limit for the maximum pressure, which is reached when the elastic limit of the ring is reached.

$$p = \frac{\frac{p_i - k_{f,i} \cdot \ln \frac{1}{Q_I}}{E_0} \cdot \left(\frac{1+Q_0^2}{1-Q_0^2}\right) + \frac{2 \cdot Q_I^2 \cdot k_{f,i} \cdot \ln Q_I}{E_I \cdot (1-Q_I^2)}}{\frac{1}{E_0} \cdot \left(\frac{1+Q_0^2}{1-Q_0^2} + v_0\right) + \frac{1}{E_1} \cdot \left(\frac{1+Q_0^2}{1-Q_0^2} + v_1\right)} \tag{1}$$

*2.2. Metallurgical joints*

Metallurgical joints can be discussed on various modelling levels. In principle, metallurgical joints are a result of thermic, electric, chemical and physical interactions of the involved surfaces. Interactions take place at contact points of both interacting surfaces. At these contact points adhesion occurs which results in an increased force to separate these contact points. Adhesion theories illustrate different behaviors of bonding which are used parallel and not exclusively.

However, in the presence of normal atmospheric conditions, free cohesive bindings will be saturated by the formation of oxide layers and the substrates' surfaces [10]. According to Buckley, these passivation layers act as a native protection inhibiting adhesion [11]. In order to establish a metallurgical joint, the joint interface is to be protected against atmospheric oxygen and non-passivated as well as contaminant free material have to get into contact with each other.

The mechanisms for the joint formation in cold pressure welding processes such as roll bonding are applicable to other conjoint forming operations and yet well understood. According to Vaidyanath et al. [12] the bond strength depends on overall reduction of the roll bonded sheets. However, a certain threshold deformation has proven to be crucial for a primarily bond formation [13]. Furthermore, the weld strength can exceed the weaker materials' strength at higher degrees of reduction. According to Wright et al. this is due to the fact that brittle covering layers on the substrates' surface exceed their formability at very low strains and juvenile material can extrude between the cracked covering layer [13]. However, Bay presented a model for cold pressure welding, which takes the surface exposure due to plastic strain as well as the state of stress at the weld interface as crucial factors into account. According to the phenomenological model two mechanisms contribute to the formation of a weld [14]. The first mechanism is characterized by the thinning of the contaminant film on the surface. Due to plastic strain on the substrate's surface the adherent deposit of fluids are thinning until the film breaks down locally [15]. In these regions, welds are formed. The second mechanism is characterized by breaking the surface covering layer. This covering layer has to be brittle which can result from a severely cold work hardening treatment like grinding or brushing as well as from the chemical composition of this layer e.g. oxides. This layer will break up at minor strain and juvenile material can be extruded between these cracks and form a weld.

Bay et al. present a theoretical model given by equation (2) to calculate the bond strength $\sigma_B$ depending on the yield stress $\sigma_0$ of the deformed material with regard to these two mechanisms [4]. The contaminants thinning mechanism contributes to the bond strength by the portion $\beta$. Furthermore, a certain threshold surface exposure $Y'$ is needed in order to exceed the contaminant's film coherence.

$$\frac{\sigma_B}{\sigma_0} = (1-\beta) Y \frac{p-p_E}{\sigma_0} + \beta \frac{Y-Y'}{1-Y'} \cdot \frac{p}{\sigma_0} \tag{2}$$

The first term on the right side corresponds to the mechanism of the covering layer's fracture. If the normal pressure $p$ exceeds the extrusion pressure $p_E$, plastic flow between the cracked covering layer sets on. By further increase of the normal pressure or decrease of the extrusion pressure due to further increase of the surface exposure, the bonding increases proportionally to the difference of the normal pressure and extrusion pressure. The presented model is further extended by Zhang [16] in order to compensate the remaining ductility of the covering layer, which has led to inaccuracies at higher superposing hydrostatic pressure [17]. Furthermore, this model is also extended for the case of welding dissimilar metals [18].

Nowadays, this numerical model has been integrated to finite element modelling by utilizing cohesive element calculations [19]. In addition to the state of stress and surface exposure, temperature and surface treatment are crucial factors for the formation of sound bonds. The weld surfaces are commonly mechanically treated prior to cold welding. In order to have a technical contaminant free and brittle surface, the combination of degreasing and wire brushing has proven to be favorable for the promotion of a weld occurrence [20]. Depending on the material combination chemical and electrochemical surface preparation methods can be even more beneficial for the weld formation [21]. Although cold pressure welding is considered as a cold state welding process, the processing temperature has a significant effect on the weld strength. Even forming at slightly elevated temperature can lead to increased bond strength and decreased threshold surface enlargement, although the temperature is considered to be far too low to promote diffusion of solids at the given processing time. In this context elevated temperatures may lead to a decrease of the surface mismatch and to a quality improvement of the weld once contact has been made. Higher temperatures may also help to break-down any absorbed water vapor or other contaminants on the samples' surfaces [22]. A subsequent heat treatment can also increase the overall bond strength [23]. However, the heat treatment of dissimilar metal joints may also lead to a decreased bond strength due to the brittleness of intermetallic phases and Kirkendall porosity at the interface [24].

## 3. Conjoint forming applications

The conjoint forming approach describes a simultaneous forming and joining process. As discussed in the previous chapter metallurgical joining differs in requirements from mechanical joining if executed simultaneously with forming. Joint requirements for the relevant joining mechanism as well as the

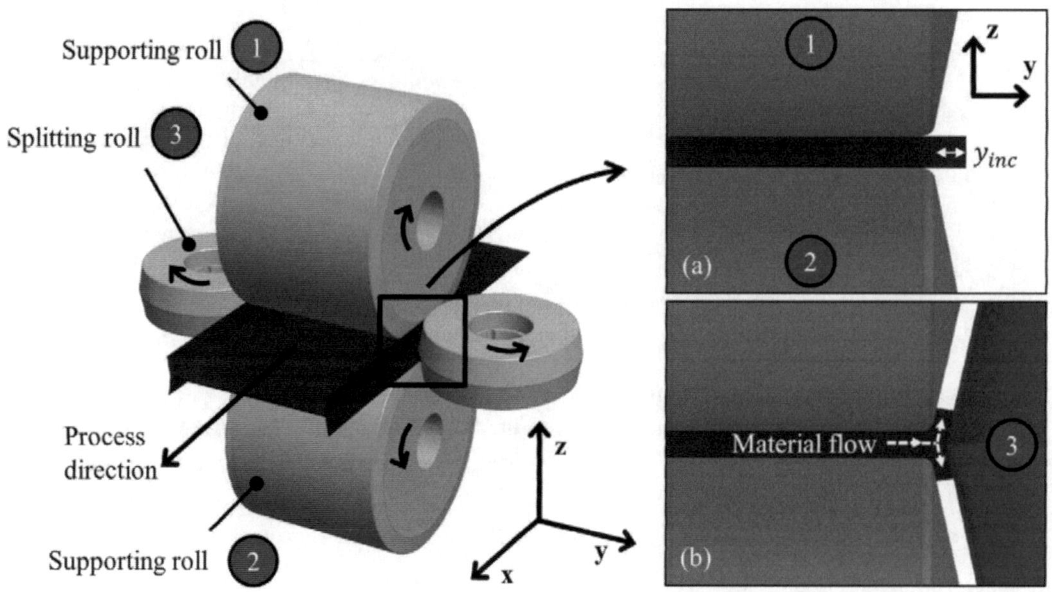


**Figure 4.** Process principle of Linear Flow Splitting [26]

conditions for the creation of the workpiece's targeted shape have to be fulfilled simultaneously. The conjoint forming approach is discussed in the following, focusing on selected conjoint forming applications.

### *3.1. Conjoint forming utilizing mechanical joining*

Linear Flow Splitting is a bulk forming process used to manufacture cold rolled profiles. The tool system for Linear Flow Splitting consists of an upper and lower supporting roll and two laterally arranged splitting rolls (Figure 4). During the forming process a sheet with a width larger than the distance between the supporting rolls is fed into the roll stand (a). The sheet metal is formed into the free flange gap between supporting and splitting roll (b) [25].

In the web of split profiles a specific distribution of longitudinal and lateral strains occurs after forming [27], which is determined through the optical measurement system GOM Argus (Figure 5). In longitudinal direction positive strains $\varepsilon_x$ appear, which are evenly distributed over the cross section. Contrary in lateral direction there is a gradient of negative strains $\varepsilon_y$ which increases in the direction to the splitting area.

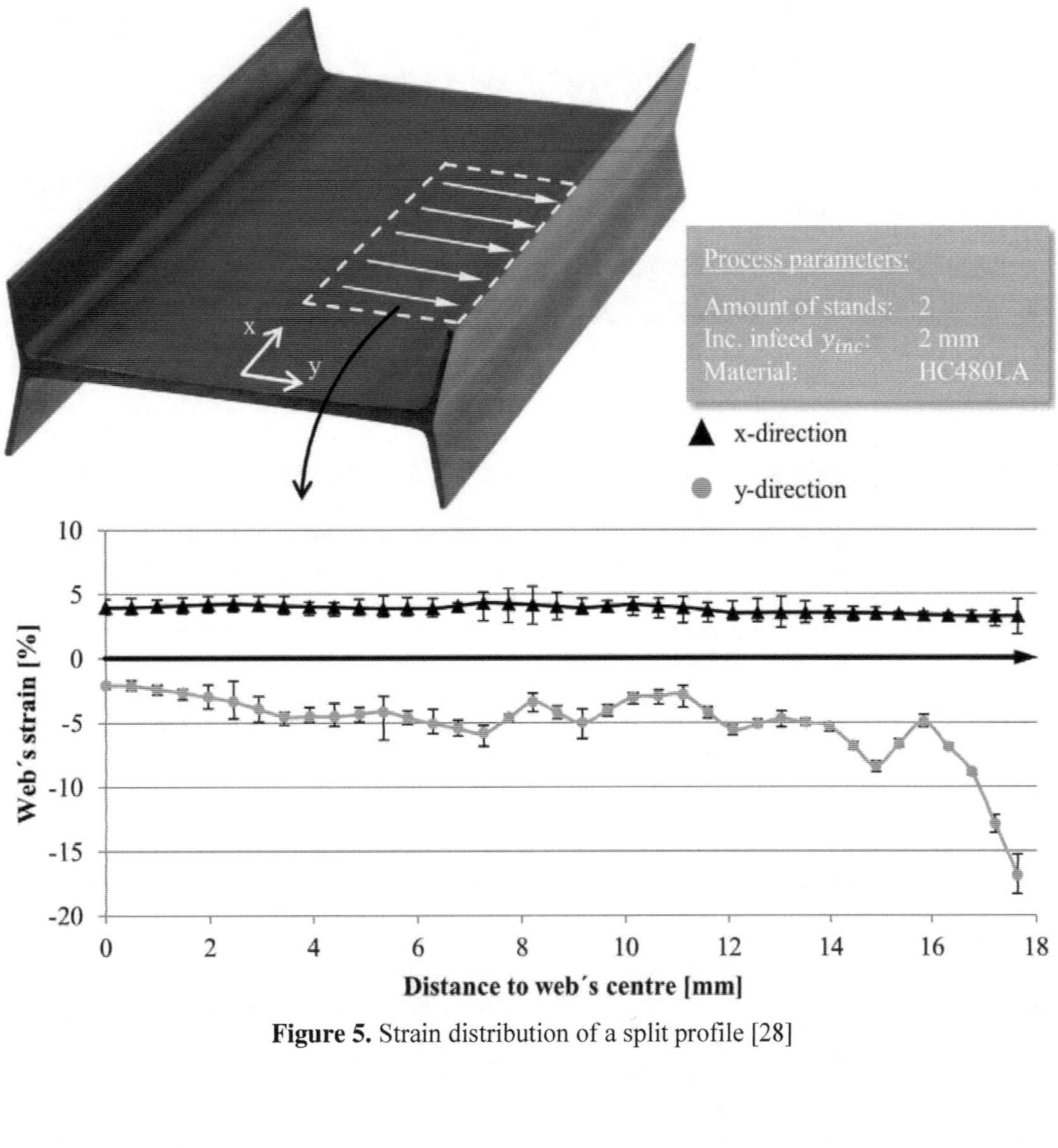


**Figure 5.** Strain distribution of a split profile [28]

Following the conjoint forming approach, mechanical joining through Linear Flow Splitting uses the energy of the forming process. Consequently, a higher efficiency by substituting additional joining processes is achieved. In order to demonstrate how process-induced strains can be utilized to create mechanical joints during the forming operation, a gear rack is joined by making use of the longitudinal stresses $\sigma_x$ (Figure 6). This integrated part can serve as a mechanical drive in linear systems [29]. Clearance holes are drilled into the split profile and a gear rack is inserted. The profile is elongated by the forming process, due to the roll out of material compressed in lateral direction. The compression results in residual stresses in the profile´s web (Figure 7). During the release a closing of the gaps between the clearance holes and the rack takes place. Subsequently, the rack is stressed with tensile stresses during the forming process. After the forming process, a partial spring-back of the elastic longitudinal stresses occurs and results in a force-fit joint between the two components. The complete spring-back of the profile is prohibited by the flanges and leads to a shifted stress level in the web, which also hinders the rack to spring back completely. In this way a remaining stress level $\Delta\sigma_{rem}$ is generated.

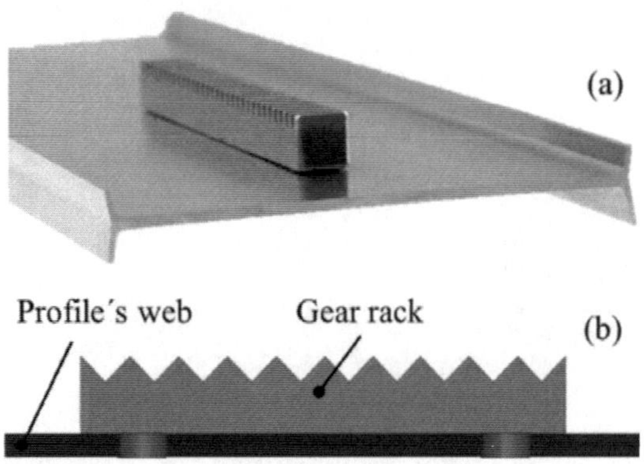


**Figure 6.** Flow split profile with joined gear rack (a) and cross section of joint (b) [29]

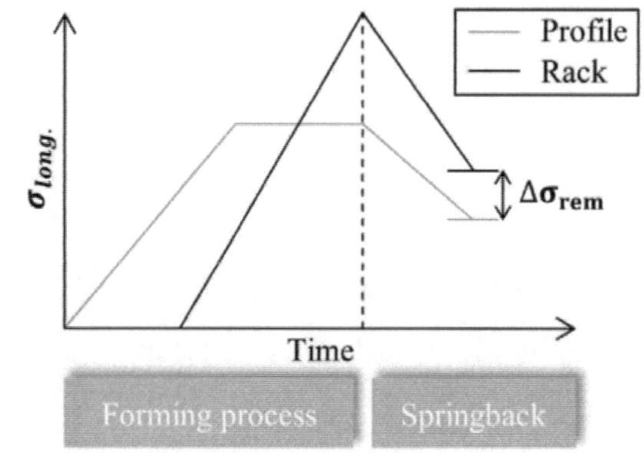


**Figure 7.** Longitudinal stresses of joint

Another example of conjoint forming is the integration of functional elements in load-bearing structures by rotary swaging. This promising approach to manufacture integrated machine elements utilizes unused internal cavities in hollow tubular work pieces. Sensors and actuators as well as the required periphery can be integrated in existing cavities of products processed by forming (Figure 8). A supplementary advantage can be achieved from the protected alignment of these elements. New product geometries are enabled which offer a high protection against moisture, stone impact or the attack of corrosive media. Compared to differentially built function-integrated structures, the conjoint forming approach offers an efficient and robust manufacturing method. Yet, common functional materials are characterized by a low ductility and sensitivity towards pressure peaks. Therefore, an adaption of existing processes is often required to enable a damage-free integration. Incremental forming processes have proven their feasibility for a damage-free joining of brittle and sensitive piezo elements in load-bearing structures [5,30,31].

Process-related stress states can be utilized in order to achieve the desired coupling of the functional elements and the structure. Only by creating a force-closure, a successful transmission of structural deformations to the sensor or an effective deflection of the structure by an actuator is possible. The damage-free joint is hereby enabled by using adapted end caps which absorb and convert high radial forming forces to axial compression. Only by redirecting the high radial forces, the piezo can withstand the joining by forming process without any damage (Figure 10). A specific preform is created by recess rotary swaging, whereby an exact length of the reduced zone is crucial to facilitate the subsequent joining step (Figure 10 a). In this second step, the functional element as well as the end caps are inserted, positioned and precompressed by a mandrel (Figure 10 b). The adaption of the

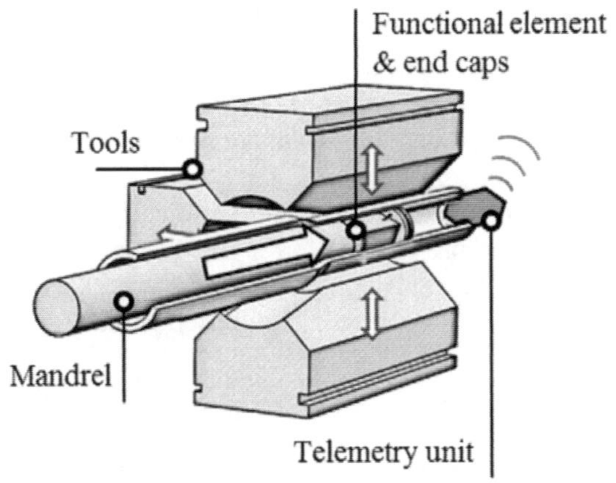


**Figure 8**. Setup of the machine and parts with attached telemetry system

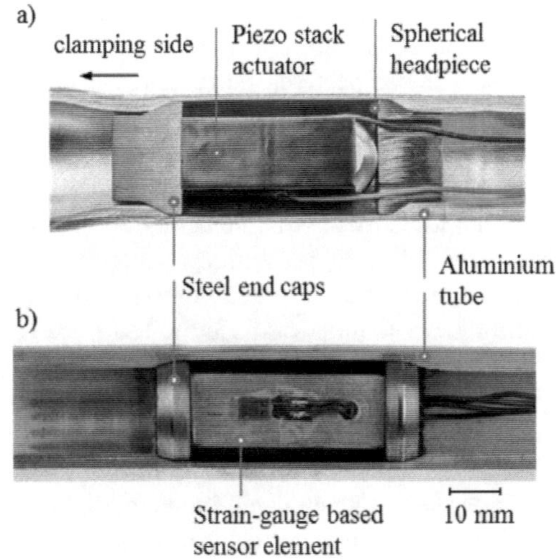


**Figure 9**. Integrated functional elements a) Piezoelectric stack actuator, b) strain-gauge based sensor element

tube's material to the end cap leads to axial forces and finally to a remaining pretension in the compound (Figure 10 c).

In Figure 11, a characteristic force curve of the joining process is illustrated, whereby a strain-gauge based sensor element is evaluated during shaping. A telemetry unit is attached to the free end of

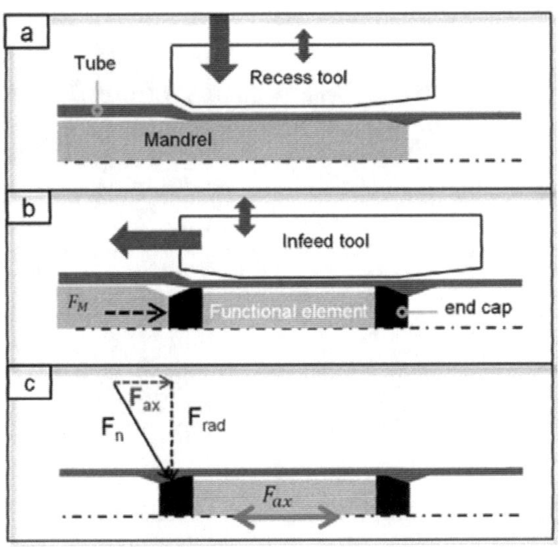


**Figure 10**. Two stage rotary swaging process for a conjoint forming of functional machine elements [31]

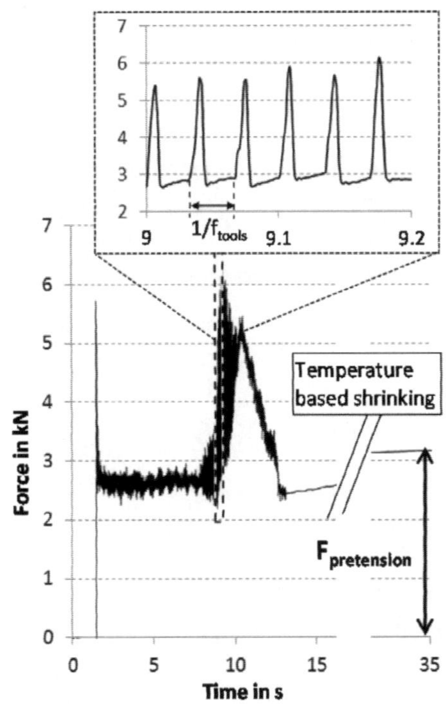


**Figure 11**. Combined formation of axial pretension in the sensor by elastic spring-back and thermal shrinking [31]

the rotating workpiece inside the rotary swaging unit. The incremental character of this forming process gets illustrated by the force peaks occurring in the same frequency as the tool strokes. From a macroscopic perspective, the force curve shows a peak in second 2, when the hydraulic mandrel force is applied. The actual forming takes place between second 8 and 12, in which considerable force peaks occur. It can be noted, that from second 8 on, each stroke leads to an increased pretension level, which can be read as the lower bound of the curve. The magnified detail view shows that a plateau is reached after each tool stroke which can be interpreted as the prevailing compression in the functional element. A reduction of the upper and lower bound of the force curve can be observed from second 11 on, when additional reduction of the remaining tube leads to reduced pretension levels. At the end of the forming process in second 13, the tools are opened and a thermal shrinking of the tube occurs as the deformation heat has partially increased the tube's temperature. After equalization of temperatures, the remaining pretension force $F_{\text{pretension}}$ amounts to 3.2 kN.

*3.2. Conjoint forming utilizing metallurgical joining*

In comparison to mechanical joints, metallurgical joints seem to be harder to obtain due to the fact that significant surface exposure and normal pressures are necessary for sound bonding. Contrary to mechanical joints, solid state welding by plastic deformation can only be achieved by the deformation of both joining partners. However, metallurgical welds can provide high bond strength, which can exceed the flow stress of the weaker material. Furthermore, no additional space is needed for the joint due to the direct welding of the contacting surface.

Figure 12 illustrates the experimental setup for cold extrusion welding of steel and aluminum. The billets are heat treated, coated with lubricant and wire brushed or polished prior to cold forging. In order to obtain a stronger bond, the flow stress of the steel and aluminum are adjusted to overcome the difference of the flow stress in the given limits [32]. Therefore, the steel billets are spheroidized and the aluminum billets are artificially aged.

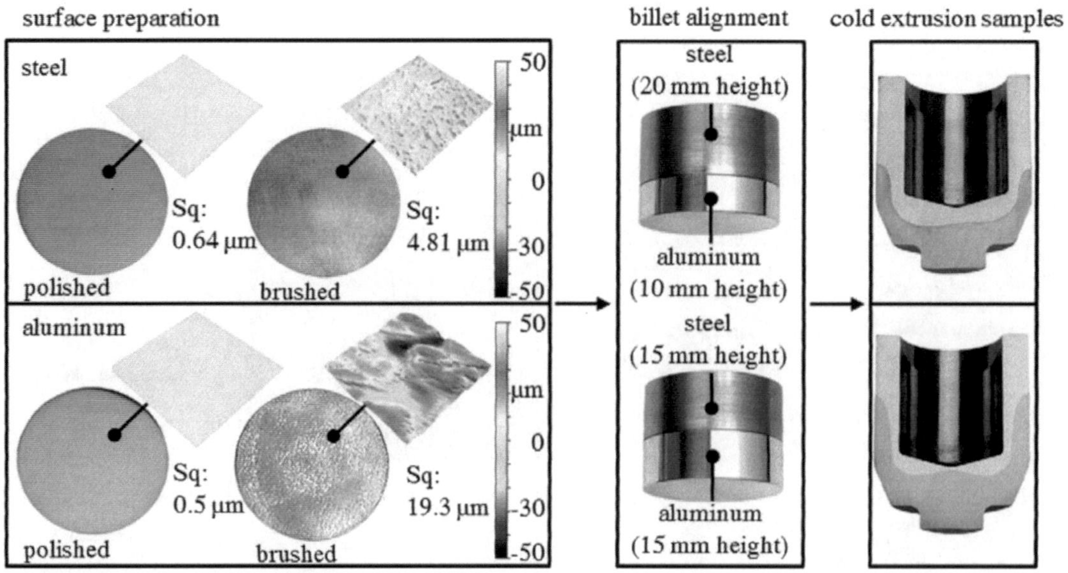


**Figure 12.** Cold extrusion welding setup [32]

Numerical modelling of the cold extrusion welding enables a deeper insight into the tribological conditions during the processing. As given by equation (3) the surface exposure $\chi$ represents the increase of the contacting surface by dividing the difference of the actual surface $A_i$ and initial surface

$A_0$ by the initial surface. Values greater than zero correspond to an increase of the interface, while values lower than zero represent a compression of the interface. The normalized pressure represents the contact pressure at the interface normalized by the flow stress. Due to a high plastic strain and an uneven strain distribution at the parts' geometry, this value expresses whether and by how far the normal pressure exceeds the materials' flow stresses.

$$\chi = \frac{A_i - A_0}{A_0}; \qquad \sigma_{np} = \frac{\sigma_n}{k_{fi}} \tag{3}$$

Figure 13 illustrates the surface exposure over the normalized pressure for the steel part at the interface prior and after coalescence of the two initial specimens is gained. It can be seen that there are

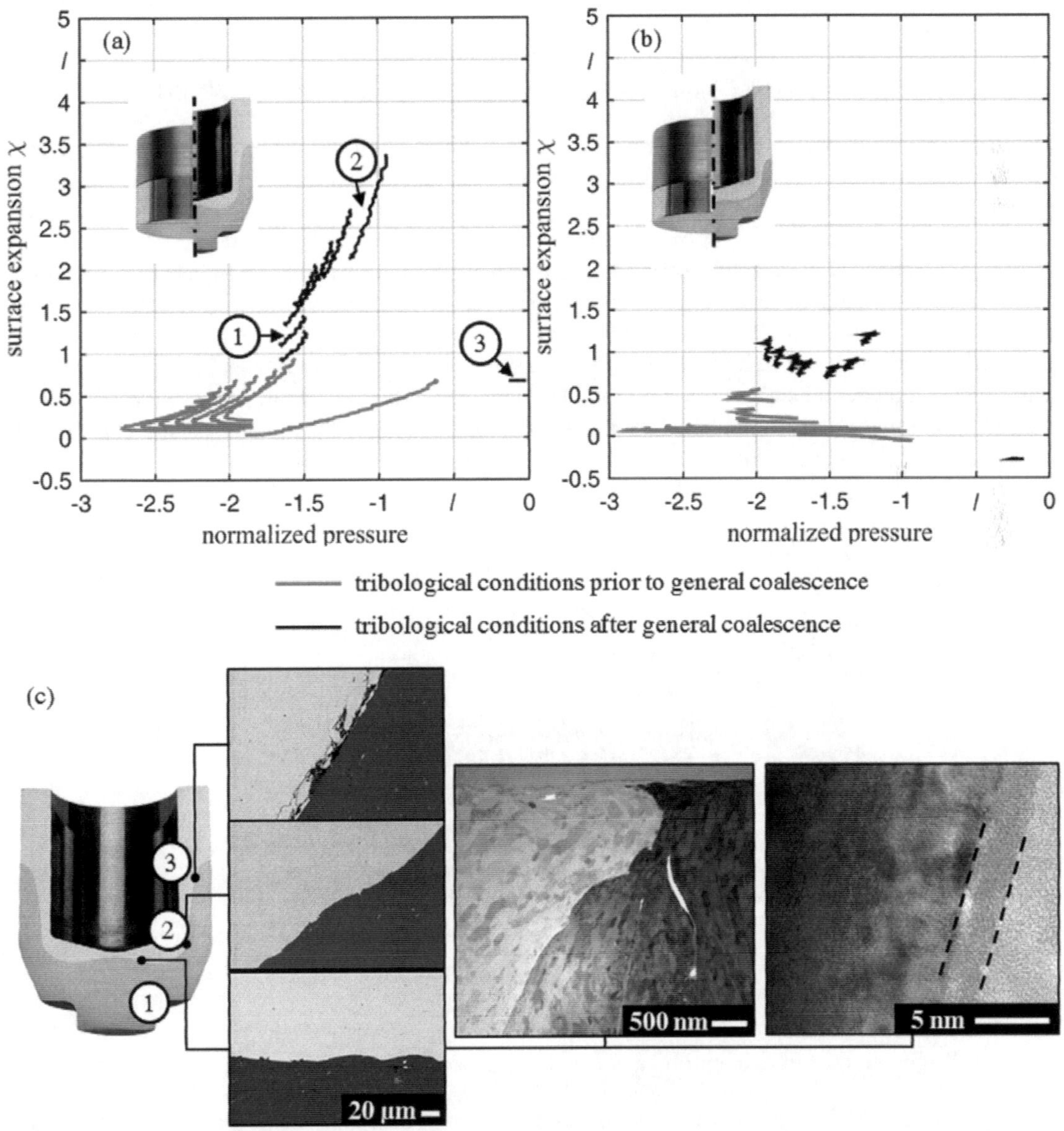


**Figure 13.** Numerical analysis of the surface expansion and normalized contact pressure for two alignments of steel and aluminum (a, b) and microscopic analysis of the weld interface (c)

mainly three regions which are characterized by different levels of surface exposure. The first region moderate values are obtained. In the second region the surface exposure reaches its maximum, whereas in the last section the interface is even compressed. By regarding the normalized pressure distribution, one can also see three distinguished regions. In the first region the pressure exceeds the material's flow stress five times. In the second section moderate values for the normalized pressure can be determined. However, a significant drop of the normalized pressure can be seen in the last region. According to the scanning electron microscopic (SEM) analysis illustrated in Figure 13 (c), no coalescences can be obtained, where the surface is compressed and the normalized pressure reaches its lowest values. However, at the first and second section a sound bond is formed. The absolute values for the surface exposure and normalized pressure also depend on the materials alignment. In case of the second alignment of 20 mm steel to 10 mm of aluminum the characteristic progression of the tribological conditions are similar (Figure 13 (c)). However, the overall values differ significantly. For the second setup the acquired values are significantly lower than in the first setup of 15 mm steel to 15 mm aluminum.

The SEM analysis enables the localization of unbonded areas. Furthermore, using the backscattering electron contrast, light materials like aluminum are illustrated darker than heavier materials like steel. Sections, which are bonded, are only characterized by a tight alignment at the given resolution. Significant intermetallic phases, which are prone to brittle fracture, are not present. As it can be seen in the transmission electron microscopic (TEM) brightfield analysis in Figure 13 (c), significant grain refinement with a mean grain diameter of about 100 nm occurs at the weld interface of the cold welded samples. This severe cold hardening on both sides of the interface contributes to the overall bond strength and is one of the reasons that the weld strength can exceed the weaker material's strength. Although, no significant intermetallic phases are formed during the cold forging, localized accumulation of intermetallic formation can be acquired. However, these formations are not evenly distributed. Further analysis in high resolution TEM shows a thin reaction layer of about 2 nm - 3 nm (Figure 13 (c) right hand side). In this reaction layer both materials' key elements iron and aluminum are present. Whether intermetallic phases are formed depends on the present temperature, the material composition and the time, in which the parts are processed at the elevated temperature. Processing at elevated temperatures can lead to the formation of intermetallic phases e.g. during lateral extrusion of aluminum and magnesium as given in [33]. Furthermore, the extractions of the cups verify the presence of a metallurgical weld.

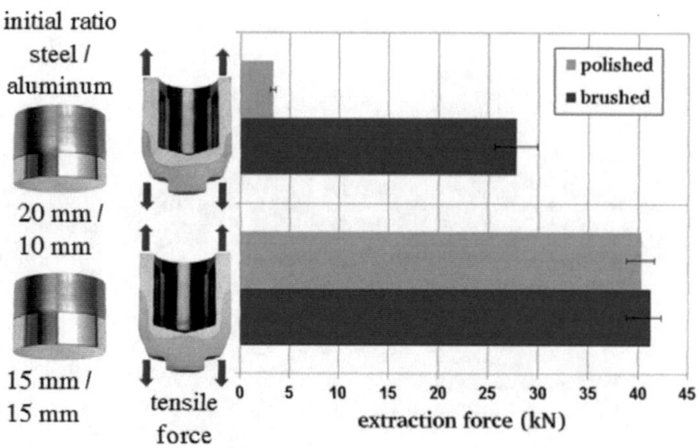


**Figure 14.** Weld interface of cold extrusion welded specimen a) and the extraction forces b) [34]

In order to measure the forces, which are needed to separate the cups, the specimens are tested in a tensile testing machine. As illustrated in Figure 14, the extraction force depends on the material distribution and on the surface preparation. The tensile forces of the joint samples with a lower aluminum ratio reveal the surface hardening effect of brushing on the bond formation. Due to that, the plastic formability of contaminated boundary layers is decreased prior to cold extrusion and the exposure of juvenile base material is promoted. A lower strain leads to lower bond strength. At a higher overall strain occurring at an initial billet height ratio of 15 mm / 15 mm the surface expansion of both parts reach significantly higher values, so that the influence of a mechanical surface pre-treatment is only of subordinate order.

In order to overcome the restrictions given by the material alignment and differences in the materials' flow stresses, the forming temperature of the parts can be adjusted. As illustrated in Figure 15 a mismatch of the materials' flow stress can lead to geometric deviations.

If in case of cup extrusion the initial temperature of the billets is not adjusted well, two kinds of failures can be observed. If the flow stress of the aluminum workpiece is too low, the extruded steel cup ruptures at the end of the punch's calibration zone due to exceeding the ultimate tensile stress. If the flow stress of the first part (e.g. the steel part in this setup) is nearly at the same level as the second part's flow stress, the first part is primarily formed and lifts off. If the flow stress of both materials is adjusted well, like in the case of the 600 °C steel billet and 200 °C aluminum billet, sound form filling and bonding occurs [32].

In addition to cold forging other forming processes, mainly massive forming processing, can be utilized for conjoint forming applications, if the afore mentioned conditions for a bond formation are met. Roll bonding sheet materials in the cold and hot stage is one of the most important conjoint forming processes, due to its widely spread industrial application. Further applications for conjoint forming by forging or bar extrusion are given in [36–38].

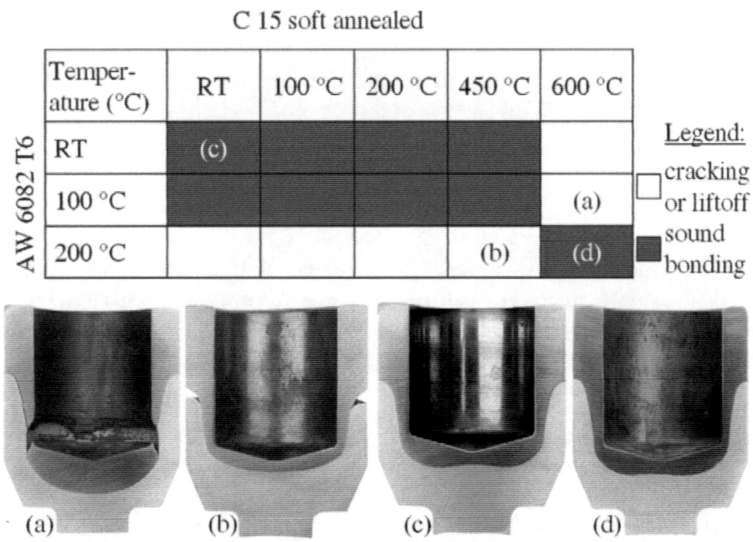


**Figure 15.** Temperature dependency by joining of dissimilar metals by cup extrusion [35]

## 4. Conclusion

This paper presents the conjoint forming approach. In the beginning the joining mechanisms for metallurgical and mechanical joints are discussed. Examples for conjoint forming applications with mechanical joining are rotary swaging and linear flow splitting. Furthermore, the weld formation of cold extrusion welding is discussed in the context of conjoint forming.

Due to the shortening in the process layout of conjoint formed parts, additional steps in between the operations are obsolete (Figure 1). Furthermore, the number of machines and tools can be reduced. However, the grade of complexity increases by following the conjoint approach, which can lead to a more demanding tool and process layout. The efforts for the preparation of the parts can increase due to the fact that the requirements for the combined forming and joining have to be met at the same time. In comparison to a sequential processing, the process time can be reduced significantly to the single, longer cycle time. Conjoint forming combines the advantages of forming technologies within a functional compound. The use of different materials enables the combination of high strength and lightweight material for the sake of weight reduction by an optimal strength to weight ratio. Furthermore, the use of costly materials in a component can be reduced by the combination of more and less costly materials in the component's structure, depending on where the specific material's properties are needed locally.The example of the metallurgical joint performed by a cold extrusion welding shows the lightweight potential of conjoint forming processes, combining steel with aluminum. Materials can be positioned and chosen by their mechanical properties to manufacture a weight optimized part.

The simultaneous forming and joining can positively affect the robustness of the manufacturing process' tolerance chain. The example of rotary swaging shows that even high geometrical deviations of the nominal shape can be compensated for producing a mechanical joint. Furthermore, the integration of functional elements allows a process monitoring of the conjoint forming process. As a future application, the process monitoring can be extended to a process control sequence. As evidenced by the linear flow splitting process, energy savings can be achieved using the process energy for the joint creation.

Thus, conjoint forming requires a high level of specific know-how with particular regard to the forming process as well as the joint mechanisms, which might also lead to a conflict of objectives. Due to the fact that forming and joining in common manufacturing process are performed sequentially, the product development process is less demanding with particular regard to the parts' geometry and positioning. For conjoint forming there is a significant gap with regard to the process design and the achievable parts' mechanical properties. For robust conjoint forming processes, there is the need for design criteria. Concerning the individual joining and forming process the mutual interactions need to be investigated in order to utilize the possible advantages of conjoint forming to the fullest.

## 5. Acknowledgment

The authors want to express their gratitude to the DFG (Deutsche Forschungsgemeinschaft) for the support of the Collaborative Research Centre 805 and 666 as well as the Priority Program 1640. The presented conjoint forming applications are investigated within the subprojects "Integration of Functional Materials", "Manufacturing of bifurcated parts by process integrated forming, cutting and joining operations" and "Investigation and enhancement on bonding by cold bulk metal forming processes" (GR 1818/48-1; ER 601/3-1)".

## References

[1] Mori K-i, Bay N, Fratini L, Micari F and Tekkaya A E 2013 *CIRP Annals - Manufacturing Technology* **62** 673.
[2] Groche P, Wohletz S, Brenneis M, Pabst C and Resch F 2014 *Journal of Materials Processing Technology* **214** 1972.
[3] Mucha J and Witkowski W 2014 *Thin-Walled Structures* **82** 55.
[4] Bay N, Clemensen C, Juelstorp O and Wanheim T 1985 *CIRP Annals - Manufacturing Technology* **34** 221.
[5] Groche P and Türk M 2011 *CIRP Annals - Manufacturing Technology* **60** 21.
[6] Marré M, Brosius A and Tekkaya A E 2008 *Int J Mater Form* **1** 1295.
[7] Matsumoto R, Hanami S, Ogura A, Yoshimura H and Osakada K 2008 *CIRP Annals - Manufacturing Technology* **57** 279.
[8] Schikorra M, Tekkaya A E and Kleiner M 2008 *CIRP Annals - Manufacturing Technology* **57** 313.
[9] Baadjou R, Shimahara H and Hirt G 2006 *SSP* **116-117** 383.

[10] Rabinowicz E 1965 *Friction and Wear of Materials* (Wiley, New York).
[11] Buckley D H 1968 *ASLE TRANSACTIONS* **11** 89.
[12] Vaidyanath L R, Nicholas M G and Milner D R 1959 *British Welding Journal* 13.
[13] Wright P K, Snow D A and Tay C K 1978 *Metals Technology* **5** 24.
[14] Bay N 1983 *Welding Journal* 137.
[15] Bay N 1981 *The Insitution of Metallurgists* 47.
[16] Zhang W 1994 *Bond Formation in Cold Welding of Metals* (Copenhagen).
[17] Zhang W and Bay N 1992 *Annals of the CIRP* **41/1**.
[18] Zhang W and Bay N 1996 *CIRP Annals - Manufacturing Technology* **45** 215.
[19] Kebriaei R, Mikloweit A, Vladimirov I N, Bambach M, Reese S and Hirt G 2014 *KEM* **622-623** 443.
[20] Vaidyanath L R and Milner D R 1960 *British Welding Journal* 1.
[21] Zhang W and Bay N 1997 *Welding Journal* 326.
[22] Cooper D R and Allwood J M 2014 *Journal of Materials Processing Technology* **214** 2576.
[23] Vaidyanath L R and Milner D R 1960 *British Welding Journal* 1.
[24] Springer H, Kostka A, Payton E, Raabe D, Kaysser-Pyzalla A and Eggeler G 2011 *Acta Materialia* **59** 1586.
[25] Groche P, Vucic D and Jöckel M 2007 *Journal of Materials Processing Technology* **183** 249.
[26] V. Monnerjahn, J. Niehuesbernd, M. Roos 2014 *Potentiale des Spaltprofilierens für die integrale Herstellung von Profilverbindungen aus Blech* (Mörfelden-Walldorf).
[27] Ludwig C, Hammen V, Groche P, Kaune V and Müller C 2013 *Mat.-wiss. u. Werkstofftech.* **44** 601.
[28] P. Groche, J. Hohmann, T. Kessler, J. Schreiner, M. Krech, A. Duschka, V. Monnerjahn, M. Ćorić 2015 *12. Umformtechnisches Kolloquium Darmstadt.*
[29] C. Wagner, S. Gramlich, V. Monnerjahn, P. Groche, H. Kloberdanz 2015 *Technology pushed process and product innovation – Joining by Linear Flow Splitting.*
[30] Brenneis M and Groche P 2012 *KEM* **504-506** 723.
[31] Krech M and Groche P 2015 *5. Wissenschaftliches Symposium PT-PIESA: Leichtbau durch Funktionsintegration 2015.*
[32] Groche P, Wohletz S, Erbe A and Altin A 2014 *Journal of Materials Processing Technology* **214** 2040.
[33] Awiszus B., Neugebauer R, Kittner K and Popp M. 2009 *UTF science. Online journal* 1.
[34] Wohletz S, Özel M and Groche P 2013 *Liewald (Hg.) 2013 – New Developments in Forging Technology* 319.
[35] Wohletz S and Groche P 2014 *Procedia Engineering* **81** 2000.
[36] Engelhardt M, Grittner N, von Senden genannt Haverkamp,Hermann, Reimche W, Bormann D and Bach F-W 2012 *Journal of Materials Processing Technology* **212** 1030.
[37] Binotsch C, Nickel D, Feuerhack A and Awiszus B 2014 *Procedia Engineering* **81** 540.
[38] Haase M, Jäger A and Tekkaya A E 2012 Influence of Extrusion Ratio on AA6060 Aluminum Alloy Chips Solid State Recycled by Integrated Extrusion and Equal Channel Angular Pressing *Metal forming 2012. Proceedings of the 14th International Conference on Metal Forming, Sept. 16 - 19, 2012, AGH Univ. of Science and Technology, Krakow, Poland* ed J Kusiak ([Düsseldorf]: Verl. Stahleisen) pp 479–82.

# Automatic optimization of localized heat treatment for Al-Si-Mg alloys

**A Ludwig[1], T Holzmann[1]**

E-mail: Tobias.Holzmann@unileoben.ac.at

**Abstract.** Material properties of aluminium alloys can usually be achieved by a heat treatment and quenching procedure. In case that only local strengthening is needed, a local heat treatment and quenching strategy could be an option to the energy intensive, time consuming and costly treatment of the whole part. One of the essential problem using a local strengthening procedure is the lack of knowledge about suitable process parameters. Therefore, a multiple criteria optimization approach with local strengthening as target function was set up, whereby the material constitution was calculated based on the precipitation evolution during local heat treatment and cooling. By automatically varying the exposure time and laser power, a series of process simulations was performed to find adequate process parameters for the sufficient local strengthening of the alloy.

## 1. Introduction

Aluminum alloy castings are often heat treated in order to improve its mechanical properties. The heat treatment is done in two steps. First, a supersaturated solid solution is obtained by heating up the casting to a certain temperature. This state is conserved by rapid quenching back to room temperature. While the supersaturated solid solution is heated up again, but now slightly (if ever), different precipitations can occur which hardens the material. In case of the AlSi7Mg0.5 alloy, which is focused on in the present work, the following precipitation sequence appears from the quenched-in supersaturated solid solution during heating: clusters with varying Mg and Si content, Guinier-Preston Zones (GP-Zones), beta'- and U1/U2-phases and the B-phase [1-3].

If only local strengthening is needed, time and energy could be saved by performing a local heat treatment. This might be quite efficient, if only one heating cycle combined with self-quenching is sufficient. Such a heat treatment is conceivable with different heat sources of which the laser is the most elegant, flexible and precise. However, when defined material improvements are desired, the power input, the beam diameter, the duration of heating and the necessary quenching strength are a priori unknown parameters.

In this paper, an automatic optimization procedure is suggested that focus on short time local laser heating with quenching of an as-cast AlSi7Mg0.5 alloy in order to improve the strength of the material so that it can resist a prescribed loading.

[1]Chair for Simulation and Modelling of Metallurgical Processes, Montanuniversitaet Leoben, Austria

## 2. Numerical Approach

### *2.1 Case of loading*

The starting point of this work is a simple rhombus-like plate casting with a centred hole. In case of loading the plate casting stresses occur in the material. Exemplary a traction around the centred hole with a load of 360 kg (area 300 mm²) is applied in this work. The task of the optimization procedure is that the material property, namely the yield strength, should locally be improved by a laser heat treatment with subsequent quenching so that plastic deformation of the material can be avoided. If ideal elasto-plastic behaviour is assumed the condition of no plastic deformation can be written as

$$\Delta = R_{p0.2}(\vec{r}) - \sigma^{eq}(\vec{r}) \cdot S \geq 0, \tag{1}$$

where S represents a security factor, $R_{p0.2}$ the local yield strength and $\sigma^{eq}$ the van Mises equivalent stress. For all positions, $\vec{r}$, in the whole casting, the condition equation (1) has to be fulfilled, especially in the vicinity of the loaded area at the centre hole. Using the rotational-symmetry condition for the considered case of loading, it is possible to consider only a segment of the plate casting for the stress analysis (and the heat treatment described later). The length of the segment was chosen in a way that the heat treatment does not affect the outer plate casting regions and therefore only local strengthening around the center hole is obtained. In Figure 1 the considered segment together with the assumed boundary conditions for the stress analysis is shown. The stress calculation is done using OpenFOAM [4], solving for the divergence of the Cauchy stress tensor to be zero [5]. The position dependent Young's modulus and yield strength were computed by a MatCalc simulation as explained below.

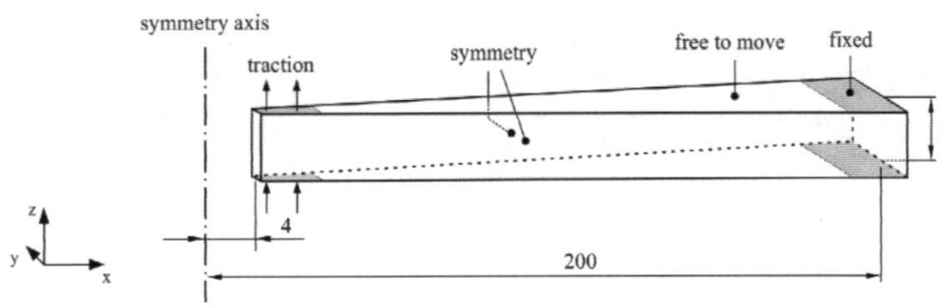


**Figure 1.** Wedge-shaped segment considered in the stress analysis. For the center hole a radius of 4 mm, for the segment a length of 200 mm and for the casting thickness 8 mm were chosen. The grey fixed area on the right side where no motion is allowed is 0.01285 m².

### *2.2 Heat treatment*

A Gaussian-shaped heat input of total power, $P$, was considered perpendicular to the plate casting for a certain time, $t$. Both, $P$ and $t$ are parameters which subject to variations during the optimization procedure. $P$ is varied between 500 and 8000 W and $t$ between 1 and 105 s. The beam diameter, also known as variance of the Gaussian shape, was taken constant with a value of 0.03 m. As initial condition, we started with the material at room temperature, used symmetry conditions at the wedge side faces and a heat transfer coefficient, $h$, with an ambient temperature, $T_{\text{amb}}$ at top, bottom, left and right surface were used [6]. During the heating period the value of the heat transfer coefficient $h$ was taken as 50 W/m²/K with $T_{\text{amb}}$ = 293 K. For the quenching period $h$ was chosen to be 15000 W/m²/K with $T_{\text{amb}}$ = 283 K. In addition the quenching period was fixed to 20 s for each simulation. A schematic draft of the heat source that is acting on the area around the center hole of the plate casting is shown in Fig. 2. For the time depending temperature evolution, the heat conductivity equation is solved with OpenFOAM [4]. The material parameters and values for the considered AlSi7Mg0.5 alloy [7] are the heat conductivity, $\lambda$ = 156 W/m/K, the heat capacity, $c_p$ = 778 J/kg/K and the density $\rho$ = 2560 kg/m³. For each of the six marked positions, which are called testing probes, the temperature history curve was further used for the phase constitutive simulation.

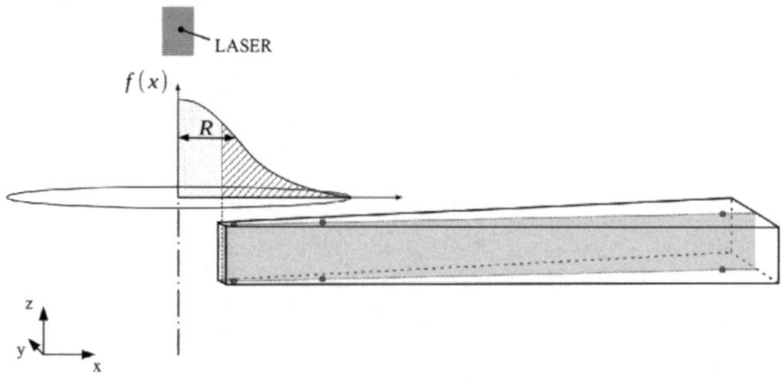


**Figure 2.** Schematic draft of the heat source. The non-hatched area below the Gauss-function defines the amount of lost energy through the centre hole. *R* denotes half of the beam diameter.

Note that the Gaussian-shaped laser beam radiates most of its energy into the centre of the hole. That can be avoided using a ring-shaped laser. Indeed, in the present work the exact shape of the laser beam is not of significant importance as the focus of the work lies on the strategy for the automated optimization rather than on specific process details.

In case of high laser power and long exposure time melting of the material can occur. This has to be avoided. If a combination of parameters leads to melting, the optimization routine will exclude the corresponding parameter set from the trials. A smarter way to prevent melting is to introduce two specific temperatures where the laser is powered off ($T_{off}$) or on ($T_{on}$). Fig. 3 shows a result of such an alternating off/on strategy. In the present numerical heat treatment this procedure is easy to implement, whereas in practice, an empirical shutter or guided power-down technique would be applied to the laser control unit.

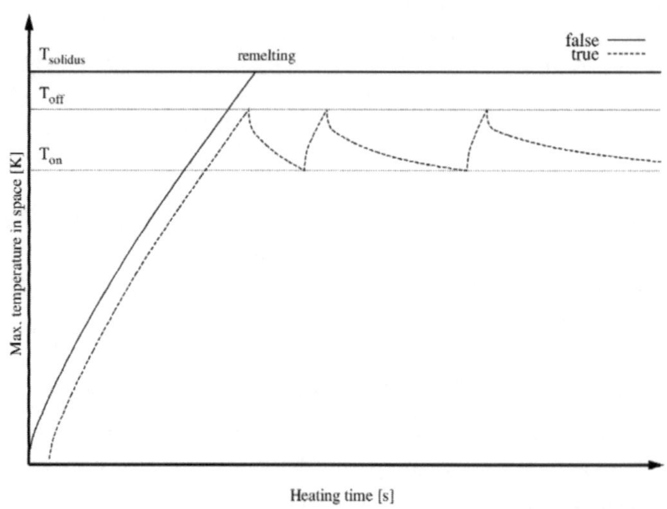


**Figure 3.** Possible temperature evolution at a surface position of the heated parted. To avoid melting an alternating off/on strategy for the laser beam was tested.

*2.3 Material constitution simulation*

During the heat treatment and quenching procedure the evolution of the material constitution of the AlSi7Mg0.5 alloy was calculated at the six testing probes shown in Fig. 2 by using the commercial software MatCalc [8]. Based on a thermodynamic database and corresponding precipitation calculations, MatCalc also provide data on the yield strength and Young's modulus. However, for the heat treatment and quenching simulation the constitution of the as-cast material has to be known as initial state. Therefore, we have performed a multicomponent Scheil solidification simulation [9] whereby MatCalc calculated the evolution of phases and precipitations together with concentration profiles. Based on these results, the temperature-time history at the six testing probes was used to calculate the evolution of the material constitution during the heat treatment and quenching step. The final Young's moduli and yield strength are then area-weighted interpolated, respectively extrapolated, onto the whole wedge-shaped segment. These position-dependent mechanical properties were then used in the stress analysis mentioned above.

*2.4 Automated optimization scheme*

The subject for the automatic optimization algorithm is finding a proper laser power and exposure time combination so that no yielding occurs. In other words, for any point in the casting, the yield strength has to be larger than the von Mises equivalent stress which occurs during loading (Eq. (1) has to be fulfilled). The requirement for this condition is that the difference between yield strength and von Mises equivalent stress is larger or equal to zero at each space point and can be expressed as

$$\delta_{\text{out}} = \min\{\Delta(\vec{r})\} \geq 0 . \tag{2}$$

From eq. (1) and (2) it is obvious that if $\delta_{\text{out}}$ is larger than zero the local strengthening of the material is stronger than necessary. Therefore, we chose $\delta_{\text{out}} = 0$ as criterion for the optimization as target function, so that a minimum of heat treatment effort results in no plastic deformation during loading.

The present study uses a gradient scheme based on the Fletcher-Reeves formula from the open source optimization toolkit Dakota [10, 11]. The optimization procedure is shown in Fig. 4. The process chain that is addressed here consists of (a) casting and solidification; (b) local heat treatment and quenching; and (c) the case of loading. Referring to the modelling point of view, the process chain has to be treated more detailed: (i) material constitution during solidification; (ii) material constitution during cooling to room temperature; (iii) temperature evolution during heat treatment and quenching; (iv) material constitution during heat treatment and cooling; (v) extraction of yield strength and Young's modulus; (vi) inter/extrapolation onto the 3D grid for stress analysis; (vii) 3D analysis in the case of loading. The steps (i) to (v) have to be executed point-wise e.g. at the selected testing probes shown in Fig. 2. All three dimensions are involved (3D) for step (vi) and (vii). If the load analysis responds $\delta_{\text{out}} = 0$, an optimum is found and the optimization procedure stops. Otherwise the process chain starts all over again using a new set of input parameters (for now the laser power, *P*, and the exposure time, $t$). The next set is suggested by Dakota based on the above mentioned gradient-based scheme.

At the moment, part (i) and (ii) could be neglected within the optimization loop because the as-cast material is the base for the local heat treatment and is identical for each loop. Currently, we are working at a process chain where the solidification conditions will influence the local material constitution. This extension will lead to new input parameters for the automated optimization toolbox like the casting temperature or cooling conditions during solidification. For this more general case, part (i) and (ii) must be considered inside the optimization loop.

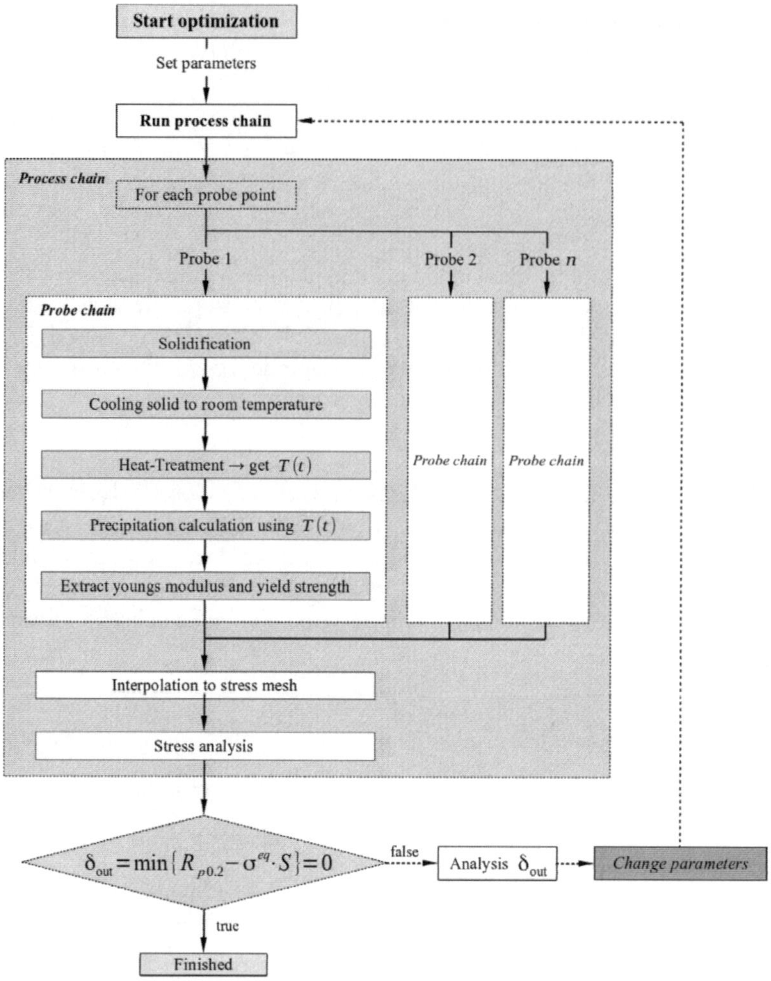


**Figure 4.** Flow chart of the automated optimization scheme.

## 3. Results and Discussion

The procedure of local heat treatment and quenching leads to the formation and growth of particular precipitations. As a results the material is strengthened by increasing the yield strength. Details on the mechanisms that cause this increase is beyond the scope of this paper and is therefore not discussed here. Fig. 5 shows the obtained yield strength (here $Rp_{0.2}$) at a position close to the surface near the centre as a function of exposure time and laser power. Lower laser power results in insufficient local strengthening because the local heating is not strong enough to reach a temperature level that favours the occurrence and growth of precipitations. Thus, the yield strength after heat treatment is the same as for the as-cast one (around 104 MPa). For higher laser powers the obtained yield strength increases rapidly to more than 130 MPa, whereby the shorter the exposure time the larger the power has to be to reach this increase. For high laser powers and longer shutter speed melting is predicted. The corresponding set of data points are not considered for the automated optimization and are thus not shown in Fig. 5.

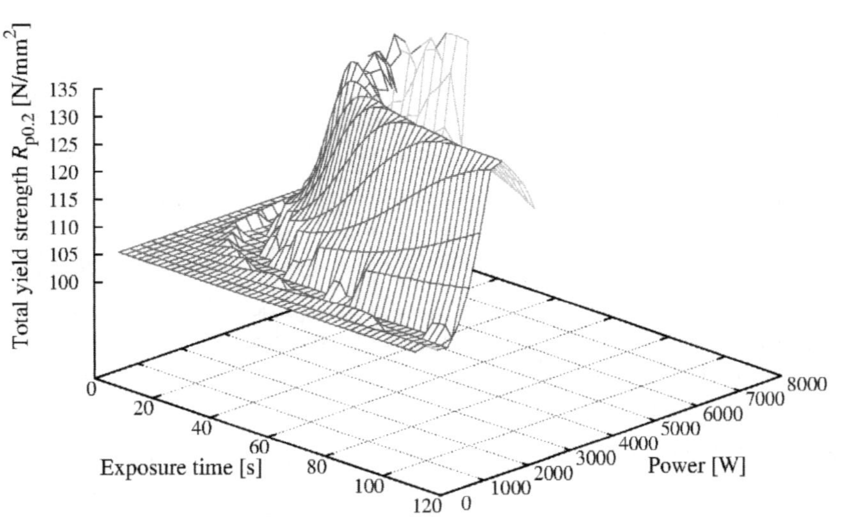


**Figure 5.** Strengthening of the material as function of laser power and heating time at the position close to the surface near the centre. For higher power values melting is predicted and therefore the corresponding power-time data set is not shown.

To avoid melting of the surface of the casting, the alternating off/on strategy mentioned above was applied with two different sets of values for $T_{\text{off}}$ and $T_{\text{on}}$. For the first set, $T_{\text{off}}$ = 793 K and $T_{\text{on}}$ = 773 K were chosen and for the second set $T_{\text{off}}$ = 593 K and $T_{\text{on}}$ = 573 K, respectively. Fig. 6 shows the corresponding obtained yield strength at the same position as for Fig. 5. As the off/on strategy avoids temperatures above the solidus temperature, higher laser powers and longer exposure times can now be accepted for the automated optimization without a risk of melting.

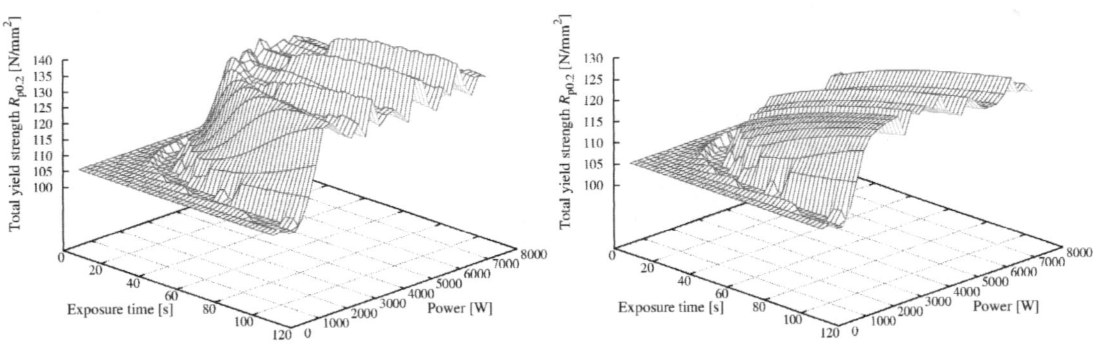


**Figure 6.** Strengthening of the material as function of laser power and heating time using the alternating off/on strategy with two different sets of values for $T_{\text{off}}$ and $T_{\text{on}}$. The left plot corresponds to higher temperatures close to solidus and the right plot to temperatures 200 K lower

While using lower temperature bounds for the off/on strategy, the yield strength surface is smoothed out (less jagged) and the absolute values are lower compared to the higher temperature set. Formation and growth of precipitations which results in a yield strength increase is favored by using higher temperatures. Thus, it is clear that an off/on interval at higher temperatures leads to higher yield strengths. For the off/on interval which is 200 K lower even extreme high laser power and long exposer time do not lead to a yield strength increase comparable to the first case. This results demonstrate the importance of the right thermal treatment.

As described above, in the present optimization scheme $\delta_{out}$ is the output quantity and *P* and *t* are the input quantities. Fig. 7 shows how $\delta_{out}$ varies on changes in *P* and *t* for the two heat treatment cases mentioned above. Note that for each $\delta_{out}(P,t)$-surface 20x20 stress simulations in 3D had been performed. Using the Fletcher-Reeves gradient-based optimization algorithm, a new input data set for *P* and *t* is evaluated based on the gradients of the respond function $\delta_{out}$ with respect to (*P*, *t*). Fig. 7 shows the absolute values of the respond quantity $\delta_{out}$ of both cases including the evaluation of the optimization strategy (black points). The optimization scheme finds the minimum rather quickly. For the first case where the heat treatment is done at elevated temperatures, the automated optimization scheme needed only 36 iterations to find the optimum $\delta_{out} \approx 0$ (Fig. 7, left). For the second case only an object function of $\delta_{out} \approx 12$ was found. That means that no parameter set (*P*, *t*) is suitable for improving the yield strength of the aluminium alloy such that plastic deformation can be avoided.

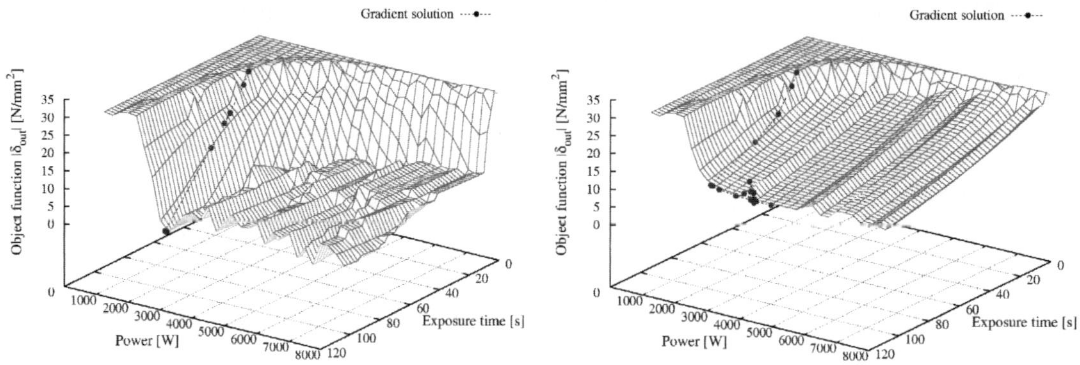


**Figure 7.** The absolute value of target function, $\delta_{out}(P,t)$, evolves corresponding to the optimization strategy. For the presented case the local minimum, $\delta_{out} \approx 0$, was found after 36 iterations in the left plot. The right plot has no set of input quantities to reach $\delta_{out} = 0$.

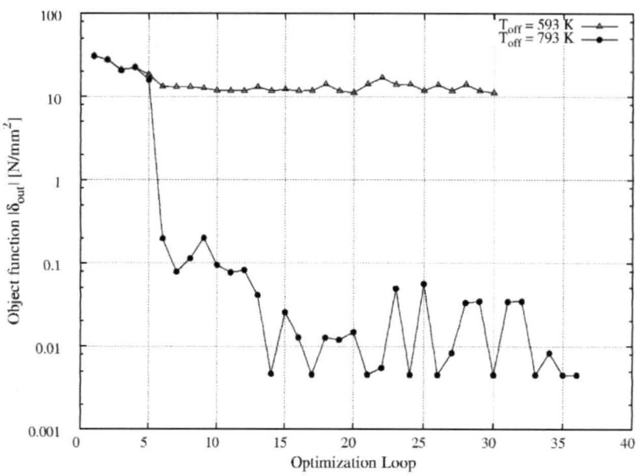


**Figure 8.** Gradient-based evolution of the target function, $|\delta_{out}(P,t)|$

Fig. 8 shows the absolute values of the target function for both cases in more detail (variations during optimization). It can be seen that for the first case after 14 iterations the minimum is already found but does not stop due to the abort criteria of the optimization. Comparing Fig. 7 and Fig. 8 it is obvious that the shape of the object function is essential for the correct choice of the optimization strategy (global or local search). Furthermore the object function and therefore the total yield strength development highly depend on the heat treatment parameters.

## 4. Summary

The paper demonstrates that a local strengthening of an as-cast AlSi7Mg0.5 alloy with a short laser heat treatment is possible. Taking the advantage of an automated optimization scheme, it is shown that the necessary laser power and exposure time could be estimated so that a local increase of the yield strength enables the material to withstand a localized traction acting without any plastic deformation. Beside a 3D heat transfer calculation and a 3D stress analysis, the procedure uses predictions from the material calculator MatCalc, which are currently under validation. The proposed procedure has great potential because it might allow to optimize features like casting temperature, cooling conditions, alloy composition, laser beam diameter and much more for the casting and heat treatment. Particularly with the increased usage of robotics, it is conceivable that local laser heat treatment, especially with self-quenching, might gain in importance. We think that the present paper is a significant step into that direction as it demonstrates how the suitable process parameters might be obtained. Energy and time saving might be the ultimate benefit of the proposed local strengthening of Al-based alloys.

## Acknowledgement

Financial support by the Austrian Federal Government (in particular from Bundesministerium für Verkehr, Innovation und Technologie and Bundesministerium für Wirtschaft, Familie und Jugend) represented by Österreichische Forschungsförderungsgesellschaft mbH and the Styrian and the Tyrolean Provincial Government, represented by Steirische Wirtschaftsförderungsgesellschaft mbH and Standortagentur Tirol, within the framework of the COMET Funding Programme is gratefully acknowledged.

## References

[1] Vissers R, van Huis M A, Jansen J, Zandbergen H W, Marioara C D and Andersen S J 2007 The crystal structure of the β′ phase in Al-Mg-Si alloys *Acta Mater.* **55** 3815–23
[2] Falahati A, Ahmadi M A, Lang P, Povoden-Karadeniz E, Warczok P and Kozeschnik E 2011 Thermo-kinetic computer simulation of precipitation and age-hardening effect in Al-Mg-Si alloys with arbitrary heat treatment *MS & T* 292–99
[3] Falahati A, Povoden-Karadeniz E, Lang P, Warczok P and Kozeschnik E 2010 Thermo-kinetic computer simulation of differential scanning calorimetry curves of AlMgSi alloys *Int. J. Mater. Res.* **101** 1089-96
[4] http://www.openfoam.org
[5] Slaugther W S 2002 *The linearized Theory of elasticity* (Birkenhauser)
[6] Vilums R 2011 Implementation of Transient Robin Boundary Conditions in OpenFOAM *Multiphy. Model. in OpenFOAM* 39-40
[7] DIN 1725: 3.2371
[8] http://matcalc.tuwien.ac.at/
[9] Dantzig J A and Rappaz M 2009 *Solidification* (Lausanne: EPFL Press) pp 396-401
[10] https://dakota.sandia.gov/
[11] Fletcher R and Reeves C M 1964 Function minimization by conjugate gradients *Computer Journal* **7** 149-54

# On the modelling of microsegregation in steels involving thermodynamic databases

**D You, C Bernhard, S Michelic, G Wieser and P Presoly**

Chair of Ferrous Metallurgy, Montanuniversität Leoben, Franz-Josef-Straße 18, 8700 Leoben, Austria

E-mail: dali.you@stud.unileoben.ac.at

**Abstract**. A microsegregation model involving thermodynamic database based on Ohnaka's model is proposed. In the model, the thermodynamic database is applied for equilibrium calculation. Multicomponent alloy effects on partition coefficients and equilibrium temperatures are accounted for. Microsegregation and partition coefficients calculated using different databases exhibit significant differences. The segregated concentrations predicted using the optimized database are in good agreement with the measured inter-dendritic concentrations.

## 1. Introduction

Microsegregation results from solute partitioning at the dendritic scale during solidification. This phenomenon can lead to the formation of defects during the casting process (e.g., hot tearing) and also negatively affect the product quality (inhomogeneous microstructure and undesirable primary precipitates). Hence, it is not surprising that this research topic has already been intensively investigated by several research groups in the past[1-6].

Existing modeling approaches consist of numerical and analytical models. Numerical solutions, depending on the magnitude of model building, are typically complex to solve. Process models are constrained by the need to save calculation time. Therefore, these models generally refer to simple analytical approaches. More elaborate software products are DICTRA®[7] and IDS®[8], which are widely applied for solving diffusion and phase transformation problems[9-11]. The present concept applies an easily solvable analytical model coupled with the thermodynamic library ChemApp®[12] and with both commercial and self-optimized databases in the background. The optimization of the databases is an ongoing activity and based on extensive DSC measurements of high-temperature phase transformations in Fe-C-Si-Mn-Al systems[13]. More details are described in the paper "*Thermodynamic optimization of individual steel databases by means of systematic DSC measurement*" which is also submitted to this conference. In future research, the proposed model will be coupled with a phase nucleation model to take the formation of non-metallic inclusions above the liquidus and during solidification into account.

## 2. Model description

In the proposed model, microsegregation was calculated using Ohnaka's model. Thermodynamic equilibrium calculation was performed using ChemApp and selected thermodynamic databases. ChemApp is an interface software developed by GTT Technologies, Herzogenrath, Germany. It may be linked with a source code written in FORTRAN, C/C++, Visual Basic® and Borland Delphi®. In this case, FORTRAN was applied as the programming language. Microsoft Visual Studio 2013 was used as a main frame provider and modern compiler. The schematic of the model is described by Figure 1.

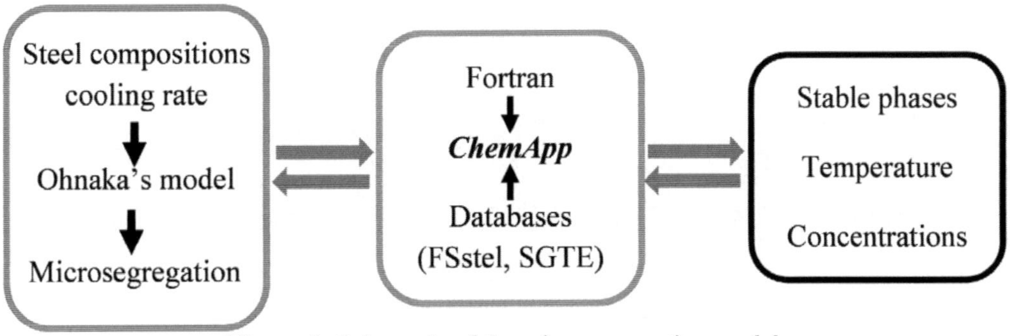


**Figure 1.** Schematic of the microsegregation model.

### *2.1. Analytical equation*

For considering the changes of the partition and diffusion coefficients, Ohnaka's model was integrated into Equation 1. Local partition coefficients and diffusion coefficients were calculated at each solidification step, but within the increase of solid fraction by $\Delta f_s$, they were assumed to be constants.

$$C_L^+ = C_L\{\frac{1-\Gamma\cdot f_s}{1-\Gamma\cdot(f_s+\Delta f_s)}\}^{\frac{1-k}{\Gamma}}, \text{ with } \Gamma = 1-\frac{4\alpha k}{1+4\alpha} \quad (1)$$

$$\alpha = \frac{4D_s t_f}{(\lambda_2)^2} \quad (2)$$

where $f_s$ represents the solid fraction; $C_L^+$ and $C_L$ are the concentrations of solutes in the residual liquid at solid fractions of $f_s$ and $f_s+\Delta f_s$, respectively; $k$ is the equilibrium partition coefficient between solid and liquid; $\alpha$ is the back diffusion coefficient, which can be calculated using Equation 2; $D_s$ is the solute diffusion coefficient in solid; $t_f$ is the local solidification time; and $\lambda_2$ is the secondary dendrite arm spacing.

### *2.2. Parameters*

In the applied model, temperatures at the solidification interface and partition coefficients ($k$) were calculated using ChemApp. At each solidification step, for a multicomponent system, the phase transformation point from liquid to solid was detected after the activity of δ-ferrite or austenite approached one. Then, the concentrations of solutes in both liquid and solid and the temperature were determined. For simplification, Figure 2 describes a binary system of the above process. The diffusion coefficients ($D_S$) applied in the calculations are listed in Table 1.

The partition and diffusion coefficients used for the calculations were the average of the values between the current and former steps. The temperatures were updated in a loop until the difference between two adjacent values was less than $10^{-3}$ K.

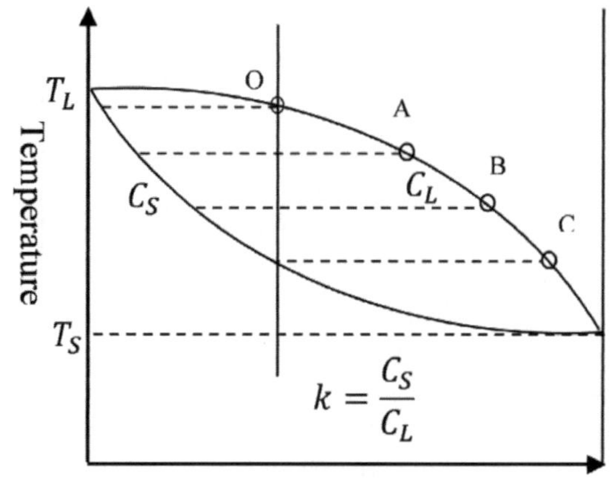


**Figure 2.** Schematic illustration of the determination of partition coefficients and temperatures.

**Table 1.** Diffusion coefficients of solutes.[14-16]

| Elements | $D_\delta$ [a] ($m^2s^{-1}$) | $D_\gamma$ [a] ($m^2s^{-1}$) |
|---|---|---|
| C | 0.0127Exp(-81301/RT) | 0.0761Exp(-134429/RT) |
| Si | 8.0Exp(-248710/RT) | 0.3Exp(-251218/RT) |
| Mn | 0.76Exp(-116935/RT) | 0.055Exp(-249128/RT) |
| P | 2.9Exp(-229900/RT) | 0.01Exp(-182666/RT) |
| S | 4.56Exp(-214434/RT) | 2.4Exp(-212232/RT) |
| Al | 5.9Exp(-241186/RT) | 5.15Exp(-245800/RT) |

[a] R: 8.314 J/(K·mol); T: temperature in Kelvin.

Secondary dendrite arm spacing was estimated using Equation 3, which is influenced by the local solidification time and initial carbon content.[17] In the first step, the local solidification time is estimated. In the second step, the local solidification time is calculated according to Equation 4 after obtaining the solidus temperature. This process was repeated until the difference between two adjacent local solidification times was less than $10^{-4}$ seconds,

$$\lambda_2 = (27.3 - 13.1C_0^{\frac{1}{3}})t_f^{\frac{1}{3}} \quad (3)$$

$$t_f = \frac{T_L - T_S}{R_c} \quad (4)$$

where $\lambda_2$ (μm) is the secondary dendrite arm spacing, $C_0$ is the initial concentration of carbon in mass percent, $t_f$ (s) is the local solidification time, $R_c$ (K/s) is the cooling rate, and $T_L$ and $T_S$ (K) are the liquidus and solidus temperatures, respectively.

In addition, the peritectic reaction was realized in a simple way similar to other analytical models. If ChemApp detects that the solid phase at the solidification interface will change from δ-ferrite to austenite, the partition and diffusion coefficients of austenite were applied.

## 3. Comparisons with characteristic temperatures

Zero strength temperature (ZST) is the critical temperature below which solidifying materials start to transmit strength inside the dendritic structure of both solid and liquid. Zero ductility temperature (ZDT) is defined as the temperature at which ductility first starts to increase during the solidification process. The schematic of ZST and ZDT is shown in Figure 3. Note that ZDT also stands for the non-equilibrium solidus temperature. Both ZST and ZDT are strongly influenced by solute enrichments and can comprehensively reflect the phenomenon. Additionally, hot tears generally form between ZST and ZDT; thus, it is also desirable to reasonably estimate ZST and ZDT through solidification modeling.

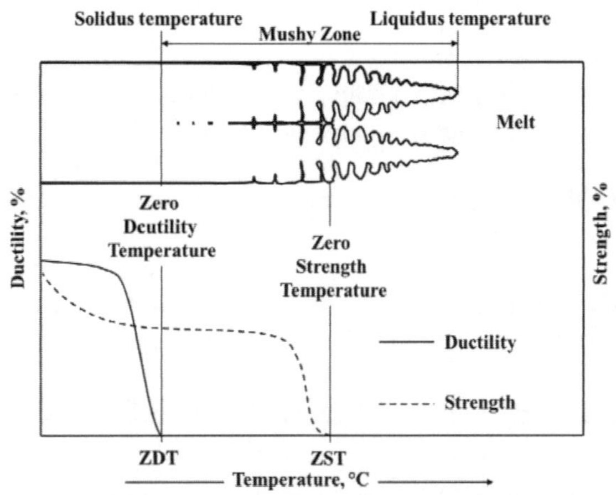


**Figure 3.** Schematic of ZST and ZDT. [18]

According to the statistical analysis and experimental results from Won[19], the temperatures at solid fractions of 0.75 and 0.99 were considered as ZST and ZDT. Note that the experimentally determined ZST values correspond with the calculated solid fractions of between 0.65 and 0.8, whereas ZDT commonly varies just between 0.98 and 1.0. Therefore, it cannot be expected that the ZST fits perfectly to the calculated solidus temperature for a solid fraction of 0.75.

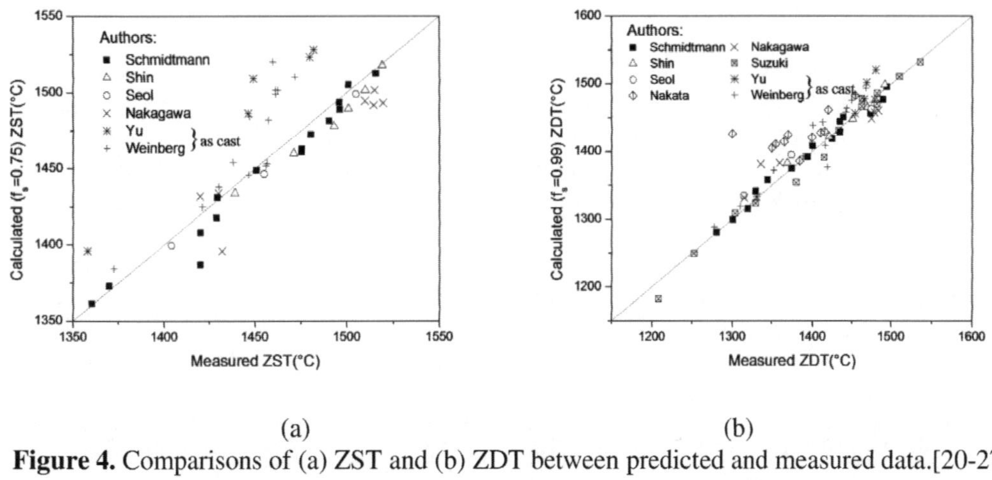


(a) (b)

**Figure 4.** Comparisons of (a) ZST and (b) ZDT between predicted and measured data.[20-27]

Figure 4 compares the ZST and ZDT from the calculated and measured data by different authors[20-27] for a certain range of steels. In this part, the calculations were carried out using FSstel database. Under the same given conditions of steel compositions and cooling rates, the calculated ZST and ZDT are in reasonable agreement with the experimental results. Note that the correspondence between the ZST from analysis and measured data is of rather poor quality. The main reason is that the ZST does not necessarily coincide with a solid fraction of 0.75 but rather depends on the steel composition; Shin et al.[21] found that the ZST varies between 0.6 and 0.7. In contrast, the correspondence of the calculated and measured ZDT is excellent. The above results, to some extent, illustrate the reasonability of the calculations. However, the comparisons of ZST and ZDT are only an inaccurate way to verify the microsegregation predictions.

## 4. Influence of different databases

In this section, three different databases, FSstel(FactSage 6.4[28]), SGTE2014 and an optimized SGTE version were applied to predict segregated concentrations of a TRIP steel. Among them, FSstel and SGTE2014 are commercial databases; the optimized SGTE is a 'private' one. An important information at this point is, that the Fe-C-Mn system is identical in all three databases and was not optimized. However, in the private database the manganese interaction with Si and Al were selectively optimized, based on model alloys and DSC measurements. The chemical composition of the selected steel is listed in Table 2. For comparing with the following measured results, secondary dendrite arm spacing was set to 18 μm and the local solidification time was calculated using Equation 3.

**Table 2.** Chemical compositions (mass%).

| C | Si | Mn | Al |
|---|---|---|---|
| 0.21 | 0.54 | 2.12 | 0.77 |

Figure 5 compares the results calculated with different databases. It shows that both solute concentrations and partition coefficients exhibit significant differences. The carbon concentration profiles (Figure 5 (a)) are similar until a solid fraction of 0.7 is reached, and then they differ; concentrations of silicon and manganese (Figure 5 (b) and Figure 5 (c)) predicted with FSstel database are close to those predicted with SGTE2014 database, while they are higher than those calculated by the private database; for aluminium, all the three calculations show its negative segregation, and the concentration in the residual liquid estimated with the private database is higher than those estimated with the other two databases.

The differences in the concentration profiles can be well explained by the differences in the corresponding partition coefficients. For positive segregation (carbon, silicon and manganese), the smaller partition coefficients indicate the higher segregation; while for negative segregation, the smaller partition coefficients lead to a lower segregation. Taking manganese as an example (Figure 5 (c)), the partition coefficient calculated using the private database is always larger than those calculated using the FSstel and SGTE2014 databases, which results in a lower concentration profile calculated with the private database. Note that the sharp changes of partition coefficients in Figure 5 are caused by the peritectic reaction. It shows that peritectic reaction times of different calculations vary with each other. At the same time, peritectic reaction leads to the decreases of back diffusions, which further increases the concentrations in residual liquid. This explains the higher concentrations calculated with the FSstel database for silicon and manganese between solid fraction 0.63 and 0.79, though their partition coefficients are larger (Figure 5 (b) and (c)). Different databases have their own parameters for the Gibbs energy minimization and thermodynamic equilibrium calculations.

Therefore, the partition coefficients and equilibrium concentrations calculated with different databases vary with each other.

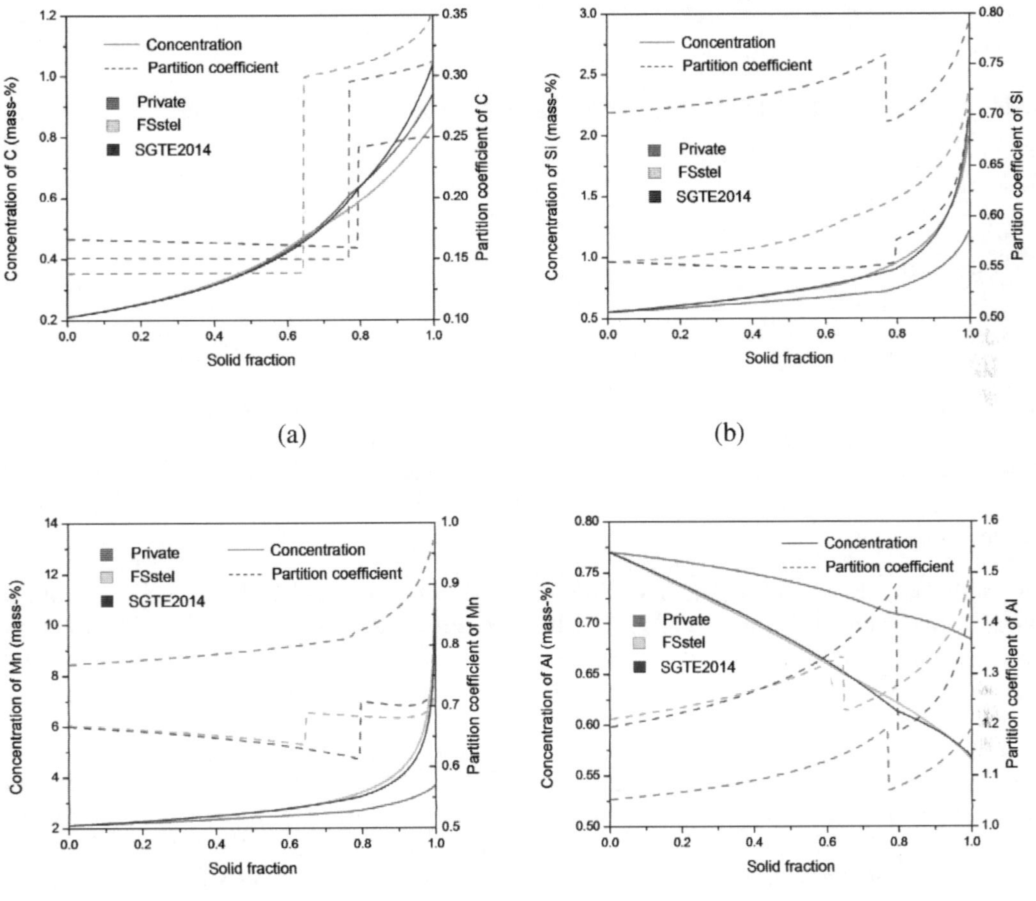


(a) (b)

(c) (d)

**Figure 5.** Predictions of segregated concentrations and partition coefficients with different databases (a) carbon, (b) silicon, (c) manganese and (d) aluminium.

For comparing with the predictions, a sample with the same compositions as listed in Table 2 was produced by means of high-frequency remelting and spin-casting. The small sample (60g) was solidified with high speed and water-quenched so that the back diffusions of solutes were minimized. Then, the microprobe analysis was performed on a selected cross section. The concentration distributions of the solutes are shown in Figure 6.

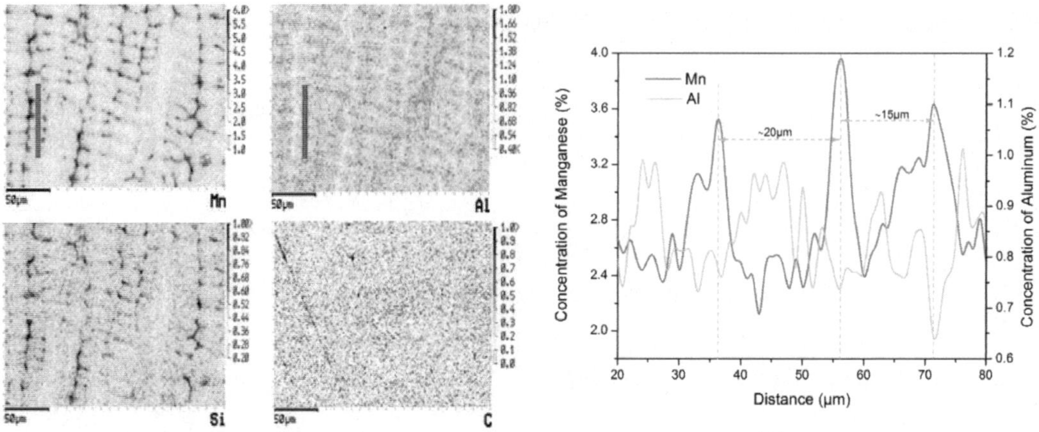


(a) (b)

**Figure 6.** Concentration distributions of (a) cross section and (b) line scan of secondary dendrites.

From Figure 6 (a), both primary and secondary structures are clearly drawn by the segregated concentrations of manganese, aluminium and silicon. The carbon concentration distributes quite evenly due to its fast diffusion. To evaluate the calculations, the concentration profile between two secondary dendrite arms was investigated in detail. A line scan was performed on the secondary dendrites as indicated by the red line in Figure 6 (a).

Figure 6 (b) describes the concentration distributions of manganese and aluminium over the line scan distance. Concentrations of manganese and aluminium reflect the microsegregation. Manganese gets enriched in inter-dendrite zones and approaches to the highest concentrations. By contrast, aluminium shows a negative segregation and reaches the lowest concentrations in inter-dendrite zones. Compared with the measurements, the segregated trends are predicted well by the proposed model. For the values, the measured inter-dendritic concentrations of manganese range from 3.4% to 4.0%. As shown in Figure 5 (c), the segregated concentration of manganese calculated with the private database is about 3.5% which fits well with the measured results. While for the predictions with the FSstel database and the SGTE2014 database, the final segregated concentrations exceed 10% which seems to be unrealistic. For aluminium, the measured inter-dendritic concentrations vary from 0.65% to 0.75%. The inter-dendritic concentration predicted with the private database of 0.68% is closer to the measured value than the results calculated with the other two databases (0.57%).

Based on the above discussion, the proposed model can better predict the microsegregation of the investigated TRIP steel with the private database. Optimization of the database is necessary to better describe the thermodynamic equilibrium of selected steels and the work is still ongoing in the current research group.

## 5. Summary

In the present study, Ohnaka's model was coupled with thermodynamic databases for predicting microsegregation of steels. The solute enrichment of a TRIP steel was calculated with two commercial and one private optimized thermodynamic databases. The results were compared with measured inter-dendritic concentrations of the selected experiment. According to the investigations, the following conclusions can be drawn:

- By coupling with a thermodynamic database, the present model offers new development space for simple analytical models.
- Using the private optimized database, solute enrichment of selected steels can be better predicted with the present model.
- The model can also function as a tool for giving feedback for the database optimization.

In future studies, using this model, the consideration of the precipitation of even very complex non-metallic phases (oxides and carbides) during solidification would be possible. Furthermore the coupling of the analytical microsegregation model with nucleation and growth kinetics for inclusions in steel is intended, opening many new areas of application.

**Acknowledgements**

The authors are grateful for the financial support from the Federal Ministry for Transport, Innovation and Technology (bmvit) and from the Austrian Science Fund (FWF): [TRP 266-N19].

**References**

[1] Scheil E 1942 *Z. Metallkd.* **34** 70
[2] Gulliver G H 1913 *J. Inst. Met.* **9** 120
[3] Brody H D and Flemings M C 1966 *Trans. TMS-AIME* **236** 615
[4] Clyne T W and Kurz W 1981 *Metall. Trans. A* **12A** 965
[5] Kobayashi S 1988 *J. Cryst. Growth* **88** 87
[6] Ohnaka I 1986 *Trans. Iron Steel Ins. Jpn.* **26** 1045
[7] Andersson O J, Helander T, Höglund L, Shi P and Sundman B 2002 *CALPHAD* **26** 273
[8] Miettinen J, Louhenkilpi S, Kytönen H and Laine J 2010 *Math. Comp. Sim.* **80** 1536
[9] Griesser S, Reid M, Pierer R, Bernhard C and Dippenaar R 2014 *Steel Res. Int.* **85** 1257
[10] Rudnizki J, Zeislmair B, Prahl U and Bleck W 2010 *Steel Res. Int.* **81** 472
[11] Röttger A, Weber S, Theisen W, Rajasekeran B and Vaßen R 2011 *Steel Res. Int.* **82** 671
[12] Petersen S, Hack K, Monheim P and Pickartz U 2007 *Int. J. Mater. Res.* **98** 946
[13] Presoly P, Pierer R and Bernhard C 2013 *Metall. Mater. Trans. A.* **44** 5377
[14] Yamada W, Matsumiya T and Ito A 1990 *Proc. 6th Int. Iron Steel Cong.* **1** 618.
[15] Ueshima Y, Mizoguchi S, Matsumiya T and Kajioka H 1986 *Metall. Trans. B* **17B** 845
[16] Bester H, Lange K W 1972 *Arch. Eisenhuttenwesen* **43** 207
[17] Pierer R and Bernhard C 2008 *J. Mater. Sci.* **43** 6938
[18] Won Y M, Yeo T J, Seol D J and Oh K H 2000 *Metall. Mater. Trans. B.* **31** 779
[19] Won Y M, Kim K, Yeo T and Oh K H 1998 *ISIJ Int.* **38** 1093
[20] Schmidtmann E and Rakoski F 1983 *Arch. Eisenhüttenwes* **54** 357
[21] Shin G, Kajitani T, Suzuki T and Umeda T 1992 *Tetsu- to- Hagane* **78** 587
[22] Seol D J, Won Y M, Oh K H, Shin Y C and Yim C H 2000 *ISIJ Int.* **40** 356
[23] Nakata H and Yasunaka H 1990 *Tetsu-to-Hagane* **76** 376
[24] Nakagawa T, Umeda T, Murata J, Kamimura Y and Niwa N 1995 *ISIJ Int.* **35** 723
[25] Suzuki G H, Nishimura S and Nakamura Y 1984 *Trans. Iron Steel Ins. Jpn.* **24** 54
[26] Yu C H, Suzuki M, Shibata H and Emi T 1996 *Mater. Trans. Japan Inst. Metals.* **37** 1251
[27] Weinberg F 1979 *Metall. Trans.* **10B** 219
[28] Bale C W, Chartrand P, Eriksson G, Jung I H, Kang Y B and Pelton A D 2009 *CALPHAD* **33** 295

# Microstructure and mechanical properties of high strength Al–Mg–Si–Cu profiles for safety parts

**J A Österreicher[1], A Schiffl[2], G Falkinger[1], G R Bourret[3]**

[1] LKR Leichtmetallkompetenzzentrum Ranshofen GmbH, Austrian Institute of Technology, Postfach 26, 5282 Ranshofen, Austria

E-mail: johannes.oesterreicher@ait.ac.at

[2] Hammerer Aluminium Industries Extrusion GmbH, Lamprechtshausener Straße 69, 5282 Ranshofen, Austria

E-mail: andreas.schiffl@hai-aluminium.at

[3] Department of Materials Science and Physics, University of Salzburg, Hellbrunner Straße 34, 5020 Salzburg, Austria

E-mail: gilles.bourret@sbg.ac.at

**Abstract.**

Aluminium extrudate used for safety parts in cars need to exhibit high yield strength and ductility, a combination that is not easily achieved. In this work, the mechanical properties and microstructure of profiles with a yield strength greater than 280MPa achieved by two different artificial ageing treatments were studied. Profiles from one of the heat treatments performed well in quasi–static compression testing while those from the other heat treatment clearly failed. The batch of profiles that failed showed higher uniform elongation in tensile testing but a lower reduction in area. However, the difference in bending angles in the three–point–bending test were not as pronounced. Microscopic investigation of polished sections and fracture surfaces revealed that failure is dominated by the fracture of intermetallic phases resulting in voids. The growth and coalescence of these voids is facilitated by another population of smaller voids within the matrix, presumably nucleating at secondary phases.

## 1. Introduction

Extruded profiles of Al–Mg–Si–Cu alloys are used for safety parts in the front section of cars to absorb kinetic energy in the case of an accident. This is ideally achieved by uniform concertina–style deformation. In a constant quest for lightweight design the automaker's specifications for strength are ever–increasing; the requirements for ductility, however, remain on a constant high level [1]. Since high strength and ductility are often mutually exclusive the production of profiles with appropriate properties is not trivial.

### *1.1. Microstructure of Al–Mg–Si–Cu extruded profiles*

For a given geometry the mechanical properties of extruded Al–Mg–Si–Cu profiles are determined by their microstructure. Characterization of microstructural features and understanding their contribution to the mechanical properties is crucial for the design of new materials with improved properties.

The grain structure of extruded high strength Al–Mg–Si–Cu profiles for safety parts is developed during extrusion and consists of a fine, fibrous core of grains elongated along the extrusion direction and a peripheral coarse grain layer formed due to dynamic recrystallization. Throughout this matrix intermetallic phases of the α–AlFeSi or α–Al(FeMn)Si type with sizes of typically a few $\mu$m are dispersed.

The peripheral coarse grain layer can lead to orange peel, bending failure and various optical defects [2]. Thus, attempts are generally made to reduce its thickness by proper alloy design and choice of manufacturing parameters, e.g. sufficient cooling. An important approach to maintain the fibrous grain structure formed during extrusion is the facilitation of the formation of secondary dispersoid phases that inhibit recrystallization and grain growth. These phases with a size under 500nm form during the homogenization treatment and only have a minor impact on strength; thus, they are usually referred to as non–hardening phases. The chemical composition and crystal structure of the dispersoids are strongly dependent on alloy composition [3].

During artificial ageing Mg and Si form nano–scale hardening precipitates, significantly increasing the hardness of the profiles by dispersion hardening. The main hardening precipitate in alloys of the 6xxx series is $\beta''$ [4]. When the alloy contains considerable amounts of Cu, $Q'$ and precursors of $Q'$ also play an important role as hardening precipitates [5, 6].

The micromechanics of ductile fracture of 6xxx Al alloys have been extensively studied by Lassance et al. [7]. They showed that the fracture of brittle intermetallics as well as particle–matrix decohesion was observed even at low strains in *in–situ* tensile tests in a scanning electron microscope. It was found that particles with the major axis oriented at an angle of less than 45° with respect to the loading direction undergo fracture while particles oriented at an angle of more than 45° lead to particle–matrix decohesion. The examination of fracture surfaces showed a ductile profile with dimples containing particles.

Mrówka–Nowotnik et al. published various papers on the mechanical properties of EN AW–6082, e.g. [8, 9]. It was found that the mechanical properties and fracture toughness of this alloy strongly depend on the artificial ageing temperature and time, or more specifically, the size and number density of $\beta''$ and $\beta'$ precipitates formed by different ageing regimes. Scanning electron micrographs of fracture surfaces of tensile samples revealed that cracks initiated by nucleation, growth and coalescence of voids. Two void populations were found through observation of the fracture surfaces. The nucleation sites for large dimples were identified as primary intermetallics, while for smaller dimples dispersion hardening phases and dispersoids were suggested as nucleation sites.

### *1.2. Assessment of crash performance*

It is generally accepted that quasi–static compression testing of whole profiles shows good correlation with high–speed crash behaviour. Thus, quasi–static compression testing is the main criterion used by automakers to assess crash performance of extruded profiles. Although rating systems for the results of quasi–static compression testing have been proposed [10], the interpretation of quasi–static compression testing remains subjective and difficult to quantify. Furthermore, the reproducibility of quasi–static compression testing is often poor for high strength profiles. These problems greatly impair comparability of results, and therefore hinder purposeful R&D efforts.

It is desirable to use small scale mechanical tests like hardness, tensile, and three–point–bending tests, and correlate their results to those of quasi–static compression testing. For example, three–point–bending of sheet specimens cut out of extruded profiles has been shown to correlate well with quasi–static compression testing for Al sheet [11]. Furthermore, stress–strain curves from small scale mechanical tests are used for the development of FEM models. For the improvement of these models, correct interpretation of the tests is important.

### *1.3. Aim of this work*

This work is part of a project aiming to understand the relationships between processing conditions, the resultant microstructure and crash properties of extruded profiles. We studied the microstructure of extruded high–strength Al–Mg–Si–Cu profiles artificially aged using different regimes and performed hardness, tensile, and three–point–bending tests. We seek to correlate parameters from these tests with the outcomes of quasi–static compression testing. The results of mechanical testing are interpreted with regard to the observed microstructure and fracture surfaces.

## 2. Materials and Methods

Extruded double chamber hollow profiles of a modification of the EN AW–6082 alloy were received as fabricated from Hammerer Aluminium Industries Extrusion GmbH (HAI) with a wall thickness of 2.7mm and outer dimensions of approximately 140mm×80mm. These were cut to a length of 300mm. Table 1 shows the composition limits of the non–standardized alloy.

**Table 1.** Composition limits of the alloy in weight percent.

| | Mg | Si | Mn | Cr | Zn | Cu | Fe | Al |
|---|---|---|---|---|---|---|---|---|
| min. | 0.6 | 0.7 | 0.4 | - | - | - | - | balance |
| max. | 1.2 | 1.3 | 1.0 | 0.25 | 0.2 | 0.5 | 0.5 | balance |

The samples were divided into two batches and proprietary artificial ageing treatments were conducted. According to their ageing conditions, the designations of the two batches are "low temperature" (LT) and "high temperature" (HT). HT was aged for a shorter time than LT. The temperatures used for ageing were in the range usually used for alloys of the 6xxx series.

Quasi–static compression testing was conducted on a 1200kN hydraulic press with a rate of 100mm/min to an end length of 100mm where possible (i.e. if no disintegration of the profile occurred).

Vickers micro–hardness testing was conducted on polished sections of the profiles using a test load of 100g.

Tensile and three–point–bending tests were conducted on a Zwick Z100 materials testing machine. Specimens were cut parallel to the extrusion direction from the outer shell of the profiles. They had a width of 5mm, a reduced section length of 14mm and a gauge length of 10mm. An initial loading rate of 8MPa $s^{-1}$ was used up to a strain of 1% and then the strain rate was increased to $0.008s^{-1}$. Using a micrometer caliper the reduction of area was measured on the failed samples.

The design of the three–point–bending test was modeled similar to VDA 238–100 [12]. Specimens had a width of 15mm and a length of 60mm, with the long side being parallel to the extrusion direction. The diameter of the supports for three–point–bending was 30mm, the distance between the supports was 5.4mm (i.e. twice the specimen thickness) and the radius of the punch was 0.4mm. Testing was conducted at a rate of 20mm/min and stopped when a drop in force of at least 15N occurred. The bending angle $\alpha$ (see Figure 1) was measured on the unloaded specimens.

For optical microscopy and scanning electron microscopy (SEM), sections were ground and mechanically polished using standard metallographic techniques. Grains were revealed by etching with Barker's reagent. SEM investigations were performed on a Zeiss Ultra Plus 55 field emission gun SEM.

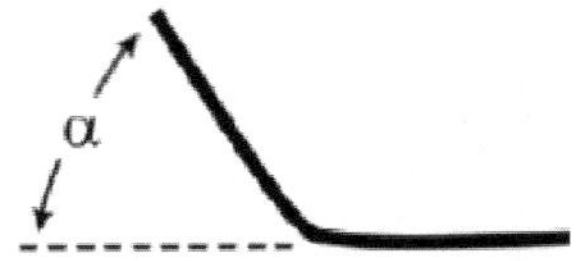


**Figure 1.** Definition of the bending angle α.

## 3. Results and Discussion

### *3.1. Microstructure*

As expected, no difference between LT and HT profiles could be observed by optical microscopy since the microstructural features visible, namely primary intermetallic phases and grain size, are not affected by the artificial ageing treatment. Figure 2 shows a polished longitudinal section of a profile. The intermetallics are mostly smaller than 10μm and exhibit rounded edges typical for α–Al(FeMn)Si –particles, indicating proper alloy design and homogenization treatment to avoid β–particles or convert them to α–particles. Most particles have an aspect ratio of approximately 1, however, some particles have aspect ratios as low as 0.2. Virtually all particles with a low aspect ratio are orientated more or less horizontally, i.e. parallel to the extrusion direction.

Figure 3 shows a longitudinal section etched with Barker's reagent to reveal individual grains. The core of the profile exhibits fine, elongated, non–recrystallized grains whereas the outer layer shows coarse, recrystallized grains that are less elongated and more equiaxed.

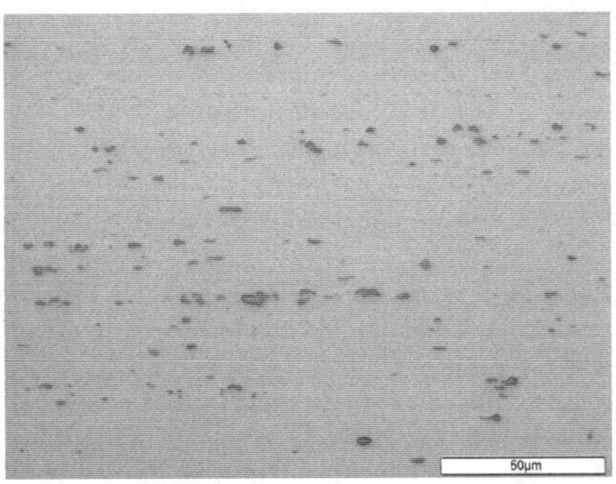


**Figure 2.** Optical micrograph of a polished longitudinal section. Intermetallic phases are dispersed in the matrix. The extrusion direction (horizontal) can be easily determined by their arrangement and orientation.

**Figure 3.** Optical micrograph of a longitudinal section etched with Barker's reagent. The microstructure consists of a fibrous core of fine non–recrystallized grains elongated along the extrusion direction and a peripheral layer of recrystallized coarse grain.

### *3.2. Mechanical properties*

The outcomes of quasi–static compression testing are illustrated in Figure 4. While the HT profile shows good folding behaviour, the test of the LT profile had to be stopped due to beginning of disintegration of the profile.

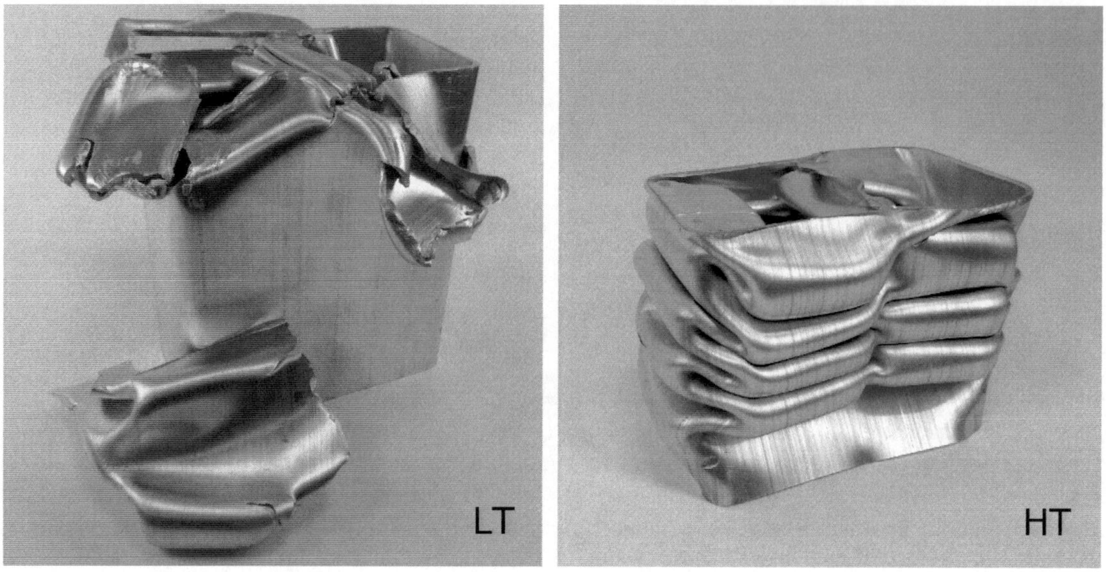


**Figure 4.** LT and HT profile after quasi–static compression test.

No difference in Vickers micro–hardness HV0.1 could be found. The mean value of five measurements was 107.8±1.5 for LT and 107.8±2.2 for HT (plus/minus values represent standard deviation).

Average stress–strain curves from the tensile tests are shown in Figure 5 and the results given in Table 2. Since the yield strength ($R_{p0.2}$) determines the *yield strength class* of a safety part it is considered the most important value by manufacturers. Both batches of profiles exhibit yield strengths over 280MPa whereat the values of HT exceed those of LT. However, LT profiles show a higher ultimate tensile strength ($R_m$) and, interestingly, a higher elongation at break ($A_{10}$). However, the reduction of area (Z) is higher for HT indicating higher localized fracture strain.

**Table 2.** Mean tensile and three–point–bending test results. $A_{10}$ is the technical elongation for a gauge length of 10mm and Z is reduction of area measured on broken specimens. $\alpha$ is the bending angle measured on unloaded specimens as defined in Figure 1. Plus/minus values represent standard deviation from three measurements (except for Z – only two measurements).

**Tensile test**

| | $R_{p0.2}$ [MPa] | $R_m$ [Mpa] | $A_{10}$ [%] | Z [%] |
|---|---|---|---|---|
| LT | 286.8±0.7 | 347.0±0.3 | 21.0±0.4 | 41.9±0.6 |
| HT | 297.5±0.4 | 330.5±0.4 | 19.7±0.0 | 51.0±2.2 |

**Three–point–bending test**

| | $F_{max}$ [N] | $\alpha$ [°] |
|---|---|---|
| LT | 1958±29.6 | 71.5±3.3 |
| HT | 1817±15.4 | 78.2±2.8 |

Although the quasi–static compression testing results of the two types of profiles were very different, the difference in bending angle ($\alpha$) in the three–point–bending test was not as pronounced. However, HT shows a lower maximum bending force ($F_{max}$) than LT (see Table 2).

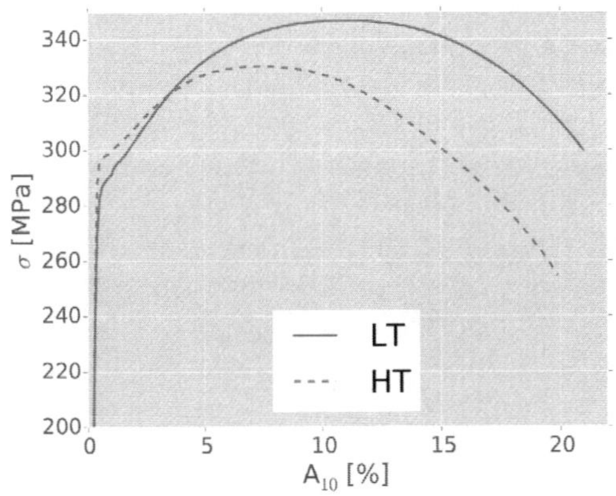


**Figure 5.** Tensile test: average stress–strain curves for the LT and HT profiles.

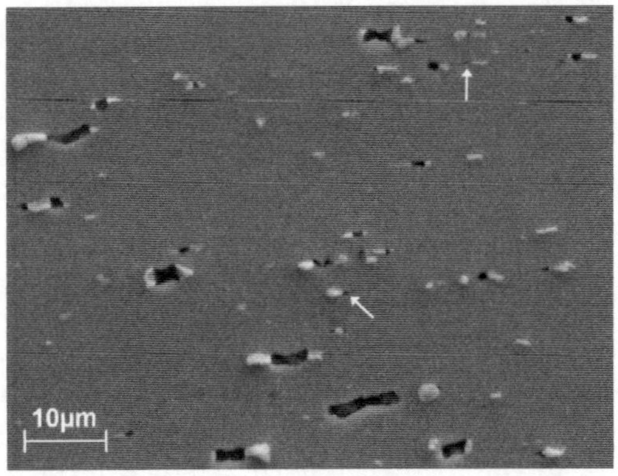


**Figure 6.** Secondary electron micrograph (20kV) of broken intermetallic phases and particle–matrix decohesion (examples marked with arrows) of intermetallic phases in a polished cross–section of an unloaded damaged tensile specimen. The original load direction was approximately horizontal.

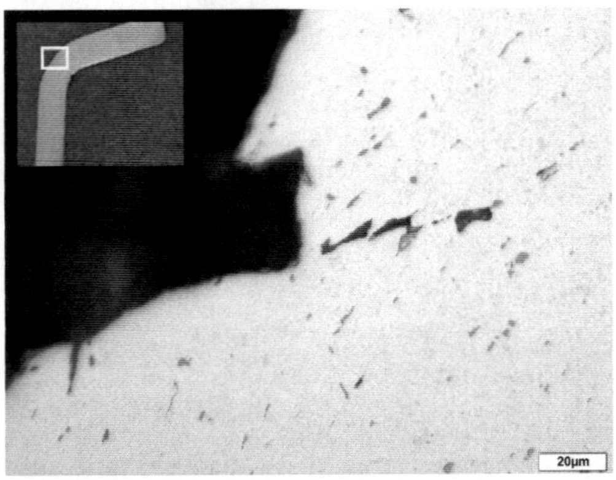


**Figure 7.** Crack propagagation in three–point–bending by void–coalescence. Optical micrograph of a polished cross–section of the tensile side of an unloaded three–point–bending specimen. The inset image shows the approximate position of the micrograph.

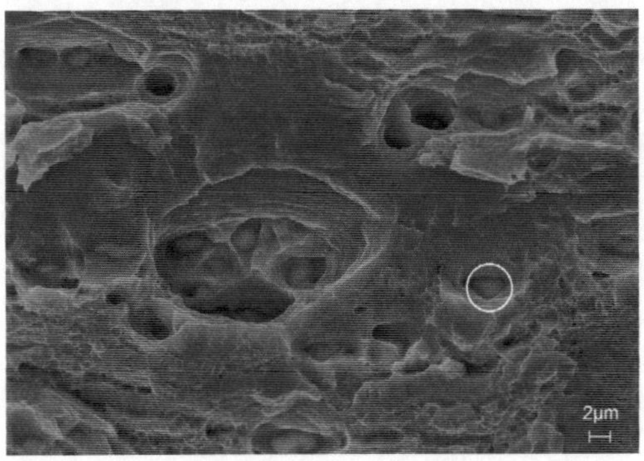


**Figure 8.** Mixed secondary/backscattered electron micrograph (15kV) of the fracture surface of a tensile specimen. The large dimples contain primary intermetallics (example marked with circle). Between the large dimples many small dimples are visible indicating a secondary population of voids.

*3.3. Microstructural investigation of failure*

Figure 6 shows breakage and particle–matrix decohesion of intermetallics due to the applied stress, resulting in the nucleation of voids. These mechanisms were observed in tensile as well as in three–point–bending specimens. In both cases breakage of intermetallics was found more frequently than particle–matrix decohesion. Since all of the particles with a low aspect ratio are oriented near parallel to the load direction, this is in accordance with the findings of Lassance et al. [7]. When, however, decohesion was observed it was not on the interface of particles oriented perpendicular to the load direction – since such particles did not exist – but rather at the sharp edges of particles oriented in line with the the extrusion direction.

The number of voids was highest near the fracture surface in tensile test and near the curved outer boundary on the tensile side of the three–point–bending specimens. In Figure 7 the coalescence of voids facilitating the propagation of a crack in the three–point–bending test can be seen.

Investigation of fracture surfaces from the tensile specimens (Figure 8) revealed large dimples. Primary intermetallic phases were found in the large dimples. Therefore, it is expected that the voids nucleated at these phases. In addition, small dimples in the matrix between the large dimples were found. These dimples suggest the existence of a secondary population of voids as described in the literature for other alloys [7, 9]. It is expected that these voids nucleate at non–hardening dispersoids due to their incoherence with the matrix or near dispersion hardening precipitates due to their effect on the matrix.

The only difference between LT and HT is the post–extrusion artificial ageing regime. However, primary intermetallic phases are hardly affected by artificial ageing. Their chemical composition, volume density, form and size are defined during casting and homogenization of the billet and their final distribution and form is created by subsequent extrusion. Therefore, no significant difference in primary intermetallics between LT and HT should exist and, in fact, no difference in intermetallics or in their breakage or decohesion behaviour could be found between LT and HT. Thus, these mechanisms cannot explain the differences in ductility.

The only factor remaining that is different between the two batches of profiles is the artificial ageing treatment. A possible explanation is that HT, which was aged at a higher temperature for a shorter time, contains fewer, larger dispersion hardening phases than LT, allowing for more dislocation mobility yielding higher ductility. Further work is needed to clarify the mechanisms involved.

## 4. Conclusions

The results show that care must be taken when interpreting small scale mechanical tests of extruded aluminium profiles for safety parts. Understanding the microstructural failure mechanisms is important for the proper design of high strength profiles for safety parts in cars.

The main findings of this work are:

- Small variations to artificial ageing can dramatically alter quasi–static compression testing performance, even though yield strength, ultimate tensile strength, and uniform elongation are similar and microhardness is the same.
- For the studied profiles, strong necking in tensile testing indicated good quasi–static compression testing behaviour.
- Higher uniform elongation did not result in better quasi–static compression testing properties.
- Differences in bending angle were rather small, however the maximum bending force was lower for profiles with good quasi–static compression testing properties.

## 5. Acknowledgments

The authors would like to thank the Austrian Research Promotion Agency (FFG), the Federal Ministry for Transport, Innovation and Technology (bmvit) and the State of Upper Austria for sponsoring this research work in the framework of COMET (project AMOREE, FFG grant no. 843537).

The authors would also like to thank their colleagues at Hammerer Aluminium Industries Extrusion GmbH and LKR Leichtmetallkompetenzzentrum Ranshofen GmbH for their assistance and Dr. Nigel Ross (LKR) for proofreading the manuscript.

## References

[1] Schiffl A and Kühlein W 2014 Hochfeste Crashlegierungen aus dem Haus Hammerer Aluminium Industries: von der Idee zur Umsetzung *Proc., 8. Ranshofner Leichtmetalltage, November 4 – 6, 2014, Geinberg, Austria* pp 37–47

[2] Sweet E, Caraher S, Danilova N and Zhang X 2004 Effects of extrusion parameters on coarse grain surface layer in 6xxx series extrusions *Proc. of the 8th Intl. Aluminum Extrusion Technology Seminar, Orlando, Florida, USA* vol 1 pp 115–126

[3] Strobel K, Sweet E, Easton M, Nie J F and Couper M 2010 Dispersoid phases in 6xxx series aluminium alloys *Materials Science Forum* vol 654 (Trans Tech Publ) pp 926–929

[4] Edwards G, Stiller K, Dunlop G and Couper M 1998 *Acta Materialia* **46** 3893–3904

[5] Chakrabarti D and Laughlin D E 2004 *Progress in Materials Science* **49** 389–410

[6] Marioara C, Andersen S, Stene T, Hasting H, Walmsley J, Van Helvoort A and Holmestad R 2007 *Philosophical Magazine* **87** 3385–3413

[7] Lassance D, Fabregue D, Delannay F and Pardoen T 2007 *Progress in Materials Science* **52** 62–129

[8] Mrówka-Nowotnik G and Sieniawski J 2005 *Journal of Materials Processing Technology* **162** 367–372

[9] Mrówka-Nowotnik G, Sieniawski J and Nowotnik A 2009 *Journal of Achievements in Materials and Manufacturing Engineering* **32** 162–170

[10] Røyset J, Tundal U, Reiso O and Furu T 2008 Al–Mg–Si alloys with improved crush properties *Proc. 9th International Aluminum Extrusion Technology Seminar, May 13–16 2008, Orlando, FL, USA, Vol. II, 1–16*

[11] Afseth A 2013 Development of high strength sheet crash alloys for BIW structures *Proc., Materials in Car Body Engineering, May 7–8, 2013, Bad Nauheim, Germany* p 253pp

[12] Neuhaus R and Borsutzki M 2013 *Materials Testing* **55** 654–659

# Semiconductor Film Grown on a Circular Substrate: Predictive Modeling of Lattice-Misfit Stresses

**E. Suhir[1,2,3], J. Nicolics[2], G. Khatibi[2], M. Lederer[2]**

[1]Portland State University, Portland, OR, USA; [2]Technical University, Vienna, Austria, and [3]ERS Co., Los Altos, CA 94024, USA, Tel. 650-969-1530, cell. 408-410-0886

suhire@aol.com

**Abstract.** An effective and physically meaningful analytical predictive model is developed for the evaluation the lattice-misfit stresses (LMS) in a semiconductor film grown on a circular substrate (wafer). The two-dimensional (plane-stress) theory-of-elasticity approximation (TEA) is employed in the analysis. The addressed stresses include the interfacial shearing stress, responsible for the occurrence and growth of dislocations, as well as for possible delaminations and the cohesive strength of a buffering material, if any. Normal radial and circumferential (tangential) stresses acting in the film cross-sections and responsible for its short- and long-term strength (fracture toughness) are also addressed. The analysis is geared to the GaN technology.

## 1. Introduction

GaN is a binary III/V direct bandgap semiconductor commonly used in bright light-emitting diodes. GaN based high-electron-mobility-transistor (HEMT) device technology is viewed as a promising one for power amplifier applications [1]. The reliability of GaN devices continues, however, to be a key factor in making promising and viable GaN technology-based devices into reliable products. The elevated lattice-misfit and thermal-mismatch stresses in GaN films are the major limitations for obtaining high-quality GaN systems on technologically important substrates, such as, e.g., Si, SiC, AlN, or diamond (C). A variety of techniques have been suggested to reduce the adverse consequence of the lattice misfit during semiconductor crystal growth (SCG) process. Based on a rather general predictive model for the evaluation of stresses in finite-size bonded joints [2], Luryi and Suhir [3] have shown that the critical thickness of an epitaxial film could be made even infinite, if a properly engineered substrate is used. The idea of employing patterned, porous or otherwise engineered substrates led to many subsequent investigations and to numerous publications exploring the use of nano-sized islands ("towers") as growth nucleation sites. Sagar et al [4] have demonstrated that a reduction in dislocation density from about $10^{10} - 10^{12}$ cm$^{-2}$ in a template prepared using molecular beam epitaxy (MBE) could be reduced to about $2.5\times10^{9}$ cm$^{-2}$, if a porous SiC substrate is employed. The relative level of the lattice-misfit and thermal-mismatch stresses in bi-material GaN assemblies was recently addressed, based on analytical (mathematical) predictive modeling, with an objective to evaluate and to compare these two types of stresses [5]. The developed models were based on the strength of material approach (SMA) and treated the GaN assembly as a bi-material elongated rectangular strip. It was determined that even if a reasonably good lattice match takes place (as, e.g., in the case of a GaN film fabricated on a SiC substrate, when the mismatch strain is only about 3%) and, in addition, the temperature change (from the fabrication temperature to the operation temperature) was significant (as high as 1000 ℃), the thermal stresses were still considerably lower than the lattice-misfit stresses. It was determined also that the interfacial shearing and peeling stresses were as important as the normal stresses acting in the cross-sections of the GaN film. While the normal stresses in the GaN film cross-sections are responsible for the fracture toughness of the film material, it is the interfacial stresses that are responsible for the ability of the assembly to resist delaminations (interfacial cracking) and for the performance of the buffering ("bonding") materials, if any. The objective of the analysis that follows is to develop a simple and physically meaningful predictive TEA based LMS model for a GaN film grown on a circular substrate. Our intent is to evaluate, using the developed model, the applicability and accuracy of the SMA (that addresses a bi-material elongated strip as a more or less suitable substitute for an actual circular assembly).

## 2. Analysis

*2.1. Normal stresses in the assembly mid-portion*

The analysis carried out in this section proceeds from the major assumption that neither the circular configuration of the assembly nor its bow affect the normal LMS in the major mid-portion of a large size bi-material assembly. Let the lattice constants for the materials of the components #1 (film) and #2 (substrate) be $a_1$ and $a_2 \leq a_1$, respectively, and, as a result of joining these components into a single bi-material assembly, the final interfacial lattice constant is $a$. Then the interfacial strains experienced by the component materials are $\varepsilon_1 = \frac{a_1 - a}{a_1} = 1 - \frac{a}{a_1}$ in compression and $\varepsilon_2 = \frac{a - a_2}{a_2} = \frac{a}{a_2} - 1$ in tension. The induced stresses, in accordance with Hooke's law for the two-dimensional state of stress, are

$$\sigma_1 = \frac{E_1}{1-\nu_1}\varepsilon_1 = \frac{E_1}{1-\nu_1}\left(1 - \frac{a}{a_1}\right), \quad \sigma_2 = \frac{E_2}{1-\nu_2}\varepsilon_2 = \frac{E_2}{1-\nu_2}\left(\frac{a}{a_2} - 1\right) \tag{1}$$

in compression and in tension, respectively. Here $E_1$ and $E_2$ are Young's moduli, and $\nu_1$ and $\nu_2$ are Poisson's ratios of the materials. The equations (1) reflect the following assumptions: these stresses are the same for all the points in the given cross-section of the given component; the assembly size (in the x-y plane) is significant and the assembly points of interest are sufficiently remote from the assembly edges. The corresponding forces are

$$T_1 = \sigma_1 h_1 = \frac{1}{\lambda_1}\left(1 - \frac{a}{a_1}\right), \quad T_2 = \sigma_2 h_2 = \frac{1}{\lambda_2}\left(\frac{a}{a_2} - 1\right), \tag{2}$$

where $\lambda_1 = \frac{1-\nu_1}{E_1 h_1}$, $\lambda_2 = \frac{1-\nu_2}{E_2 h_2}$, are the axial compliances of the assembly components, and $h_1$ and $h_2$ are their thicknesses. The condition $T_1 = T_2$ of equilibrium yields: $a = \frac{\lambda a_1 a_2}{\lambda_1 a_1 + \lambda_2 a_2}$, where $\lambda = \lambda_1 + \lambda_2$.

The normal strains in the cross-sections of the assembly components are $\varepsilon_1 = \lambda_1 T_0$, $\varepsilon_2 = \lambda_2 T_0$, where $T_0 = \frac{a_1 - a_2}{\lambda_1 a_1 + \lambda_2 a_2}$ is the longitudinal force (whether $T_1$ or $T_2$). The corresponding normal stresses acting in the cross-sections of the assembly components are $\sigma_1 = \frac{T_0}{h_1} = \frac{1}{h_1}\frac{a_1 - a_2}{\lambda_1 a_1 + \lambda_2 a_2}$, $\sigma_2 = \frac{T_0}{h_2} = \frac{1}{h_2}\frac{a_1 - a_2}{\lambda_1 a_1 + \lambda_2 a_2}$ in compression and in tension, respectively. When the component #2 is a thick substrate, and the component #1 is a thin film, the axial compliance $\lambda_2$ of the substrate becomes significantly smaller than the compliance $\lambda_1$ of the film, and, if the Young's moduli of the component materials are in the same order of magnitude such as GaN and SiC, the above formulas can be simplified as follows:

$$T_0 = \frac{1}{\lambda_1}\left(1 - \frac{a_2}{a_1}\right) = \frac{E_1 h_1}{1-\nu_1}\left(1 - \frac{a_2}{a_1}\right), \quad \sigma_1 = \frac{E_1}{1-\nu_1}\left(1 - \frac{a_2}{a_1}\right), \tag{3}$$

These formulas indicate that the normal compressive stress in the mid-portion of the component #1 (thin film), as long as it is thin enough, is independent of its thickness, and that the normal tensile stress in the substrate is proportional to the thickness ratio and is very low.

### *2.2. Interfacial shearing stress*

#### *2.2.1. Assumptions*

The following major assumptions are used in our analysis:

- The assembly components (the film, and the substrate) can be treated as thin circular plates experiencing small deflections, and the engineering theory of bending of thin plates can be used to predict their physical behavior;
- The peeling stresses do not affect the interfacial shearing stresses and need not be accounted for when evaluating the shearing stresses;
- The interfacial compliances of the assembly in its plane is due to the joint interfacial compliances $\kappa_1 = \frac{h_1}{3G_1}$ and $\kappa_2 = \frac{h_2}{3G_2}$ of the film and the substrate [2];
- The interfacial radial displacements, $u_1(r)$, of the component #1 (film) can be evaluated as the sum of the radial displacements, $u(r)$, caused by the lattice-misfit-induced forces, and additional displacements, $\kappa_1\tau_0(r)$, of the interfacial point at the given radius $r$, with respect to the displacements $u(r)$ of the inner points of the cross-section: $u_1(r) = u(r) + \kappa_1\tau_0(r)$. In this formula, $\tau_0(r)$ is the interfacial shearing stress in the given cross-section, and $\kappa_1$ is the interfacial compliance of the film layer. The displacements $u(r)$ can be evaluated based on the Hooke's law, and are considered the same for all the points of the given (circumferential) cross-section. The second term in this relationship is, in effect, a correction that considers the deviation of the given cross-section from planarity;
- The interfacial radial displacements,. $u_2(r)$, of the substrate can be evaluated as $u_2(r) = -\kappa_2\tau_0(r)$.
- Assembly bow has a small effect on the state of stress in the film and need not be accounted for.
- The interfacial shearing stress $\tau_0(r)$ increases with an increase in the film thickness and with an increase in the shearing stress gradient $\frac{\partial\tau_{xz}(r)}{\partial z}$ in the through-thickness direction; in an approximate analysis the interfacial shearing stress $\tau_0(r)$ can be sought as a product $h_1\frac{\partial\tau_{xz}(r)}{\partial z}$ of the film thickness $h_1$ and the through-thickness gradient $\frac{\partial\tau_{xz}(r)}{\partial z}$ of the shearing stress at the given radius. This means that the through-thickness gradient $\frac{\partial\tau_{xz}(r)}{\partial z}$ of the shearing stress can be sought in the approximate form: $\frac{\partial\tau_{xz}(r)}{\partial z} \approx \frac{\tau_0(r)}{h_1}$.

#### *2.2.2. Basic equation and its solution*

The taken assumption and the condition $u_1(r) = u_2(r)$ of the displacement compatibility result in the following formula for the radial interfacial displacements of the film:

$$u(r) = -\kappa\tau_0(r), \tag{4}$$

where $\kappa = \kappa_1 + \kappa_2$ is the total interfacial compliance of the assembly. The formula (4) and the Cauchy formulas [7] $\varepsilon_r = u'(r)$, $\varepsilon_\theta = \frac{u(r)}{r}$ for the normal radial, $\varepsilon_r$, and the normal circumferential (tangential), $\varepsilon_\theta$, strains yield: $\varepsilon_r = -\kappa\tau_0'(r)$, $\varepsilon_\theta = -\kappa\frac{\tau_0(r)}{r}$. The corresponding normal stresses in the film can be evaluated, using Hooke's law equations $\sigma_r = \frac{E_1}{1-\nu_1^2}(\varepsilon_r + \nu_1\varepsilon_\theta)$, $\sigma_\theta = \frac{E_1}{1-\nu_1^2}(\varepsilon_\theta + \nu_1\varepsilon_r)$, as

$$\sigma_r = -\frac{E_1}{1-\nu_1^2}\kappa\left[\tau_0'(r) + \nu_1\frac{\tau_0(r)}{r}\right], \quad \sigma_\theta = -\frac{E_1}{1-\nu_1^2}\kappa\left[\frac{\tau_0(r)}{r} + \nu_1\tau_0'(r)\right]. \tag{5}$$

Introducing the obtained formulas into the equilibrium equation $\frac{\partial\sigma_r}{\partial r} + \frac{\partial\tau_{rz}}{\partial z} + \frac{\sigma_r - \sigma_\theta}{r} = 0$, the following basic equation of Bessel type for the shearing stress function, $\tau_0(r)$ can be obtained:

$$\tau_0''(r) + \frac{\tau_0'(r)}{r} - \frac{\tau_0(r)}{r^2} - k^2\tau_0(r) = 0, \tag{6}$$

where $k = \sqrt{(1+\nu_1)\frac{\lambda_1}{\kappa}}$ is the parameter of the interfacial shearing stress, and $\lambda_1$ is the radial compliance of the film. Note that when the SMA is used and the film is significantly thinner than the the substrate, the longitudinal axial compliance $\lambda$ of the assembly is due primarily to the compliance $\lambda_1$ of the film, and the parameter of the shearing stress is $k = k_* = \sqrt{\frac{\lambda_1}{\kappa}}$.The difference should be attributed to the circumferential loading in circular assemblies.
The equation (13) has the following solution:

$$\tau_0(kr) = C_1 I_1(kr). \tag{7}$$

where $C_1$ is the constant of integration, $k$ is the parameter of the interfacial shearing stress, and $I_1(kr)$ is the modified Bessel function of the first kind of the first order [8]. The Bessel function in (7) obeys the following rules of differentiation: $I_1'(kr) = kI_0(kr) - \frac{I_1(kr)}{r}$, $I_0'(kr) = kI_1(kr)$, where $I_0(kr)$ is the modified Bessel function of the first kind of zero order. The function $I_1(kr)$ is anti-symmetric with respect to the origin and plays the same role as the hyperbolic sine plays in the SMA solution for an elongated strip [2]. Introducing the solution (7) into the formulas (5), we have:

$$\sigma_r = -\frac{E_1}{1-\nu_1^2}\kappa C_1\left[kI_0(kr) - (1-\nu_1)\frac{I_1(kr)}{r}\right], \quad \sigma_\theta = -\frac{E_1}{1-\nu_1^2}\kappa C_1\left[\nu_1 kI_0(kr) + (1-\nu_1)\frac{I_1(kr)}{r}\right] \tag{8}$$

Then the equilibrium equation, with consideration of the the solution (7), yields:

$$\frac{\partial\sigma_r}{\partial r} = C_1\left[\frac{E_1}{1+\nu_1}\kappa\left(k\frac{I_0(kr)}{r} - 2\frac{I_1(kr)}{r^2}\right) - \frac{I_1(kr)}{h_1}\right] \tag{9}$$

This equation can be also written as

$$\frac{\partial\sigma_r}{\partial r} = \frac{C_1}{h_1}\left[\frac{1-\nu_1}{k^2}\frac{d}{dr}\left(\frac{I_1(kr)}{r}\right) - \frac{1}{k}\frac{dI_0(kr)}{dr}\right]. \tag{10}$$

By integration, we find

$$\sigma_r = \frac{C_1}{h_1}\left[\frac{1-\nu_1}{k^2}\frac{I_1(kr)}{r} - \frac{1}{k}I_0(kr)\right] + f(z), \tag{11}$$

where $f(z)$ is an arbitrary function of the coordinate $z$. Since the radial normal stress in the film does not change in its through-thickness direction, one should put $(z) = C_0$, where $C_0$ is thus far unknown constant of integration. Thus,

$$\sigma_r = \frac{C_1}{h_1}\left[\frac{1-\nu_1}{k^2}\frac{I_1(kr)}{r} - \frac{1}{k}I_0(kr)\right] + C_0. \tag{12}$$

There are no external loads acting on the assembly edges, and therefore the boundary condition $\sigma_r(r_0) = 0$ should be fulfilled. This condition and the expression (12) yield:

$$C_0 = -\frac{C_1}{h_1}\left[\frac{1-\nu_1}{k^2}\frac{I_1(kr_0)}{r_0} - \frac{1}{k}I_0(kr_0)\right], \tag{13}$$

so that

$$\sigma_r = \frac{C_1}{kh_1}\left[\frac{1-\nu_1}{k}\left(\frac{I_1(kr)}{r} - \frac{I_1(kr_0)}{r_0}\right) + I_0(kr_0) - I_0(kr)\right]. \tag{14}$$

For small radii $r$ the following approximations can be used [8]: $I_0(kr) \approx 1,\ I_1(kr) \approx kr$. Then the expression (14) yields:

$$\sigma_r = \frac{C_1}{kh_1}\left[(1-\nu_1)\left(1 - \frac{I_1(kr_0)}{kr_0}\right) + I_0(kr_0) - 1\right]. \tag{15}$$

Comparing this formula with the first formula in (8), we conclude that the constant $C_1$ is expressed as

$$C_1 = \frac{1}{2}\frac{E_1}{1-\nu_1}\frac{kh_1\left(1-\frac{a_2}{a_1}\right)}{(1-\nu_1)\left(1-\frac{I_1(kr_0)}{kr_0}\right) + I_0(kr_0) - 1}. \tag{16}$$

This formula indicates particularly that while the radial normal stress is proportional to the $\frac{C_1}{kh_1}$ ratio and is, hence, film thickness independent, the interfacial shearing stress is proportional to the $C_1$ value and increases linearly with an increase in the film thickness.

Using the solution (7) and the formula (16) for the constant of integration $C_1$, the following formula for the interfacial shearing stress can be obtained:

$$\tau_0(r) = \tau_{\max}\frac{I_1(kr)}{I_1(kr_0)}, \tag{17}$$

where $\tau_{\max} = \tau_0(r_0) = \frac{E_1}{1-\nu_1}kh_1\left(1-\frac{a_2}{a_1}\right)\chi(kr_0)$ is the maximum shearing stress that occurs at the assembly edge, and the function

$$\chi(kr_0) = \frac{I_1(kr_0)}{(1-\nu_1)\left(1-\frac{I_1(kr_0)}{kr_0}\right) + I_0(kr_0) - 1} \tag{18}$$

considers the effect of the product $kr_0$ of the parameter $k$ of the interfacial shearing stress and the assembly size (radius) $r_0$ on the maximum interfacial shearing stress. Thus, the maximum interfacial shearing stress at the assembly edge increases with an increase in the effective Young's modulus of the material of the film, with an increase in the parameter $k$ of the interfacial shearing stress and with the increase in the thickness $h_1$of the film. The stress increases, of course, with an increase in the lattice misfit $1-\frac{a_2}{a_1}$ that plays the role of the "external loading". The maximum shearing stress is inversely proportional to the longitudinal (axial) compliance $\lambda_1 = \frac{1-\nu_1}{E_1h_1}$ of the film. The function (18) is tabulated in the second line of Table 1 for $\nu_1 = 0.25$. As evident from the calculated data, the maximum shearing stress increases with an increase in the parameter $kr_0$ when this parameter changes from zero to about $kr_0 \approx 10.0$, and then remains constant, i.e., assembly size independent.

*2.2.3. Theory-of-elasticity (TEA) vs. strength-of-materials (SMA) solutions*

Let us compare the TEA solution with the SMA solution for the interfacial shearing stress. One can write the SMA solution as [2]

$$\tau(x) = \tau_{\max} \frac{\sinh kx}{\sinh kl}, \tag{19}$$

where the maximum value of the interfacial shearing stress is at the edge of the assembly and is

$$\tau_{\max} = \tau_0(l) = \frac{E_1}{1-\nu_1} kh_1 \left(1 - \frac{a_2}{a_1}\right) \coth \frac{kl}{2}. \tag{20}$$

**Table 1.** Tabulated function $\chi(kr_0)$ that considers the effect of the product $kr_0$ on the maximum interfacial shearing stress and the function $\chi_1(kr_0)$ that considers the effect of this product on the ratio of the maximum interfacial stresses computed based on the theory-of-elasticity approach to the maximum interfacial stress calculated using the strength-of-materials approach

| $kr_0$ | 0 | 0.1 | 0.5 | 1.0 | 2.0 | 3.0 | 5.0 | 10.0 | $\infty$ |
|---|---|---|---|---|---|---|---|---|---|
| $\chi$ | 0 | 0.133 | 0.605 | 0.955 | 1.110 | 1.085 | 1.043 | 1.021 | 1.000 |
| $\chi_1$ | 0 | 0.00665 | 0.1512 | 0.4622 | 0.9823 | 1.2972 | 1.5152 | 1.6085 | 1.5811 |

Comparing the TEA solution with the SMA solution we conclude that the distribution of the interfacial shearing stress along the assembly radius (length) is governed, in the case of the TEA solution, by the ratio $\frac{I_1(kr)}{I_1(kr_0)}$ of the modified Bessel functions, and by the ratio $\frac{\sinh kx}{\sinh kl}$ of the hyperbolic sines, in the case of the SMA solution. Another, more important, difference is due to the considerably higher value of the parameter $k$ of the interfacial shearing stress in the TEA solution. The maximum TEA shearing stress $\tau_{\max}^{TEA}$ can be determined from the maximum SMA shearing stress $\tau_{\max}^{SMA}$ as

$$\tau_{\max}^{TEA} = \frac{k_{TEA}}{k_{SMA}} \chi(k_{TEA} r_0) \tanh\left(\frac{k_{SMA} l}{2}\right) \tau_{\max}^{SMA} = \sqrt{(1+\nu_1)}\, \chi(kr_0) \tanh\left(\frac{kr_0}{2\sqrt{2(1+\nu_1)}}\right) \tau_{\max}^{SMA} = \chi_1(kr_0)\, \tau_{\max}^{SMA} \tag{21}$$

where the function $\chi_1(kr_0) = \sqrt{(1+\nu_1)}\, \chi(kr_0) \tanh\left(\frac{kr_0}{2\sqrt{(1+\nu_1)}}\right)$ is the maximum interfacial stress ratio.

The half-assembly-length $l$ is replaced here with the radius $r_0$ value, and the notation $k = k_{TEA}$ is used. The function $\chi(kr_0)$ changes from zero to one, when the product $kr_0$ changes from zero to infinity. For small $kr_0$ values this function can be computed as $\chi(kr_0) = \frac{kr_0}{1-\nu_1}$. The function $\chi_1(kr_0)$ changes from

$\chi_{1,\min} = kr_0 \chi_{\min} = \frac{(kr_0)^2}{1-\nu_1}$ to $\chi_{1\max} = \sqrt{1+\nu_1}\, \chi_{\max}$, when the product $kr_0$ changes from zero to infinity (actually, from a small number smaller than, say, 2, to a large number exceeding 10). The calculated values of the function $\chi_1(kr_0)$ are shown in the bottom line of Table 1. The computed data indicate that the SMA underestimates the maximum shearing stress that takes place for large $kr_0$ values $(kr_0 \succ 3)$ and overestimates this stress for small $kr_0$ values $(kr_0 \prec 2)$. This is due primarily to the different values of the parameter $k$ of the interfacial stress: this parameter is by the factor of $\sqrt{1+\nu_1}$, greater in the case of the TEA based solution. If this parameter were the same in the two approaches, then the shearing stress predicted for large enough $(kr_0 \succ 10)$ assemblies on the basis of the TEA and SMA would be exactly the same. This means that, in an approximate stress analysis of large size and/or stiff assemblies $(kr_0 \succ 10)$, one can compute the parameter $k$ using the TEA formula and then calculate the maximum interfacial shearing stress on the basis of the simpler SMA formula. The functions $\frac{I_1(kr)}{I_1(kr_0)}$ and $\frac{\sinh(kr)}{\sinh(kr_0)}$ reflect the distributions of the interfacial stress along the assembly in the cases of TEA and SMA based solutions, respectively. The parameter $k$ of the interfacial shearing stress is assumed to be the same in the TEA and in the SMA based solutions. This means that TEA predicts somewhat higher stress concentration at the assembly ends than the SMA (for the same maximum stress at the assembly end). The predicted relative ordinates of the interfacial shearing stress are rather close though.

For large arguments $z$ ( $z \succ 10$ ), the modified Bessel function of the order $n$ can be evaluated by the approximate formula [8]: $I_n(z) = \frac{e^z}{\sqrt{2\pi z}}$. Then, with

$$I_0(kr_0) \cong I_1(kr_0) \cong \frac{e^{kr_0}}{\sqrt{2\pi kr_0}}, \quad I_0(kr) \cong I_1(kr) \cong \frac{e^{kr}}{\sqrt{2\pi kr}} \tag{22}$$

we obtain:

$$\tau_0(r) = \tau_{\max}\sqrt{\frac{r_0}{r}}e^{-k(r_0-r)}, \tag{23}$$

where the maximum value of the shearing stress may be derived from the formulas (20) and (21). As evident from the formula (23), the interfacial shearing stress, $\tau_0(r)$, concentrates along a narrow peripheral ring, and is next to zero for the inner radii of the assembly $r \prec\prec r_0$.

### *2.3. Normal stresses*

Introducing the expression (16) for the constant $C_1$ into the formula (15) for the normal stress $\sigma_r(r)$ in the film, we obtain:

$$\sigma_r = \sigma_1\left[1 - \frac{(1-\nu_1)\left(1 - \frac{I_1(kr)}{kr}\right) + I_0(kr) - 1}{(1-\nu_1)\left(1 - \frac{I_1(kr_0)}{kr_0}\right) + I_0(kr_0) - 1}\right] \tag{24}$$

where $\sigma_1$ is the normal stress in the mid-portion of the assembly. The expression in the brackets is, in effect, a “correction” that considers the role of the finite radius $r_0$ of the assembly. By differentiation we find:

$$\frac{d\sigma_r}{dr} = -\sigma_1\left[\frac{\frac{1-\nu_1}{kr}\left(2\frac{I_1(kr)}{r} - kI_0(kr)\right) + kI_1(kr)}{(1-\nu_1)\left(1 - \frac{I_1(kr_0)}{kr_0}\right) + I_0(kr_0) - 1}\right] \tag{25}$$

Then the circumferential stress $\sigma_\theta$ can be found from the equilibrium equation as follows:

$$\sigma_\theta = \sigma_r + r\frac{d\sigma_r}{dr} + 2\frac{\tau_0(r)}{h_1}r = \sigma_1\left[1 + (1-\nu_1)\frac{I_0(kr) - \frac{I_1(kr)}{kr}}{(1-\nu_1)\left(1 - \frac{I_1(kr_0)}{kr_0}\right) + I_0(kr_0) - 1}\right] \tag{26}$$

This stress changes from $\sigma_1$ in the midportion of the assembly, where it is not different from the normal radial stress, to the

$$\sigma_\theta(r_0) = \sigma_1\frac{(1-\nu_1)\left(I_0(kr_0) - 2\frac{I_1(kr_0)}{kr_0}\right)}{(1-\nu_1)\left(1 - \frac{I_1(kr_0)}{kr_0}\right) + I_0(kr_0) - 1} \tag{27}$$

value at the assembly end. Clearly, this stress cannot be found on the SMA basis.

In the case of a large size (large $r_0$ values) and/or stiff (large $k$ values) assemblies, the formulas for the normal stresses can be simplified:

$$\sigma_r = \sigma_1\left[1-\sqrt{\frac{r_0}{r}}e^{-k(r_0-r)}\right],\quad \sigma_\theta = \sigma_1\left[1+(1-\nu_1)\sqrt{\frac{r_0}{r}}e^{-k(r_0-r)}\right]. \tag{28}$$

These formulas indicate that the normal stresses, $\sigma_r$ and $\sigma_\theta$, in the film are uniformly distributed over the inner portion of the assembly ($r \prec\prec r_0$). At the assembly end the radial stress $\sigma_r$ is zero, and the circumferential stress is

$$\sigma_\theta = (2-\nu_1)\sigma_1, \tag{29}$$

i.e., by the factor of $2-\nu_1$ higher than the normal stresses, $\sigma_1$, in the mid-portion of the film.

## 3. Numerical example

### *3.1. Input Data*

| Component | Film, GaN (Component #1) | Substrate,SiC (Component #2) |
|---|---|---|
| Young's modulus | 181000 MPa | 461070 MPa |
| Poisson's ratio | 0.352 | 0.175 |
| Lattice constant | 3.18A | 3.08A |
| Thickness | 0.002mm | 0.100mm |
| Assembly radius $r_0$ = 25.4 mm | | |

### *3.2. Calculated plots*

The calculated stresses for the example of 2μm GaN film on a SiC substrate are presented in figures 1 and 2. Figure 1 displays plots of interfacial shearing stress calculated by using TEA and SMA, respectively and the plots of radial and tangential stresses based on TEA are shown in figure 2. Experimental studies show that thermally induced stresses in epitaxial grown GaN layers may relax by cracking or occurrence of high dislocation densities during the cooling down from deposition temperatures. Still biaxial stresses with values exceeding 1 GPa have been observed as a consequence of lattice-misfit and thermal-mismatch strains on GaN epitaxial films deposited on SiC substrates [4,9].

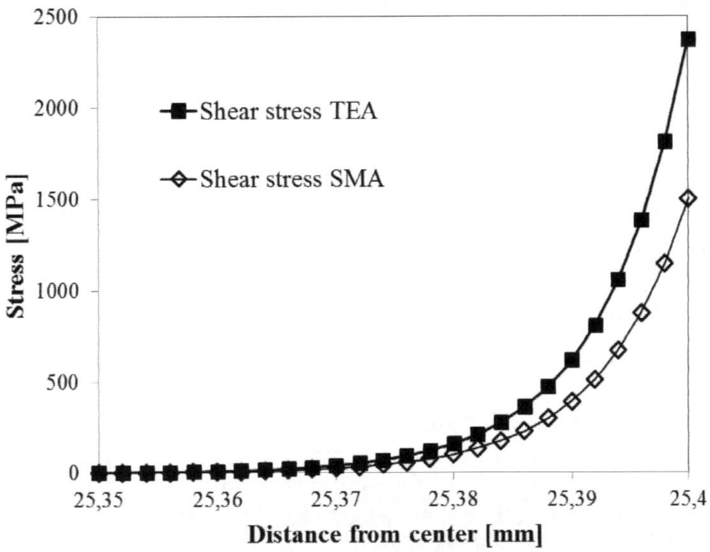


**Figure 1.** Plots of interfacial shearing stress at the assembly end.

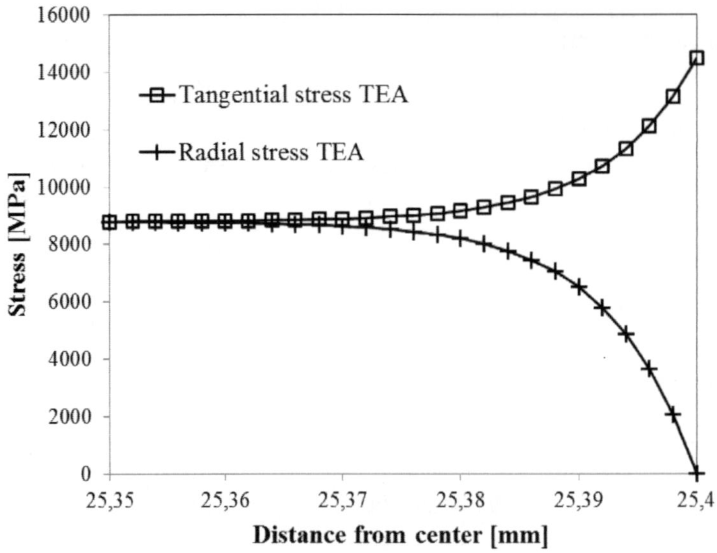


**Figure 2.** Plots of tangential and radial stresses at the assembly end.

## 4. Conclusions

- The obtained theory-of-elasticity solution enables one to estimate the state of stress normal in a GaN film.
- The computed peripheral shearing stresses and circumferential stresses are significant. It is envisioned that the occurrence of dislocations in thin semiconductor films starts at their peripheral portions and then rapidly propagates inwards the assembly, until the relief in the shearing stresses in the inner portion of the assembly is sufficient to "arrest" the further penetration of dislocations.
- The theory-of-elasticity solution predicts that the state of stress in the assembly in question becomes assembly size independent, when the product $kr_0 = \sqrt{3\frac{1-\nu_1^2}{1+\nu_2}\frac{E_2}{E_1}}\frac{r_0}{\sqrt{h_1 h_2}}$ of the parameter $k$ of the interfacial shearing stress and the assembly radius $r_0$ exceeds 10. This takes place when the modified Bessel functions of the first kind of the first and the second order become sufficiently large and equal to each other at a sufficiently high current radius r value. In the above formula, $E_1$ and $E_2$ are Young's moduli of the film and the substrate materials, $\nu_1$ and $\nu_2$ are their Poisson's ratios, $h_1$ and $h_2$ are the thicknesses of the film and the substrate, and $r_0$ is the assembly radius.
- The theory-of-elasticity and the strength-of-materials predictions lead to the same $kr_0$ criterion if the parameter $k$ of the interfacial shearing stress is evaluated based on the theory-of-elasticity formula, which predicts the parameter $k$ value that is by the factor of $\sqrt{1+\nu_1}$ larger than the strength-of-material prediction.
- The following simple formulas can be used to evaluate the maximum elastic stresses in a bi-material circular semiconductor thin film large size assembly:

$$\sigma_{r,\max} = \sigma_1 = \frac{E_1}{1-\nu_1}\varepsilon, \; \varepsilon = 1 - \frac{a_2}{a_1}, \quad \tau_{\max} = kh_1\sigma_1, \; \sigma_{\theta,\max} = (2-\nu_1)\sigma_1 .$$

Here $\sigma_1$ is the maximum normal radial and normal circumferential (tangential) stress in the film's midportion, $\varepsilon$ is the lattice-misfit strain, $a_1$ and $a_2$ are the lattice constants of the substrate and the film materials (the film is in compression if the strain $\varepsilon$ is positive), $\tau_{\max}$ is the maximum interfacial shearing stress at the assembly end, $k$ is the parameter of the interfacial shearing stress, $h_1$ is the film thickness, and $\sigma_{\theta,\max}$ is the maximum normal circumferential stress at the assembly end.

- The developed predictive model can be used, in addition to the routine finite-element analyses (FEA), for the assessment of the merits and shortcomings of a particular semiconductor-crystal growth technology, as far as the expected lattice-misfit stresses are concerned, before the actual experimentation and/or fabrication is considered and conducted. The models can be used particularly to determine, from the observed critical film thickness, the material, design and technological factors that lead to elevated dislocation densities.

**Acknowledgement**
M. Lederer and G. Khatibi wish to thank Christian Doppler Gesellschaft, Vienna Austria (CD labor RELAB).

**References**
[1] Lee J et al, *Wafer bonding of SiC and GaN*, MRS Symp. Proc., vol. 681E, MRS, 2001
[2] Suhir E, *Stresses in bi-metal thermostats*, ASME J. of Appl. Mech., vol.55, No.4, 1986
[3] Luryi S, Suhir E, *New approach to the high quality epitaxial growth of lattice-mismatched materials*, Appl. Phys. Lett., vol.49, No. 140, 1986
[4] Sagar A, Feenstra R M, *Growth of GaN on porous SiC by molecular beam epitaxy* in Porous Silicon Carbide and Gallium Nitride: Epitaxy, Catalysis, and Biotechnology Applications, ed Feenstra R M, Wood C E C, Wiley, West Sussex, 2008
[5] Suhir E, *Stresses in Bi-Material GaN Assemblies*, J. Appl. Phys. 110, 074506, 2011
[6] Suhir E, *Predicted Stresses in a Circular Substrate/Thin-Film System Subjected to the Change in Temperature,* Journal of Applied Physics, vol.88, No.5, 2000
[7] Timoshenko S, and Goodier, J N, *Theory of Elasticity*, 3rd ed., McGraw-Hill, New York, 1970
[8] Sneddon I N, *Special Functions of Mathematical Physics and Chemistry*, Oliver and Boyd, Edinburgh, 1956
[9] Ahmad I, Holtz M, Faleev N N, Temkin H, *Dependence of the stress–temperature coefficient on dislocation density in epitaxial GaN grown on α–$Al_2O_3$ and 6H–SiC substrates*, J. Appl. Phys. vol.95, No.4, 2004

# Nanostructural haemocompatible coatings for the internal side of artificial blood vessels

**K Trembecka-Wojciga[1], R Major[1*], J M Lackner[2], B Butruk-Raszeja[3], M Sanak[4] and B. Major[1]**

1. Institute of Metallurgy and Materials Science, Polish Academy of Sciences, 25 Reymonta Street, 30-059 Krakow, PL
2. Joanneum Research Forschungsges mbH, Institute of Surface Technologies and Photonics, Functional Surfaces, Leobner Strasse 94, A-8712 Niklasdorf, A
3. Biomedical Engineering Laboratory, Faculty of Chemical and Process Engineering, Warsaw University of Technology, Waryńskiego 1,00-645 Warszawa, PL
4. Department of Medicine, Jagiellonian University Medical College, Cracow, PL

*corresponding author E-mail:

**Abstract.** The main goal of the work was to elaborate low thrombogenicity of surface inside tube-like elements for cardiovascular system support by combination of low-temperature glow discharge and hydrogel coatings to inhibit blood-clotting cascade activation. A large share of amorphous phase silicon was observed in the microstructure analysis. The crystalline elements were uniformly distributed in the amorphous structure. Combination of low thickness, the proper microstructure and density of the coatings provided a highly flexible nature of the whole system. The blood-material interaction was analyzed in vitro in dynamic conditions by using a designed and fabricated novel blood flow simulator. Coatings deposited by the glow discharge expressed good hemocopatibile properties. The use of hydrogel coatings did not reduce coagulation parameter. Hydrogel coatings did not improve the hemocompatibility of the surface modified with carbon based coatings. Modification of surface with hydrogel resulted in further increased risk of hemolysis.

## 1. Introduction

Non-thrombogenic blood-contacting surfaces and appropriate blood flow characteristics are essential for the clinical application of artificial blood vessels. Similarly, providing various implants in the soft tissue decisively improves the quality of patients' lives. Tube-like devices (cannulas, connectors, and guiding drains) are currently used in many cardiovascular applications, and their blood-contacting surfaces are often modified with heparin, which protects against blood clots. This modification lowers the thrombogenicity of the tube materials. The main disadvantage of heparinisation is the risk of bleeding. Thus, heparinised blood surfaces could be replaced by biomimetic materials and surfaces (regarding pseudointima formation), which better prevent thrombosis and minimise the bleeding risk [1].

Despite the constant development of materials science dedicated to the development of biomaterials, progress in this area is limited. This is because of the need for an individual approach to the patient. Organs and systems that are being replaced, repaired or supported could can react differently in different people for the introduction of an artificial material. A great body of data [2] describing the material and biological aspects has been accumulated. Unfortunately, these data are not only scattered throughout different types of publications but are also obtained from experimental and clinical studies that vary greatly in accuracy and precision.

The guidelines of the International Organization for Standardization (ISO) ISO-10933-4 determine the degree of hemobiocompatibility defined as the absence of adverse effects of biomaterials with circulating blood [3]. These effects concern cell activation, damage to blood cells, activation of the coagulation system, changes in the composition of the electrolytes and blood proteins, inflammation and immune responses of the body. In the body, clots are formed in response to damage to the vascular wall. The first reaction in this process is platelet adhesion. On the surface, an enzymatic thrombin complex is formed [4]. Commonly used methods for dynamic testing of whole blood mimic the effects of biomaterials in terms of blood flow. The work, that has been done in the context of the present issue, exhibits aspects of the mechanics and biology to study the impact of the artificial surfaces on blood. The work considered several approaches, i.e., hemocompatible coating design and deposition for the internal side of artificial vessels, preparing protocols for the blood-material interaction analysis and determining the appropriate test performance. Important parameters determined in the analysis of activated platelets are procaspase activating compound-1 (PAC-1), evaluation of platelet activation indicator P-selectin and microparticles procoagulant activity in human plasma, in vitro.

PAC-1 is monoclonal anti-platelet antibody, designated PAC-1, which binds to activated platelets but not to unstimulated platelets [5-9]. It is a pentameric IgM that binds to agonist-stimulated platelets. PAC-1 recognises an epitope on the glycoprotein IIb/IIIa (gpIIb/IIIa, αIIbβ3) complex of activated platelets at or near the platelet fibrinogen receptor.

P-selectin functions as a cell adhesion molecule (CAM) on the surfaces of activated endothelial cells, which line the inner surface of blood vessels, and activated platelets [10-14]. P-selectin belongs to adhesion molecules located in the alpha-granules of platelets and Weibel-Pallade'a bodies of endothelial cells. P-selectin mediats processes related to the blood cell rolling on the endothelial surface connecting blood leukocytes to platelets, endothelial cells and other leukocytes to parts of tissue damage or inflammation.

Microparticles (MPs) are small membrane vesicles that are released by activated or apoptotic cells. Microparticles from various origins are present in plasma. They are generated by the activation of blood cells (platelets, endothelium, leucocytes, monocytes). According to their origin, these microparticles can expose a procoagulant phospholipids surface and thrombogenic proteins such as Tissue Factor. Cell membrane microparticles are also called microbubbles. Their size is 0.2-2.0 microns. Microparticles are released into the plasma membrane of all cellular components of the blood. MP release from eukaryotic cells is a normal physiological process that occurs during maturation and aging of the cell [15, 16]. Increased release of MP can occur upon activation of cells, for example, by binding to complement components or immune complexes; cytokines; chemokines; stress factors such as temperature, change in the osmotic pressure, tension generated during flow in vessels; or under the influence of the factors such as improperly prepared material introduced directly into the bloodstream leading to apoptosis. It is now known that increased numbers of MPs are detected in the plasma of patients in the course of many diseases (e.g., deep vein thrombosis, cardiovascular diseases, diabetes and various types of infections, as well as in patients supported with artificial organs).

The release of cytoplasmic granules and platelet mechanical failure results in platelet microparticles. These include fragments of cell membranes in the form of circulating bubbles having a diameter on the scale of fractions of micrometres. Phosphatidylserine-rich phospholipids appear on the surface of microparticles as a result of the enzymatic activity. This phenomenon is a characteristic marker of the functional importance of the microparticles [17, 18]. In addition to the expression of phosphatidylserine, platelet microparticles retain all of the markers that are characteristic of the platelets, such as P-selectin and CD61. Phospholipids of the microparticles form a platform on which the tissue factor complex and the activated coagulation factor X complex are formed with the participation of calcium ions. The complex is known as the prothrombinase complex. As the consequence of prothrombinase complex formation, thrombin is generated. Thrombin is an enzyme causing fibrin precipitation and clot formation. Clinical studies confirmed that the concentration of circulating microparticles in the healthy organism is less than 10 nM of phosphatidylserine [16,17]. The research hypothesis of the described work concerns the surface modification by the biopassive hydrogel coating stabilised by the physical vapour deposited thin films can effectively influence the minimization of the coagulation processes.

## 2. Materials and methods

### *2.1. Thin coating preparation*

The surfaces of the clinically used polyurethane AR/LT was modified in two stages. A first phase considered hemocompatibile coating, based on carbon deposited by magnetron sputtering. These coatings were supposed to protect the substrate, improve hemocompatibility and stabilize final hydrogel coating. The hydrogel coatings were deposited by dip coating. They were supposed to be the most effective surface for the direct contact with blood.

#### *2.1.1. Headers and footers*

The applied coating setup for inner surfaces of tubes is basically comparable to gas discharge lamps (Figure 1). However, the glass tube is replaced by a polycarbonate (PC) polymer tube of 12 mm inner diameter and 30 cm length, which was itself substrate and vacuum chamber together. For easier characterization of the deposited films, flat PC and (100) silicon substrates were placed inside in the center of this tube for some experiments. The setup (Diener Elektronik GmbH, Nagold, Germany), described elsewhere [4] (Figure 1), is afterwards closed on both sides with vacuum flanges, acting as electrodes (cathode and anode). After the evacuation of the tube to 10 Pa base pressure by a fine vacuum pump through the mesh anode, the process gases are delivered to the polymer tube through the mesh cathode.

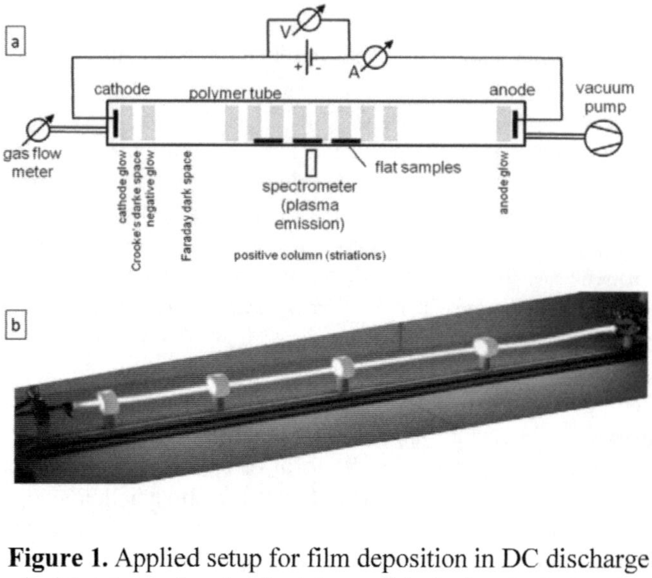


**Figure 1.** Applied setup for film deposition in DC discharge inside tubes, whereby the tube wall is the barrier between vacuum and atmosphere (a) schematics (b) image taken during.

The following gas atmospheres were chosen to obtain different types of film materials, consisting always of a base and a top layer. As base layer, a SiO2 coating of about 20 nm thickness was obtained by using 80 Pa HMDSO (hexamethyl disiloxane, Sigma Aldrich, 99.7%) and 20 Pa O2 gas mixtures. This layer provides the required adhesion on the tube substrates for the following top layer deposition, where 3 types of different C (and Si) containing plasmapolymers were applied:

- A: 60 Pa C2H2
- B: 110 Pa C2H2
- C: 35 Pa HMDSO + 60 Pa C2H2

The plasma modification and coating starts in the pulsed 535 Hz DC discharge between the cathode and the anode with a duty cycle of 15% (= percentage of active plasma discharge in pulsed plasma). The supplied voltage signal has a rectangular shape with slightly decreasing voltage in the end of the discharge. The control of the gas supply (O2, C2H2) occurs by gas mass flow meters, the control of liquid HMDSO vaporization by temperature control (25 ± 1°C) and a liquid gas flow meter. Pressure measurement was performed by a Pirani cell and correlated to the applied gas and the precursor vapor types. This pressure was used for the control of the deposition process, allowing a reproducibility of film thickness of ±8%. A reproducible modification with nearly constant chemical film composition and contact angles over the tube length is reached in the so called "positive column" of the discharge, which covers > 90% of the discharge length for 50 cm long tubes [19, 20].
Thin coatings were deposited on a substrate made of clinically applicable in heart assist system, polyvinyl chloride (PCV). The goal of applying amorphous carbon (a-C:H) and silicon doped amorphous carbon (a-C:H:Si) coatings was to separate the polymer from the tissue environment and stabilise the final hydrogel layer. The detailed deposition parameters are listed in Table 1.

**Table 1.** The parameters of coatings deposition.

| Parameter | **Pre-treatment (for all Versions)** | **Undercoating (for all Versions)** | **Top coating Version A** | **Top coating Version B** | **Top coating Version C** |
|---|---|---|---|---|---|
| Starting pressure | 0.15 mbar | 0.15 mbar | 0.15 mbar | 0.15 mbar | 0.15 mbar |
| Gas-flow | 6 sccm Ar | HMDSO (counter = 340); 3 sccm O2 | 4 sccm C2H2 | 10 sccm C2H2 | HMDSO (counter=340); 4 sccm C2H2 |
| Coating pressure | 0.27 mbar | - | 0.60 mbar | 1.10 mbar | 0.95 mbar |
| Coating time | 3 min | 0.5 min | 2 min | 2 min | 2 min |

The industrially-scaled equipment for plasma polymerization of nanoparticles (Diener Electronik, Ebhausen, Germany) is based on a cubic chamber, which is pumped to start pressure of 5 Pa by a dry vacuum pump. The PCV substrates are mounted on a vertical rotating cage between square, vertically positioned electrodes. Gases (Ar, C2H2) and precursors (hexamethyldisiloxane HMDSO purchased in Sigma Aldrich), are introduced in the chamber from the top and their pressure is adjusted before igniting plasma by the pulsed DC discharge (40 kHz frequency). Gas pressure measurements occurred by a Pirani gauge. Thin coating deposition occurred in the "volume polymerization mode" at pressures of around 50 Pa in the capacitive pulsed DC discharge, whereby gas mixtures of 20 Pa hexamethyldisiloxane (HMDSO, 99.5%, Sigma Aldrich, Austria) as liquid precursor and 25 Pa argon (99.999%, Linde Gas, Stadl-Paura, Austria) were applied. In contrast to the commonly used surface polymerization mode, the higher gas pressure (applying of non-reactive Ar as process gas) leads to higher probability of collisions between the ionic species in the plasma and chemical reactions, which initiates nanoparticle formation [19, 20]. The whole deposition process was performed at a temperature of 32 °C for controlled precursor vaporization conditions. The temperature of the substrate cage was not changing significantly (<2°C) during nanoparticle deposition, which was measured by temperature indicator strips.

*2.1.2. Biopassive hydrogel coatings*

The best way for the final surface modification are hydrophilic passive coatings [21-23]. Biopassive surfaces are characterized by low interfacial energy, which results in a decreased number of surface-adsorbed proteins and surface-adhered cells. In addition, the strength of interaction between the protein and the surface is small, which limits the conformational changes occurring in the protein molecules and enables maintenance of their biological activity. The final hydrogel coatings were prepared from polyvinylpyrrolidone (PVP) macro-molecules, which were bonded to polyurethane (PU) substrate, previously coated with carbon based thin films. The coatings was designed to improve the surface hemocompatibility of blood-contacting medical devices. The hydrogel coatings were deposited by dip coating. The hydrogel coating preparation on the luminal side of the tube-like element made of polyvinylchloride. The coating was carried out in 2 stages using dip coating. Two steps meant in this case two immersion in two different solutions. The exact concentrations of the reagents used in the preparation of the hydrogel coatings are listed in table 2. Each process was preceded by flushing in a 5% aqueous alcohol solution and then with water. They were dried to constant weight at 40 °C. In the first stage, the covered tubes were immersed in a solution of hexane containing 5% ethylene glycol dimethacrylate (EGDMA) and 3% cumene hydroperoxide (CHP) for 5 min at 25 °C. In a second step, the samples were placed in an aqueous solution containing 0.1% FeCl2, 1% ascorbic acid (AA) and 5% polyvinylpyrrolidone (PVP) for 15 min at 25 °C. PVP is highly hydrophilic, biocompatible polymer with low cytotoxicity. Therefore, this material is appropriate for many medical applications such as an artificial pancreas, wound dressings, artificial skin, and cardiovascular devices [24-27]. The tubes were immersed in an aqueous polymer solution containing 0.1% sodium dodecyl sulphate (SDS) and were placed on a shaker for 5 min to remove unbound PVP. The samples were washed in water then transferred to phosphate buffered saline (PBS) and incubated at 37 °C overnight. The films were chemically cross-linked by ethylene glycol dimethacrylate (EGDMA).

**Table 2.** The exact concentrations of reagents used in the preparation of hydrogel coatings. The process consists of two steps: the first step involves reagent diffusion to the interface; during the second step macroradical formation along with their recombination occurs.

| Steps in hydrogel layer formation by dip coating | **Chemical compound of solution** | **Content %** |
|---|---|---|
| | PVP | 5.0 |
| STEP 1 | | |
| Reagents diffusion to interphase | FeCl2 | 0.1 |
| | AA | 1.0 |
| STEP 2 Radical polymerization | CHP | 3.0 |
| | EGDMA | 5.0 |

*2.2. Microstructural analysis*
The microstructure of the deposited coatings was characterised using a TECNAI G2 FEG (200kV) transmission electron microscope (TEM) equipped with a field emission gun (FEG). Cross-sectional analysis of the microstructure was performed for the samples with the crystallised thin buffer titanium carbonitride coating. Thin foils for TEM observation were prepared directly from the location of interest using the focused ion beam (FIB) technique. A detailed description of the TEM microstructure analysis is presented in [28]. A Quanta 200 3D DualBeam was used for FIB preparation. Phase analysis was performed using electron diffraction patterns and was confirmed through identification of high-resolution images (HR-TEM).

*2.3. Validation of the blood flow chamber*
As part of this work, a dynamic test on blood was designed and manufactured (Figure 2). The original layout of the test was dedicated to analysing the dynamic haemocompatibility of the coating applied on the internal parts of the tube-like elements. The system was prepared in accordance with the ISO-10933-4 standard. The concept and the design were based on assumptions of the commercially available dynamic test on blood (Impact-R) [29]. The Impact-R test is clinically used to evaluate the quality of blood flow in hydrodynamic conditions. Important parameters that have to be transferred from the Impact-R test to the aortic flow simulator are the surface roughness of the rotor, the distance between the analysed surface and the rotor, the rotor rotational speed and the duration of the single experiment [30, 31].

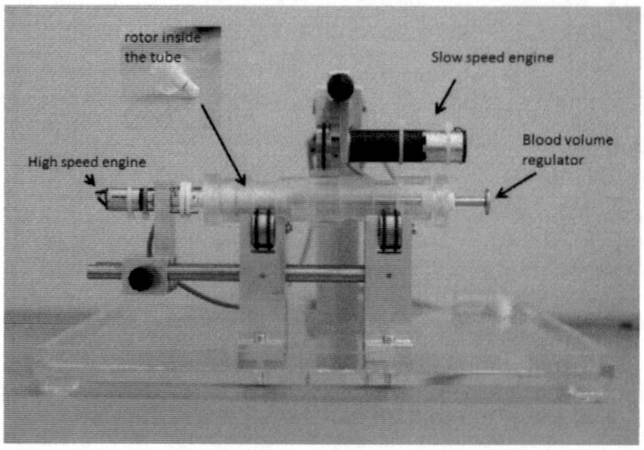


**Figure 2.** Self elaborated and validated to ISO 10993 blood flow chamber

Table 3 shows the parameters of the individual elements of a commercial assay and their implementation to the aortic flow simulator.

**Table 3.** Comparison of the most important parameters of a commercial test Impact-R with parameters of the aortic flow analyzer.

| | **Impact-R** | **Aortic fow analyzer** |
|---|---|---|
| The shape of the sample | Flat disc d=14.4 mm | Tube, internal diameter 7 mm |
| The shape of the rotor | Conical | Flat |
| The rotor speed | 720 rpm | 72 rpm |
| The rotor drive | The rotor rotates in the magnetic field | The rotor attached to an electric motor |
| Time of analysis | 2 min | 20 min |

The newly designed testing apparatus was equipped with two motors that were responsible for two different and important aspects resulting from the features of the blood. The low rotation motor was introduced to prevent sedimentation of the blood before the test. The high rotation motor was applied to simulate arterial flow conditions in the internal (luminal) side of the tube. This flow caused shear stress between the blood cells and the analysed luminal surface, which was modified with the thin coatings. The dedicated control unit allowed precise control of the rotation rate, yielding shear stresses in the range of 0 to 100 Pa. The operational and hydrodynamic parameters, such as the rotation rate and inner diameter of the tube, were recorded by developed software (the rotational rate was automatically recorded on a computer connected to device).

The effect of the shear forces on inner surfaces of the tubular elements was studied. The number of active platelets and leukocytes on the surface and in the blood collected above the tested samples were evaluated after exposure to flow.

The protocol for analysing blood-material interactions was prepared according to the ISO-10933-4 standard. The test procedure using the aortic flow simulator allows studying the degree of platelet activation and the amount of circulating monocyte-platelet aggregates in the blood taken from above the test material. For this purpose, the corresponding antibodies were used to communicate with the cellular membrane protein receptors. Measurements of active platelets were performed by analysing the expression of PAC-1 and P-selectin antigens bound to the platelet fibrinogen receptor and the expression of active markers. The analysis focused on the activated platelets. A mixture of antibodies and adenosine diphosphate (ADP) solutions were prepared with different concentrations to perform the positive control (Table 4).

**Table 4.** Antibodies composition.

| Name | Composition | Application |
|---|---|---|
| M | CD61-PerCP, CD62P, PAC-1 | Estimate platelet activation upon contact with the material (+ control) |
| A | CD61-PerCP, CD62P, PAC-1 | Estimate of platelet activation by ADP |
| ADP 1 | ADP 0.4mM | Positive control |
| ADP 2 | ADP 40mM | Negative control |
| B | CD61-PerCP, CD621-FITC | Estimate of the amount monocyte-platelet aggregates |

The tube-like element having a diameter of 7 mm and a length of approximately 6 cm was placed the inside flow chamber. The active volume was determined using an appropriate holder. The tube-like element was filled with 2 ml of blood for each specimen. Before performing the actual test, the blood was gently mixed for 2 min. Then, the test began automatically, as controlled by the software. The applied shear stress can be expressed according to equation [32]:

$$\tau=M/(2\pi R2l) \quad (1)$$

where M is the torque input by the engine (M= 2.7 dyne-cm), R is the radius (cm; Ra≤R≤Rb; Ra-radius of rotor; Rb-radius of tube) and l is the effective length of the rotor (cm).

The values of shear stress decreased along the radius (Figure 3).

The shear rate can be calculated by using equation (2).

$$\gamma= \tau /\eta \quad (2)$$

where $\eta$ is the viscosity of blood in 37 °C.

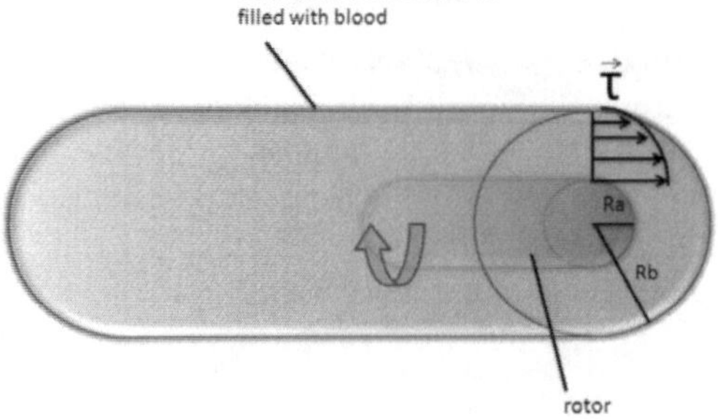


**Figure 3.** Layout of the shear stress distribution inside the tube-like element during the dynamic blood-material interaction test

*2.3.1. Blood-material interaction analysis*

The blood used in this test was collected from a healthy male volunteer. Aliquots of blood (0.4 □L) were activated with adenosine diphosphate (ADP, 20 μM final concentration) for 5 minutes. These aliquots served as the positive controls that were prepared to simulate conditions of maximum platelet activation. Blood samples stored under static conditions were used as a negative control (bas). The blood collected from above the sample was mixed with the antibodies CD61-PerCP, CD62P, and PAC-1 to estimate the platelet activation upon contact with the material. The platelet activation was analysed using flow cytometry.

PAC-1 recognises an epitope on the glycoprotein IIb/IIIa (gpIIb/IIIa, αIIbβ3) complex of activated platelets at the platelet fibrinogen receptor. PAC-1 is a receptor for fibrinogen [33] and von Willebrand factor and occurs in platelet activation. P-selectin (P-SEL) is a cell adhesion molecule (CAM) that accumulates in thrombocite granules α of unactivated platelets. When there platelet activation occurs, rapid translocation of P-selectin from granules to the cellular membrane occurs [34, 35]. P-SEL promotes platelet aggregation by binding platelets with other platelets and with fibrin.

The inner surfaces of the tube-like elements were imaged using confocal microscopy (CLSM; Carl Zeiss, Exciter 5). Before image acquisition, the materials were divided into 10x15 mm sections. Cells adhered to material surfaces were fixed and permeabilised for 10 min with 4% paraformaldehyde (PFA). After washing the samples three times for 5 min with PBS, they were incubated with anti-human CD62P P-selectin antibody conjugated with fluorescein isothiocyanate (FITC), which is a dye that marks platelets with the P-selectin active receptor. After 30 minutes of incubation, the samples were washed three times with PBS and were covered by the human monoclonal antibody CD45 PE conjugated with Texas Red for 10 minutes. This secondary antibody marks active leukocytes on the material surfaces. Finally, the samples were rinsed three times in PBS and were immediately imaged. The cell densities on the surfaces were measured using the "colocalisation" tool of the microscope software (Carl Zeiss Exciter 5 equipped with ZEN 2008). Colocalization analysis is done on basis considering pixel by pixel analysis.

*2.3.2. Microparticle analysis*

The ZYMUPHEN MP-Activity kit was proposed for the measurement of the in vitro microparticle procoagulant activity in human plasma using automated or manual methods [36-38].

The diluted assayed plasma sample supplemented with calcium, Factor Xa and thrombin inhibitors was introduced into one of the microplate wells coated with streptavidin and biotinylated annexin V and then incubated. Following a washing step, the Factor Xa-Va mixture containing calcium and purified prothrombin was introduced. When present in the tested sample, microparticles bind to annexin V and expose the phospholipid surface, thus allowing FXa-FVa, in the presence of calcium, to activate prothrombin into thrombin. Therefore, the phospholipid concentration is the limiting factor. There is a direct relationship between the phospholipid concentration and the amount of thrombin generation, which is measured via its specific activity on the thrombin substrate. Blood plasma was collected through a frank venipuncture on 0.109 M (or 0.129 M) citrate anticoagulant. The plasma supernatant was rapidly decanted (within 2 hours) following a 15 min centrifugation at 1,500 g and room temperature. The plasma supernatant was then again rapidly centrifuged for 2 min at 13,000 g and room temperature. The plasma was obtained by collecting the supernatant, avoiding any contact with the platelet pellet. The plasma was tested within 4 hours or stored frozen at -80 °C or below for up to 6 months and thawed for 15 min at 37 °C prior to use.

## 3. Results and discussion

*3.1. Microstructure*

Microstructural characterisation was performed on the cross-sections of the coatings in bright field observation mode (Figure 4). The phase analysis was performed using selected area electron diffraction patterns (SAED).

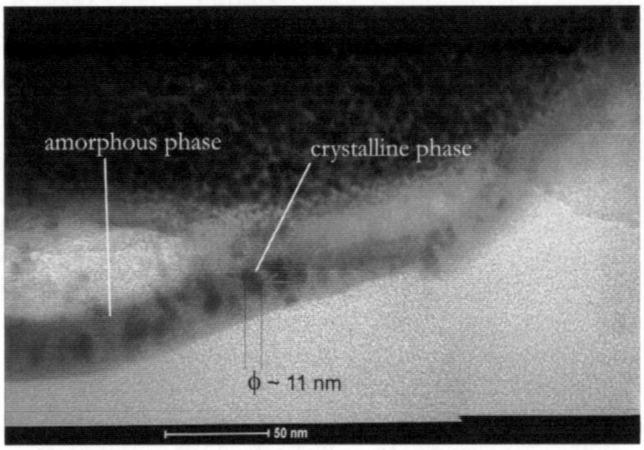


**Figure 4.** TEM microstructure of the ceramic coating directly deposited on the polymer substrate with discrenible two different phases; crystalline and amorphous

The microstructural characterisation confirmed the high elastic properties of the coating. The nano-crystalline characteristic of the coating was confirmed by TEM BF imaging at high magnifications and high-resolution observations (HRTEM) (Figure 5). The crystallite diameter according to the bright field imaging was approximately 11 nm. The high-resolution TEM analysis of the crystalline areas observed in bright field mode showed crystallite agglomerates. The actual crystallite diameter was 7 nm (Figure 4). The high-resolution analysis showed the presence of crystallised elements in the amorphous matrix, which was associated with the initial mechanism of thin film nucleation from the gas phase. Two-dimensional thin film nucleation allows unique mechanical properties, which are unusual for ceramic coatings. The coating that had elastic properties was directly applied to the polymer. Crystalline elements were uniformly arranged along the coating (Figure 5). To fabricate biologically functional coatings, a final top hydrogel layer was applied to the surface of the partially amorphous layer, introducing functional molecular groups. Crystalline components irregularly distributed in the hydrogel were observed.

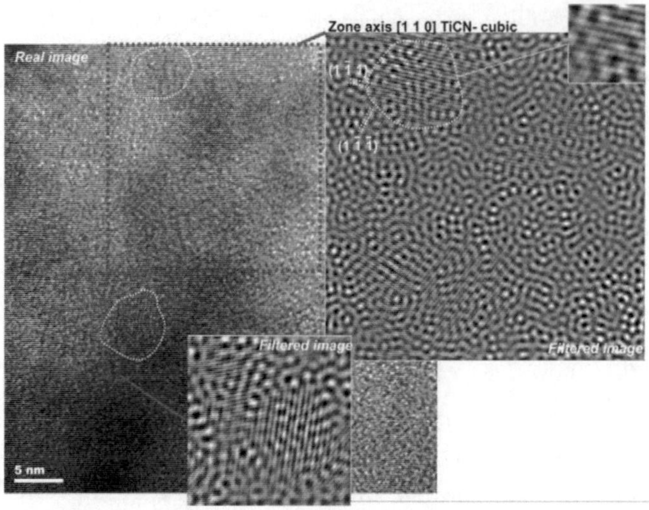


**Figure 5.** Microstructure and phase characterisation of the coating at the cross section done by HRTEM technique.

### *3.2. Blood-material interactions in the dynamic condition*
When exposed to physiological-like conditions, blood components are activated, which may result in platelet and leukocyte adhesion to the biomaterial surface. This process is preceded by coagulation of blood above the surface. Blood coagulation, followed by increasing platelet aggregate density in the blood could initiate thrombus formation. Physiological shear stress promotes all of these processes and induces the immune response of the organism. Shear forces can also destroy platelets, resulting in microparticle formation.

#### *3.2.1. Blood cell response to shear forces and material contact*
The cell density on a surface is a relevant parameter determining blood-material interactions. The results of the number of active blood components in particular leukocytes and platelets attached to the surfaces, designated as option "A", "B", or "C" (Table 1), are shown in figure 6. The figure shows a qualitative (N/N0) analysis of the impact of blood cells on the tested surfaces. For this case, subjected active platelets and leukocytes were studied. As the control substrate, the tubular element currently used for regeneration of the circulatory system was used. The comparison was made between the control substrate and tubular implants coated with haemocompatible coatings. N0 is the number of active blood elements on the surface of the control tube substrate, and N is the number of active blood elements on the surface of the tube substrate with modified luminal side. Among the coatings designated as "A", "B", and "C" (Table 1), the smallest number of platelets and leukocytes were observed for sample "B"; therefore, sample "B" appears to be the best solution.
Decreased the immune response and decreased activation of the coagulation cascade was visible for sample "B". Conversion of the immune response instead of activation of the coagulation system was observed. For variant "C", the number of active leukocytes and platelets increased. The predomination of the immune response over activation was retained.

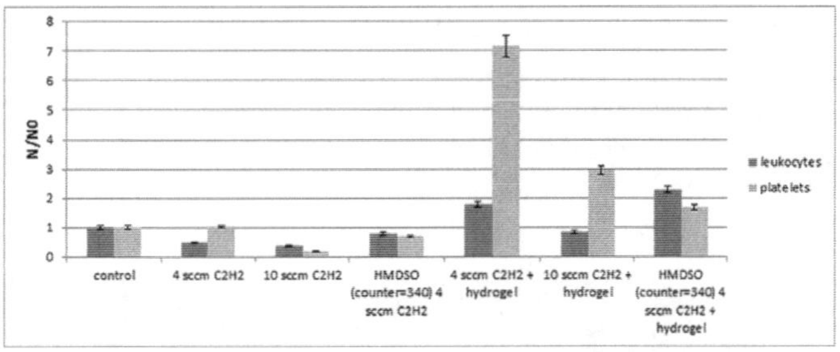


**Figure 6.** The relative amount of activated blood cell counts in the function of the coating design.

The final modification of the surfaces involved hydrogel coatings. The final surface modification caused increased activation of platelets and leukocytes (Figure 6). The layers with hydrogel coatings showed worse response in dynamic contact with blood compared with the coatings "A", "B", and "C" without hydrogel coatings. Comparing option "B" without a hydrogel layer with the coated option "B" with the hydrogel, a change in the response of the cellular components of blood can be clearly observed. The activation analysis of circulating blood above the analysed surfaces was performed using a flow cytometer. Platelet activation of the PAC-1 active receptor above the surface of option "A", "B", and "C" is shown in figure 7. The average amount (MNF) of active platelets with active PAC-1 and P-SEL receptors and an indicator of platelet activation (IPA) were measured. In unactivated endothelial cells, P-selectin is stored in granules called Weibel-Palade bodies, and α-granules are stored in unactivated platelets. Other names for P-selectin include CD62P, granule membrane protein 140 (GMP-140), and platelet activation-dependent granule to external membrane protein (PADGEM). The expression level of CD62 was specified as the binding parameter to determine the ratio of platelet activation. This analysis indicated that the IPA is determined as a percentage (%). The value of P-SEL MNF was the smallest for the sample designated as version "B" among all of the tested coatings. The MNF value for PAC-1 for this coating was the greatest. A hydrogel layer applied on the inner side of the tubular element resulted in a constant level MNF and IPA for PAC-1 receptor (Figure 7), regardless of the coating. The second receptor, whose activity was analysed using flow cytometry, was P-selectin (P-SEL). The result of the analysis of P-SEL shows a similar dependence to the analysis of PAC-1. A hydrogel layer applied on the inner side of the tubular element resulted in a constant level MNF and IPA of the receptor P-SEL (Figure 7), regardless of the coating.

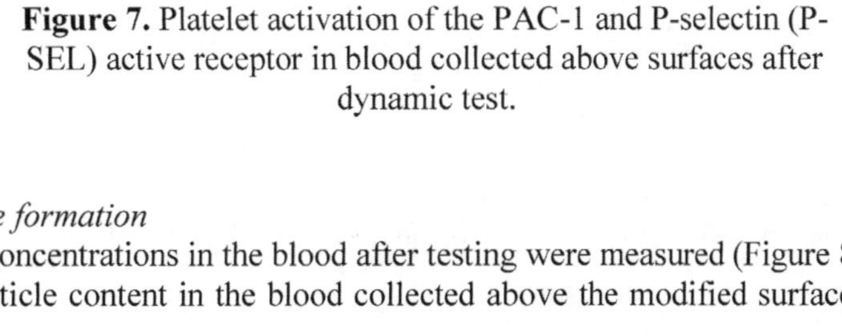


**Figure 7.** Platelet activation of the PAC-1 and P-selectin (P-SEL) active receptor in blood collected above surfaces after dynamic test.

### *3.2.2. Microparticle formation*

The microparticle concentrations in the blood after testing were measured (Figure 8). The results show decreased microparticle content in the blood collected above the modified surfaces of "A", "B", and

"C" compared with the control sample (control PCV). The highest concentration of membrane fragments was observed for the coating "B". For the materials with hydrogels, the microparticle concentration was higher compared with the coatings "A", "B", and "C". This result indicates high influence of the hydrogel layer on the platelet destruction process.

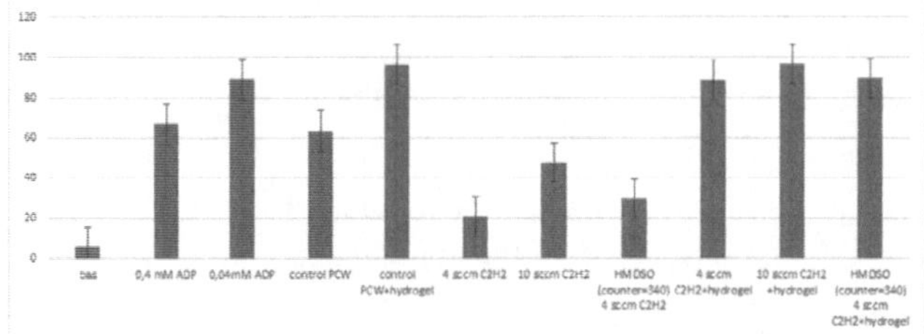


**Figure 8.** Microparticles concentration on the function of the coating design.

## 4. Conclusions

- The ceramic coatings showed elastic properties. This result was due to the amorphous structure with uniformly distributed crystalline particles. The elastic properties of the ceramic coating depended on the appropriate mechanism of thin film nucleation from the gas phase. Coatings that are utilised for coating implants should not change their mechanical properties and would improve biological properties. This property has been achieved in the present study.
- The coating "B" (110 Pa C2H2) comprising the greatest amount of carbon in the structure, showed high haemocompatible properties. Based on the dynamic analysis of the blood, decreased coagulation processes and immune response were observed compared with "A" (60 Pa C2H2), and "C" (35 Pa HMDSO + 60 Pa C2H2). It is apparent from the carbon content in the structure. As shown in table 1, The coating indicated as "A" has the highest carbon content. The deposition process was performed in the strongest flow of acetylene. Hydrogel coatings are highly hydrophobic. On the surface, an additional layer of water is created, which protects the surface against aggregation processes. Hydrogel coatings are also very highly liquid absorbent, which affects the swelling of the material. In the dynamic tests on the blood, strong haemolysis was revealed, therefore it was found to abandon the last modification of the surface which was hydrogel coating.
- Based on extensive investigations, thin coating were selected to modify the internal parts of the tubular elements for applications in regenerative cardiac surgery

## 5. References

[1] Ratner B D 2007 *Biomaterials*, **28**, 5144–7.
[2] Anderson J 2001 *Annu.Rev. Mater. Res.*, **31**, 81–110.
[3] Seyfert U T, Biehl V and Schenk J 2002 *Biomol. Eng.*, **19**, 91-96.
[4] Major R, Lackner J M, Gorka K, Wilczek P and Major B 2013 J. *R. Soc. Chem.*, **3**, 11283–11291.
[5] Abrams C, Deng Y J, Steiner B, O'Toole T and Shattil S J 1994 *J. Biol.l Chem.***269**, 18781-18788.
[6] Abrams C and Shattil S J 1991 *Thromb. Haemostasis*, **65**, 467-473.
[7] Shattil S J 1990 Blood, **75**, 128-138.
[8] Abrams C, Deng J, Steiner B and Shattil S J 1994 *J. Biol. Chem.*, **269**, 18781–18788.
[9] Bennett J S, Shattil S J, Power J W and Gartner T K 1988 *J. Biol. Chem.*, **263**, 12948-12953.
[10] Monzavi-Karbassi B, Stanley J S, Hennings L, Jousheghany F, Artaud C, Shaaf S, Kieber-Emmons T 2007 *Int. J.Cancer*, **120 (6)**, 1179–91.
[11] Garcia j, Callewaert N and Borsig L 2007 *Glycobiology*, **17 (2)**, 185–96.

[12] Bar-Ner M, Eldor A, Wasserman L, Matzner Y, Cohen I R, Fuks Z and Vlodavsky I 1987 *Blood*, **70 (2)**, 551–7.
[13] Lazo-Langner A, Goss G D, Spaans J N and Rodger M A 2007 *Thromb. Haemostasis*, **5 (4)**, 729–37.
[14] Borsig L, Vlodavsky I, Ishai-Michaeli R, Torri G and Vismara E 2011 Neoplasia, **13 (5)**, 445–52.
[15] Gelderman M P and Simak J 2008. *J. Mol. Biol.,* **484**, 79–93.
[16] Piccin A, Murphy W G and Smith O P 2007 *Blood*, **21**, 157–171.
[17] Gemmel C H 1995 *J. Lab. Clin. Med.*,**125 (2)**, 276–287.
[18] Gemmel C H 2001 *J. Biomat. Sci. Polymer*, **12 (8)**, 933–934.
[19] Lackner J M, Kahn M and Waldhauser W 2011 Vacuum, **86(2)**, 144-150.
[20] Lackner J M, Wiesinger M, Kaindl R, Waldhauser W, Heim D and Hartmann P 2014 *Plasma Chem. Plasma P.*, **34(2)**, 259-269.
[21] Butruk-Raszeja B A, Łojszczyka I, Ciach T, Koscielniak-Ziemniak M, Janiczak K, Kustosz R and Gonsior M 2015 *Colloid Surface* B, **130**, 192–198
[22] Krishnan S, Weinman C J and Ober C K 2008 *J. Mater. Chem.*, **18**, 3405.
[23] Kasemo B 2002 *Surf. Sci.,* **500**, 656–677.
[24] Peppas N A 1987 *Hydrogels in Medicine and Pharmacy*, Vol. 3, CRC Press.
[25] Hickey A S and Peppas N A 1995 *J.Membrane Sci.*, **107**, 229-237.
[26] Kaźmierska K A, Kuc K and Ciach T 2008 *Acta Pol. Pharm.*, **65**, 763-766.
[27] Nicolson P C and Vogh J 2001 Biomaterials, **22**, 3273–83.
[28] Major L, Tirry W and Van Tendeloo G 2008. *Surf. Coat. Tech.,* **202**, 6075–6080.
[29] Sanak M, Jakiela B and Wegrzyn W 2010 *B. Pol. Acad. Sci. - Tech.*, **58**, 317-322.
[30] Shenkman B, Inbal A, Tamarin I, Lubetsky A, Savion N and Varon D 2003 *Brit. J. Haematol.*, **120**, 597–604.
[31] Wang X, Dorsam R T, Lauver A, Wang H, Barbera F A, Gibbs S, Varon D, Savion N, Friedman S M and Feuerstein G Z 2002 *J. Pharmatol. Exp. Ther.,* **303**, 1114–1120.
[32] De Hoog F R and Anderssen R S 2005 *B. Aust. Mat. Soc.*, **72**, 461.
[33] Abrams C, Deng Y J, Steiner B, O'Toole T and Shattil S J 1994 *J. Biol. Chem.,* **269**, 18781-18788.
[34] Hattori R, Hamilton K K, Fugate R D and McEver RP 1989 *J. Biol. Chem.*, **264**, 7768-71.
[35] Hahne M, Jäger U, Isenmann S, Hallmann R and Vestweber D 1993 *J. Cell Biol.,* **121 (3)**, 655-64.
[36] Morel O, Toti F, Hugel B and Freyssinet J M 2004 *Curr. Opin. Hematol.,* **11**, 156-164.
[37] Freyssinet J M 2003 *J Thromb Haemost*, **1**, 1655-62.
[38] www.aniara.com; ZYMUPHEN MP-Activity (# A521096) Last revision: 10/06/2014

# Silver nanoparticles influence on the blood activation process and their release to blood plasma from synthetic polymer scaffold

**R Major[1], J M Lackner[2], M Sanak[3], B Major[1]**
1 Institute of Metallurgy and Materials Science, Polish Academy of Sciences, Reymonta St. 25, Crakow, PL.
2 Joanneum Research Forschungs-GmbH, Materials – Functional Surfaces, Leoben, A.
3 Department of Medicine, Jagiellonian University Medical College, 8 Skawinska Street, 31-066 Cracow, PL.

E-mail: b.major@imim.pl

**Abstract.** In the present work, blood and blood plasma interaction to silver stabilised polyelectrolytes was investigated in vitro. The designed materials are dedicated for regeneration of the cardiovascular system. Silver nanoparticles were introduced into the polyelectrolyte structure in order to reduce the risk of bacterial biofilm formation. The introduction of Ag nanoparticles occurred by deposition at high vacuum by magnetron sputtering. The analysis of blood-materials interactions were performed by using commercially available tester, Impact-R (Diamed). The assessment of silver ion nanoparticles release into the plasma consisted in determining the Prothrombin Time (PT) and Activated Partial Thromboplastin Time (APTT). Unmodified surface of polyelectrolytes is a strong activator for blood elements. The introduction of silver nanoparticles resulted in a significant reduction in the probability of clotting. The extrinsic pathway of coagulation determined on the basis of the PT and the intrinsic and common pathways of coagulation measured by the APTT did not indicate the danger out of range. Microstructure was studied using TEM on thin foils prepared from the cross-section of samples subjected to biomedical treatments. The observations revealed hetero- interface between two different crystalline solids.

## 1. Introduction

In the frame of the work, biologically mimicking coatings were considered. The key issue was to express the minimal impact on blood activation process. The designed coatings form the scaffolding of cell repopulation. Tissue analogs allow better and longer use in the treatment of cardiovascular diseases. The work concerns the implementation of porous coatings and surface nano-functionalization.

A fundamental problem that affects the field of cardiovascular surgery is the paucity of autologous tissue available for surgical reconstructive procedures [1]. Although the best results are obtained when an individual's own tissues are used for surgical repair, this is often not possible as a result of pathology of autologous tissues or lack of a compatible replacement source from the body.

[1] To whom any correspondence should be addressed.

Extracellular matrix (ECM) structure can be easily mimic by the synthetic, degradable polymers [2,3]. Degradable scaffold materials are advantageous in tissue engineering applications because they provide initial biomechanical integrity as cells adhere, mature to the desired phenotype, and secrete natural ECM, and, most important, they are designed to be replaced by the neo-tissue. Polymers that have been extensively used in tissue engineering applications are poly glycolid acid (PGA), polylactic acid (PLA), and poly-E-caprolactone (PCL) [4-6]. They are widely used because of their biocompatible degradation properties. Hydrolysis of ester bonds in situ is thought to release nontoxic monomers that are excreted or absorbed by surrounding tissues. A release of acidic degradation products can be a source of toxicity that causes adverse local and systemic reactions that could result in implant rejection.

An introduction of the nanoparticles of precious metals created an antibacterial function [7].

Binding anions and proteins in biologic systems, silver ion ($Ag+$) avidly binds to cell surface receptors of bacteria, yeasts and fungi [8]. Silver cations also strongly bind to electron donor groups of biologic molecules containing sulphur, oxygen and nitrogen. The binding of silver ion to sulfhydryl groups and proteins on cell membranes appears to be critical to the antimicrobial action. The ionizing capacity of various silver compounds is critical in comparing the antimicrobial activities.

Nanotechnologies allow an efficient exploitation of the antimicrobial properties of silver by using silver in the form of nanoparticles [9].These are used in applications such as preservatives in cosmetics, textiles, water purification systems, coatings in catheters and wound dressings. The anticipated widespread exposure to silver nanoparticles in the near future has prompted governmental bodies and the public to raise questions about the safety of such applications.

An understanding of protein adsorption and macromolecular interactions is crucial for the control of surface initiated coagulation, the humoral immune response, cell binding, activation of inflammation, and wound healing [10-12]. Such processes are of large interest for currently expanding technological areas in biosensors and the controlled drug release, but may be of even larger significance for the in vivo behaviour of artificial surfaces in the cardiovascular system during short to medium long contact times (minutes to hours) [13, 14]. The dynamics of plasma protein adsorption, exchange and degradation at solid interfaces are still obscure phenomena during physiological conditions. The time course of adsorption and displacement of fibrinogen in plasma is called the Vroman effect [15]. Its kinetics depends on surface chemical properties, where negatively charged hydrophilic surfaces are suggested to facilitate the displacement of adsorbed proteins and hydrophobic surfaces to retain them over extended times in plasma [16]. Positively charged hydrophilic surfaces (with point of zero charge (p.z.c.) on the other hand, are not at all understood in this context. The research activity was focused on studies of displacement of pre-adsorbed human serum albumin (HSA) by buffer and blood plasma, and the activation of the intrinsic pathway of coagulation and complement on silver nanoparticles introduces.

## 2. Materials and methods

### *2.1. Polyelectrolyte and Ag nanoparticle preparation*

Polyelectrolyte multi-layered films were deposited by a "layer by layer" technique [17]. Substrates were activated by 10 M NaOH and washed with pure Milli-Q water. PLL and HA were dissolved in 400mM HEPES/ 0.15M NaCl solution with concentrations of 0.5 and 1mg/ml, respectively and pH of the solutions was set at pH 7.4 by adding 0.5M NaOH. Films were manufactured by an automatic dipping machine using an alternately immersing substrate in solutions of PLL and HA for 8 min each. After each deposition step, the substrate was rinsed in 0.15 M NaCl solution buffered at pH 7.4 to remove excess polyelectrolyte. The process was repeated until a desired number of 12 bilayers were obtained. Multilayers were further processed for the cross-linking. Then additional layer of PLL was adsorbed. Afterwards, the samples were subjected to final rinsing steps and stored at 4 °C in 400 mM HEPES/0.15M NaCl solution buffered at pH 7.4 until measurements were performed.

After the evacuation of the tube to 10 Pa base pressure by a fine vacuum pump through the mesh anode, the process gases are delivered to the polymer tube through the mesh cathode. The introduction of Ag nanoparticles occurred by deposition at high vacuum by magnetron sputtering from pure Ag targets (99.9%) at extremely high gas pressure. Therefore, the dried polyelectrolyte samples were introduced to the vacuum chamber, which was afterwards pumped down to 3x10-5 mbar. Afterwards, sputtering occurred at 1x10-2 mbar Ar pressure at 150 mm distance between the Ag sputter target and the substrates. The deposition was stopped after 30 s.

*2.2. Microstructure*

The microstructure characterization of the deposited coatings was performed using a TECNAI G2 FEG (200kV) transmission electron microscope (TEM) equipped with a field emission gun (FEG). Thin foils for TEM observation were prepared for one sample in every group of materials using the focused ion beam (FIB) technique. For FIB preparation a Quanta 200 3D DualBeam was used. A detailed description of the TEM microstructure analysis is presented elsewhere [18]. Phase analysis was performed using electron diffraction patterns and was confirmed through identification of high-resolution images (HR-TEM). All thin foils were cut as a cross- section of investigated coatings.

*2.3. Blood-material interaction*

The analysis of blood-materials interactions were performed by using commercially available tester, Impact-R (Diamed) [18]. Expression of platelet activation was evaluated based on glycoprotein IIb/IIIa, and using CD62P for P-selectin.

Receptor IIb/IIIa is the characteristic receptor to assess the degree of activation. It is an integrin complex found on platelets. It is a receptor for fibrinogen and von Willebrand factor and aids in platelet activation. The complex is formed via calcium-dependent association of gpIIb and gpIIIa, a required step in normal platelet aggregation and endothelial adherence [19, 20]. Platelet activation by ADP leads to the aforementioned conformational change in platelet gpIIb/IIIa receptors that induces binding to fibrinogen.

P-selectin is a protein that in humans is encoded by the SELP gene [21]. P-selectin functions as a cell adhesion molecule (CAM) on the surfaces of activated endothelial cells, which line the inner surface of blood vessels, and activated platelets. In inactivated endothelial cells, it is stored in granules called Weibel-Palade bodies. Platelet activation (through agonists such as thrombin, Type II collagen and ADP) results in "membrane flipping" where the platelet releases α- and dense granules and the inner walls of the granules are exposed on the outside of the cell. The P-selectin then promotes platelet aggregation through platelet-fibrin and platelet-platelet binding [22, 23].

Integrated fluorescence of the activation marker is calculated as a multiplication total of geometric mean fluorescence by percentage of marker-positive objects. Aggregates of platelets are analysed after erythrocyte lisps by mixing 25 μL of blood with 0.4 mL FLS and subsequent fixation by addition 3.5 mL 1% paraformaldehyde in PBS. The cellular material is recovered by centrifugation (1 000 g, 7 minutes) and immunostained (25 μL aliquots) with 4 μL PerCP-CD14 and 5 μL FITC-CD61 or 5 μL FITC- CD61 alone for 30 minutes at room temperature. Samples are then washed in PBS, and analyzed by the flow cytometer. The percentage of granulocyte-platelet aggregates (leukocytes stained with platelet marker CD61) is calculated using granulocyte forward/side scatter gate and additional monocyte CD14+ gate. Absolute number of platelets is calculated as the number of CD61 positive objects in reference to total granulocytes count. Small and big platelet aggregates are counted using forward/side scatter gates for CD61 positive objects in the flow cytometer analysis. Samples were washed in PBS solution three times and analyzed by using confocal laser scanning microscopy Carl Zeiss Exciter 5. The specific monoclonal antibodies were used in order to analyze the adhesion of active platelets by using two antibodies; P-selectin conjugated with fluorescein isothiocyanate (FITC) and (platelets with the active receptor selectin P), human monoclonal antibodies CD45 PE conjugated with PE-Texas Red (active leucocytes).

### *2.4. Silver release into blood plasma*

In the assessment of nanoparticles of silver ion release into the plasma, the main parameters for their evaluation were taken under consideration, Prothrombin Time (PT) and Activated Partial Thromboplastin Time (APTT).

PT is used to assess the extrinsic coagulation. Its value is dependent on the concentration of blood plasma coagulation factors such as factor II, factor V, factor VII, factor X, and fibrinogen. Tissue factor (TF) is added to the citrated plasma to activate factor VII. The binding of calcium ions by the citric acid causes that the activation of factor X does not take place by VIIa (activated factor VII) [24]. After incubation, the time when calcium ions are introduced is measured.

APTT measure of the activity of plasma coagulation factors. Kaolin-activated partial thromboplastin time- one of the indicator of blood clotting, since it is a measure of the activity of plasma clotting factors XII, XI, IX and VIII. They form the intrinsic system of prothrombin activation. Kaolin-activated partial thromboplastin time should remain in the level of 26.0 to 36.0 seconds.

For the analysis 22.5 mL of blood was taken on citrate (Vacutainer, 5x 4.5 ml). Blood was poured into one tube of 50 ml Falcon to ensure homogeneity of the experiment. 1 ml of blood was taken to the eppendorf and centrifuged 4000g/8min./4 °C. After centrifugation the plasma was transferred into 20 ul of appropriately signed eppendorf (0 bass) and frozen - 80 °C to determine the

concentration of input value of micro-particles. The rest of the plasma (~370 ul) was transferred to a new tube (0 bass) and measured for prothrombin time (PT) and partial activated tromboplastin time (APTT). Three plate wells were selected with the appropriate substrates coated with silver and having discs on the bottom of the test which was rinsed once with sterile physiological saline solution. For each testing well 3.5 ml of whole anticoagulated blood was added with sodium citrate and mixed in the incubation conditions at 37°C with shaking (120 rev./ Min.) for 1, 3 and 6 hours respectively. After appropriate incubation the wells were collected aseptically, taking 1 ml of blood and centrifuged at 4000g/8 min/ 4 °C. Then taken after 20 ul of plasma was taken of each sample (to measure micro-particles) and frozen at - 80 °C. 370 ul of plasma was transferred to a new tube for the determination of PT and APTT using BCS XP Siemens.

## 3. Results

### *3.1. Microstructure characterization*

Analysis of the coating microstructure with incorporated silver nanoparticles is shown in figure 1. The investigated coating was uniform, without the effect of delamination. Silver nanoparticles were located in an orderly manner in the amorphous matrix. It was ascertained on the basis of high resolution transmission electron microscopy (Figure 2a,b). Interplanar distances were measured in pictures obtained by inverse fast Fourier transform (IFFT) of atomic planes to obtain better contrast (Figure 2c). Figure 3a shows hetero- interface between two different crystalline solids. The lattice mismatch was possibly adjusted by edge dislocation or by strain (Figure 3b).

**Figure 1.** TEM bright field mode microstructure analysis from the cross section of the investigated coating..

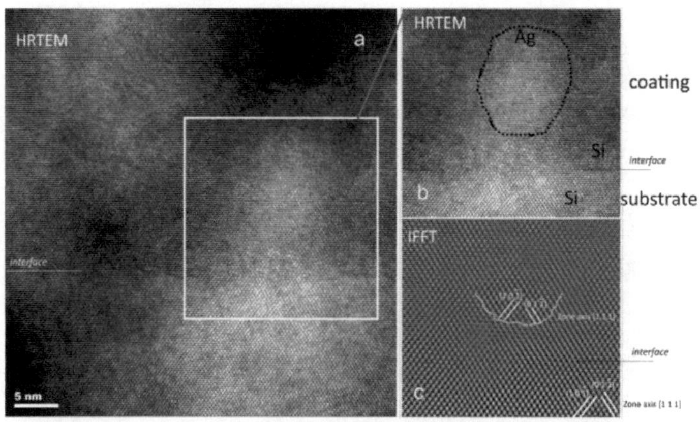


**Figure 2.** High resolution transmission electron microscopy analysis of the Ag nanoparticles introduced into porous synthetic coating a.) HRTEM, the general characteristics; b.) HRTEM, region of interest; c.) Inverse fast Fourier transform (IFFT) analysis of atomic planes.

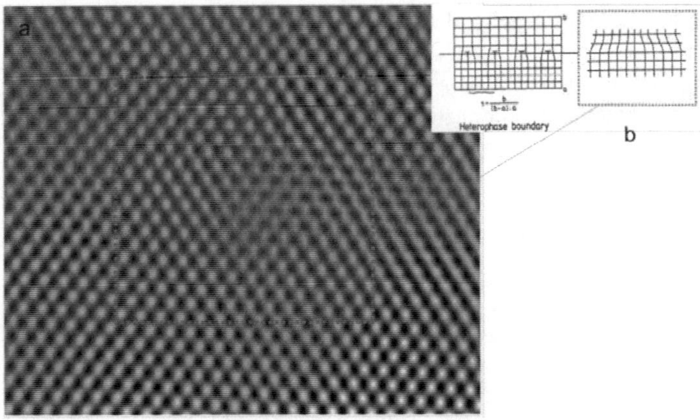


**Figure 3.** Hetero- interface between two different crystalline solids; b.) Theoretical interpretation of the lattice mismatch caused by edge dislocation.

*3.2. Blood-material interaction in the dynamic conditions*

Activation of platelets and its aggregation in blood collected after dynamic test and adhesion of active platelets were analysed. Analysis of the quality of blood through the techniques of flow cytometry is presented in figure 4.

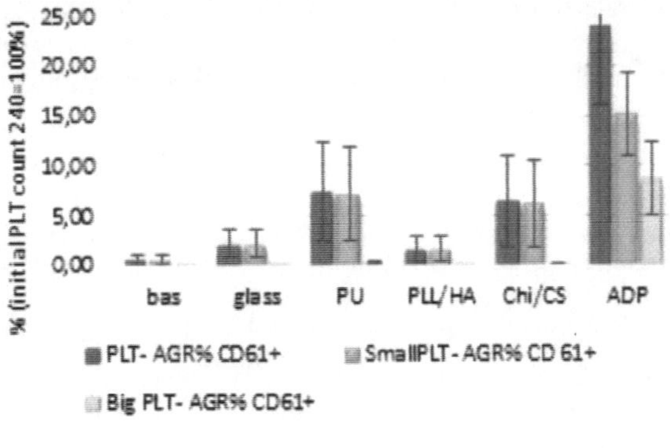


**Figure 4.** The probability of the platelet aggregates formation (CD61+) in a function of the tested material; bas- negative control, ADP- adenosine-di- phosphate activated platelets- positive control.

In the case of platelet aggregates, aggregates were distinguished as small and large once. Small units defined as two plates connected together. Big aggregates were defined as a conglomeration of more than two platelets.

The reference substrate was clinically used polyurethane. In the case of surface modification using polyelectrolyte-modified with silver nanoparticles, for all cases, hemocompatibility was improved. Better blood- material interaction was observed for PLL/HA modified with silver both for small and big platelets aggregates.

Analysis of platelet activation related to the analysis of membrane receptors, P-selectin and IIb / IIIa receptor.
Expression of platelet activation markers is measured on CD61 gated objects using PAC-1 antibody for conformational change of glycoprotein IIb/IIIa, and using CD62P for P-selectin (Figure 5).

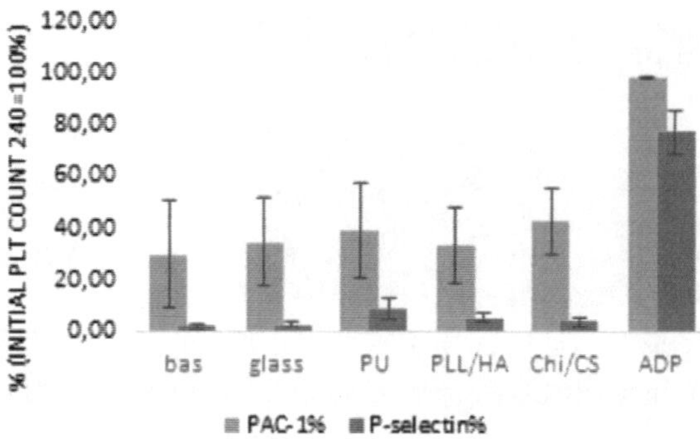


**Figure 5.** The amount of platelets with active receptors IIb/IIIa and selectin- P respectively in a function of the tested material; bas- negative control, ADP- adenosine-di- phosphate activated platelets- positive control.

In this case, considerable differences between the modified polyurethane with a polyelectrolyte with silver and a non-modified polyurethane were not observed. All materials showed the activation on the negative control level which means heparinized blood. However, in the case of pure, non-modified polyelectrolytes PLL/HA and Chi/CS without silver ions, the followed reaction concerns a very fast and strong activation and aggregation of blood on the surface (Figure 4).

*3.3. Silver release into blood plasma*

The coatings tested did not affect the blood clotting factors. Prothrombin time (PT) was marked. It increased with a decrease in fibrinogen, prothrombin, factors V, VII, and X (extrinsic coagulation pathway) (Figure 6).

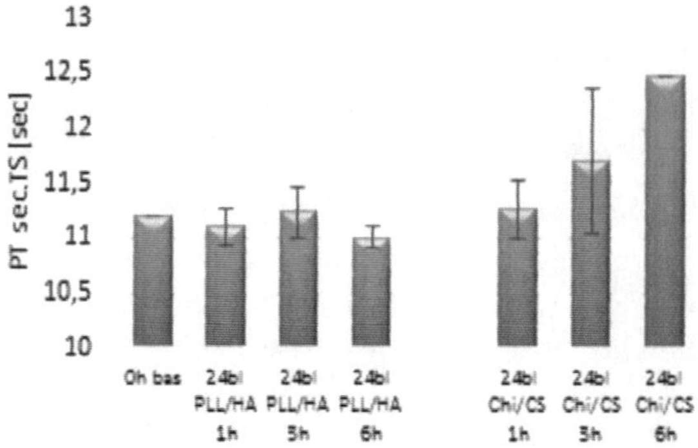


**Figure 6.** Prothrombin time (PT), analysis of extrinsic coagulation pathway V, VII and X.

APTT (partial activatet tromboplastin time), measures the extrinsic path and evaluates shortage activation of the contact pathway. It is able to further demonstrate the usage of factor VIII, IX, XI and XII. There are the factors form the intrinsic system of the activated prothrombin (Figure 7).

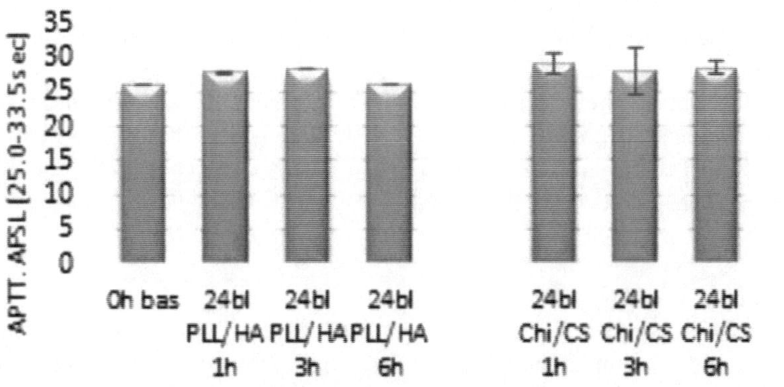


**Figure 7.** APTT (partial activatet tromboplastin time), demonstrating the usage of factor VIII, IX, XI and XII.

The second coagulation factor, it is a glycoprotein which is contained in the plasma, produced in the liver with the support of vitamin K. Prothrombin is a protein soluble in the serum, and is converted to insoluble thrombin. It is built of about 600 amino acids containing four oligosaccharide chains. Glycosylation of prothrombin occurs in the liver and is determined by the participation of vitamin K. APTT is also dependent on factors involved in the thrombin formation (prothrombin, factor X and V) and the conversion of fibrinogen to fibrin. Kaolin-activated partial thromboplastin time should remain in the range of 26.0 to 36.0 seconds. During our experiment the incubation time had to be reduced up to 6 hours because the mixed blood showed hemolysis.

## 4. Conclusions

The best hemocompatibility express materials mimicking tissue structure and its properties. For this reason extracellular like polymer structures were prepared.

Extracellular membrane is produced by the cells, filling the space between them, which is assembling part of the tissues.

Under normal conditions, the surface of polyelectrolytes causes severe thrombosis on the surface. A similar phenomenon is observed on the surface of pure extracellular matrix, excluding the vascular endothelial cell in the blood vessels.

Introduction of silver nanoparticles into the structure of the synthetic coating in order to stabilize the polyelectrolyte film, reduced the activation processes of blood without the adverse ion release into the body.

A specific type of intercellular substance is blood plasma. In this case, considerable differences between the modified polyurethane with the polyelectrolyte comprising silver and the non-modified polyurethane were not observed. It can be considered that the polyurethane surface modification is not necessary. There is another phenomenon that must be taken into account. The polymers in tissue environment change characteristics over their physical and mechanical properties. They become stiffer what could effect on the increased risk of clotting on the surface. Synthetic materials have a fundamental drawback. In a tissue environment they are biodegradable. This is not a significant problem when the materials dissolve into simple compounds. The real problem arises when synthetic

materials decompose into the complex compounds. Then a toxic effect on surrounding tissue is observed. Therefore, first of all, the appropriate polyelectrolyte selection is necessary.
Another important point was an influence on time of biodegradation and maintain the control over the process. For this purpose, the silver nanoparticles were proposed, introduced into the porous structure of the synthetic, porous, tissue-like coating. Silver is commonly known antibacterial material. It protects perfectly against the formation of biofilm. Another factor that was investigated was an influence of nanoparticles to processes of thrombosis and the release into the blood plasma.
It has been found that nanoparticles introduced into the porous polymer structure reduces coagulation processes and their release into the plasma falls within the limits of acceptability. The overall conclusion is as follows, the silver nanoparticles may be used to control the biodegradation of extracellular matrix- mimicking coatings designed for the cardiovascular therapy.

**Acknowledgements**
This publication was prepared under the project: "Development of innovative bioactive prosthetic heart valve". This project is implemented under the Programme for Applied Research in the path A, according to the agreement PBS3 / A7 / 17/2015, funded by the National Centre for Research and Development. The topic of polyelectrolyte was developed in Grenoble Institute National Polytechnique Minatec Phelma LMGP

**5. References**

1. Lee Y A, Mahler N, Best C, LEE Y U, and Breuer C K 2014 *Translational Research*, **163(4)**, 321-341.
2. Chung S, King M W 2011 *Biotechnol Appl Biochem*, **58**, 423–38.
3. Naderi H, Matin M M, Bahrami A R, 2011 *J Biomater Appl.*, **26**, 383–417.
4. Richert L, Boulmedais F, Lavalle P, Mutterer J, Ferreux E, Decher G, Schaaf P, Voegel J C, Picart C, 2004 *Biomacromolecules*, **5(2)**, 284–94.
5. Major R, 2013 *Journal of Material Science Materials in Medicine; Springer*, **24**, 725-733.
6. Mzyk A, Major R, Kot M, Gostek J, Wilczek P, Major B, 2014 *Archives of civil and mechanical engineering*, **14**, 262–268.
7. Marx D E, Barillo D J, 2014 Burns **40S**, 9 –18.
8. Lansdown G, 2006 *Curr Probl Dermatol.*, **33**, 17–34.
9. Parka M., Neigh A M, Vermeulen J P,de la Fonteyne L, Verharen H W, Briedé J J, van Loveren H, de Jong W H, 2011 *Biomaterials* **32**, 9810-9817.
10. Arvidsson S, Askendal A, Tengvall P, 2007 *Biomaterials* **28**, 1346–1354.
11. Horbett T A, The role of adsorbed proteins in tissue response to biomaterials. In: Ratner B D, Hoffman A S, Schoen F J, Lemons J E, editors. Biomaterials science, an introduction to materials in medicine. second ed. London: Elsevier Academic Press; 2004. p. 237–46.
12. Brash J L, Ten Hove P, 1984 *Thromb Hemostas*, **51**, 326–30.
13. Merrill E W, Distinctions and correspondences among surfaces contacting blood. In: Leonard E F, Turitto V T, Vroman L, editors. Blood in contact with natural and artificial surfaces, vol. 516. New York: Annals of the New York Academy of Sciences; 1987. p. 196–203.
14. Hoffman A S, Modification of material surfaces to affect how they interact with blood. In: Leonard E F, Turitto V T, Vroman L, editors. Blood in contact with natural and artificial surfaces, vol. 516. New York: Annals of the New York Academy of Sciences; 1987. p. 96–101.
15. Wojciechowski P, Ten Hove P, Brash J L, 1986 *J Colloid Interface Sci*, **111**, 455–65.
16. Elwing H, Askendal A, Lundstrom I, 1987 *J Biomed Mater Res*, **21**, 1023–1028.
17. Picart C, Lavalle P, Hubert P, Cuisinier F J G, Decher G and Schaaf P, 2001 *Langmuir*, **17**, 12531–12535.
18. Major L, Lackner J M and Major B, 2014 *RSC Adv.*, **4**, 21108–21114.

19. Sanak M, Jakiela B, Wegrzyn W, 2010 *Bull Pol Acad Sci Tech Sci.*, **58(2)**, 317–22.
20. Calvete J J, 1995 *Proc. Soc. Exp. Biol. Med.,* **208 (4)**, 346–60.
21. Shattil S J, 1999 *Thromb. Haemost.* **82 (2)**, 318–325.
22. Vickers J D, 1999 *Platelets*, **10(4)**, 228–237.
23. Cleator J H, Zhu W Q, Vaughan D E, Hamm H E, 2006 *Blood*, **107(7)**, 2736–2744.
24. Wein M, Sterbinsky S A, Bickel C A, Schleimer R P, Bochner B S, 1995 *Am. J. Respir. Cell Mol. Biol.*, **12(3)**, 315–319.